The 3rd International Conference on Green Civil and Environmental Engineering (GCEE2021)

IOP Conference Series: Earth and Environmental Science
Volume 847

Online
12 August 2021

Editors:

M. Mirza Abdillah Pratama
Cynthia Permata Dewi
Apif Miptahul Hajji
Buntara Sthenly Gan
N. Altinay Perendeci
Ta-Chih Hsiao

ISBN: 978-1-7138-4811-0
ISSN: 1755-1307

Printed from e-media with permission by:

Curran Associates, Inc.
57 Morehouse Lane
Red Hook, NY 12571

Some format issues inherent in the e-media version may also appear in this print version.

Additional copies of this publication are available from:

Curran Associates, Inc.
57 Morehouse Lane
Red Hook, NY 12571 USA
Phone: 845-758-0400
Fax: 845-758-2633
Email: curran@proceedings.com
Web: www.proceedings.com

TABLE OF CONTENTS

IOP Conf. Series: Earth and Environmental Science **847** (2021) 011001 doi:10.1088/1755-1315/847/1/011001

PREFACE

The 3rd International Conference on Green Civil and Environmental Engineering (GCEE2021) has been scheduled and successfully taken place in a virtual meeting room by Zoom in Malang, Indonesia on 12 August 2021. This conference, at first, was originally set to be hosted and held offline at Universitas Negeri Malang, Malang, East Java, Indonesia, but then was held virtually due to COVID-19 and the government rules in travel restriction, especially for visitors from abroad. Since GCEE has been proclaimed as a biannual event and preparations have started since mid-2020 then this event is still held instead of being postponed. Moreover, the publication of the call of paper was induced great enthusiasm from academics and researchers. But then, the second wave of COVID-19 forced the committee to hold the event virtually so that it could run on time.

GCEE2021 was organised by the Department of Civil Engineering, Faculty of Engineering, Universitas Negeri Malang, Indonesia. The theme of this international conference was **"Energy-Responsible Building Technology and Life Cycle for Sustainable Built Environment"** with topic areas including sustainable construction and building materials, sustainable technology in building and infrastructure, the forensic process in structures and infrastructure: assessment, repair, and rehabilitation, disaster-resistant structures designs, technologies for safety and health management in construction, supply chain management, green building and materials, sustainable cities and architecture, renewable energy for the built environment, waste treatment and management, energy and nutrient recovery from waste and wastewater, sustainable technology for water and wastewater treatment, the resilience of water resources management in a global challenge, challenges of integrated water resources management, and GIS application in civil engineering.

The plenary session was presenting four keynote speakers from different countries and topics, which were: 1) Assoc. Prof. Henri Siswanto from Universitas Negeri Malang, Indonesia with speech: "Strategy for The Improvement of Regency Road Performance (Case Study on Indonesian Regency Roads)"; 2) Assoc. Prof. Mohd Syahrul Hisyam Mohd Sani from Universiti Teknologi MARA, Malaysia with speech: "Experimental Investigation of Built-up Cold-formed Steel Stub Column with and without Concrete-filled"; 3) Assoc. Prof. Ta-Chih Hsiao from National Taiwan University, ROC Taiwan with speech: "Fine and Ultrafine Particles in Taiwan Urban Area"; and 4) Prof. Nuriye Altınay Perendeci from Akdeniz Üniversitesi, Turkey with speech: "Boosting Biogas Production and Process Stability by Pretreatment". Each keynote speaker delivered their speech for 20 minutes and 30 minutes for a discussion session for every two speakers. In the afternoon, the conference was continued with parallel sessions to present 65 papers from authors. Each presenter has 10 minutes for paper presentation and 5 minutes for discussion with participants. This conference involves 298 participants from Indonesia and overseas, such as Malaysia, Taiwan, Turkey, and Japan. The virtual conference was using the Zoom platform. Overall, the conference was conducted smoothly without any technical issues problems.

We would like to acknowledge all of those who supported GCEE 2021:

- The entire steering, scientific, and organising committee members of GCEE 2021 for their help and support.
- Faculty of Engineering, Universitas Negeri Malang for the support for GCEE 2021.
- University partners and institutions
- The distinguished keynote speakers for their acceptance to give speeches and intellectual contributions on their respective fields of expertise.
- The participants of GCEE 2021 for their contribution and sharing knowledge.

IOP Conf. Series: Earth and Environmental Science **847** (2021) 011001 doi:10.1088/1755-1315/847/1/011001

LIST OF COMMITTEES

Host:
Department of Civil Engineering, Faculty of Engineering Universitas Negeri Malang, Indonesia

Steering Committee
- A.H. Rofiuddin – Universitas Negeri Malang, Indonesia
- Marji – Universitas Negeri Malang, Indonesia
- Arif Nur Afandi – Universitas Negeri Malang, Indonesia
- Eddy Sutadji– Universitas Negeri Malang, Indonesia
- Karyadi – Universitas Negeri Malang, Indonesia
- B. Sri Umniati – Universitas Negeri Malang, Indonesia

International Scientific Committee
- Davide Pogio – University of Sheffield, UK
- Wen Hsiang Hsieh – Formosa University, Taiwan
- Philip Van Den Heede – Ghent University, Belgium Julius Marpaung – Northeastern University, USA
- Faimun – Institut Teknologi Sepuluh Nopember Surabaya, Indonesia
- Buan Anshari – Universitas Mataram, Indonesia
- Yuli Panca Asmara– INTI International University, Malaysia
- Eng. Achfas Zacoeb – Universitas Brawijaya, Indonesia
- Ni Nyoman Nepi Marleni – Universitas Gadjah Mada, Indonesia
- Vera Agustriana Noorhidana – Universitas Lampung, Indonesia
- Akmaluddin – Universitas Mataram, Indonesia
- Jojok Widodo Soetjipto– Universitas Jember, Indonesia
- Heri Sulistiyono – Universitas Mataram, Indonesia
- Bobby Rio Indriyantho – University of Technology Dresden, Germany
- Agung Murti Nugroho – Universitas Brawijaya, Indonesia
- Nur Aini Iswati Hasanah – Kementrian Pekerjaan Umum dan Perumahan Rakyat, Indonesia
- Christina Kristiyani – Universitas Sanata Dharma, Indonesia
- Abd. Qadir Muslim – Universitas Brawijaya, Indonesia
- Nurina Fitriani – Universitas Airlangga, Indonesia
- Sugeng Dwi Hartantyo– Universitas Islam Lamongan, Indonesia
- Muslich Hartadi Sutanto – Universiti Teknologi Petronas, Malaysia
- Husna binti Takaijudin – Universiti Teknologi Petronas, Malaysia
- Radin Maya Saphira Binti Radin Mohamed – Universiti Tun Hussein Onn Malaysia
- Maranatha Wijayaningtyas – Institut Teknologi Nasional Malang, Indonesia
- Nurhafizah Abd Khalid – Universiti Teknologi Malaysia, Malaysia
- Henri Siswanto– Universitas Negeri Malang, Indonesia
- Dian Ariestadi – Universitas Negeri Malang, Indonesia
- Monita Olivia – Universitas Riau, Indonesia
- Poppy Puspitasari – Universitas Negeri Malang, Indonesia
- Aisyah Larasati – Universitas Negeri Malang, Indonesia

Chairman
Anie Yulistyorini – Universitas Negeri Malang, Indonesia

Editor
- M. Mirza Abdillah Pratama – Universitas Negeri Malang, Indonesia
- Cynthia Permata Dewi – Universitas Negeri Malang, Indonesia
- Apif Miptahul Hajji – Universitas Negeri Malang, Indonesia
- Buntara Sthenly Gan – Nihon University, Japan
- N. Altınay Perendeci - Akdeniz University, Turkey
- Ta-Chih Hsiao - National Taiwan University, Taiwan

GCEE
CONFERENCE

IOP Conf. Series: Earth and Environmental Science **847** (2021) 011003 doi:10.1088/1755-1315/847/1/011003

Peer review declaration

All conference organisers/editors are required to declare details about their peer review. Therefore, please provide the following information:

- **Type of peer review: Double-blind**
  The conference papers were double-blind reviewed, where the identity of both author and reviewer on the papers will be hidden. The process is started by manuscript submission to the online conference management system (CMS). The editorial team then doing plagiarism check using Turnitin on submitted papers. The plagiarism process is under no-repository mode. If the plagiarism result exceeds 15%, the paper will be returned to the author for revision and request for new submission. For papers passed the similarity checks, the editorial member will then assign the papers to two reviewers for topic-related content review within designated time limits. All reviewers will have access link to the papers to evaluate and judge substantial indicators, such as: abstract sufficiently informative; clarity in the presentation of findings; methodology appropriate to study; results or conclusion supported by data analysis; and originality/ novelty of finding. Clarity of figures and texts are also the main concern. The reviewers can also write their constructive critics directly in the CMS and put their judgement for the papers to be: (1) accepted and continued for camera-ready version, (2) accepted with minor revision, (3) accepted with major revision, or (4) rejected. Editors will notify all authors about the review results by email.

- **Conference submission management system:**
  ConfBay Conference Management System
  (https://submit.confbay.com/conf/icgcee2021)

- **Number of submissions received:**
  150 papers

- **Number of submissions sent for review:**
  116 papers

- **Number of submissions accepted:**
  36 papers

- **Acceptance Rate (Number of Submissions Accepted / Number of Submissions Received X 100):**
  36 / 150 x 100 = 24%

- **Average number of reviews per paper:**
  2 reviews

- **Total number of reviewers involved:**
  25 reviewers

IOP Conf. Series: Earth and Environmental Science **847** (2021) 011003 doi:10.1088/1755-1315/847/1/011003

- **Any additional info on review process:**
  -
- **Contact person for queries:**
  Mr. M. Mirza Abdillah Pratama (mirza.abdillah.ft@um.ac.id)
  Universitas Negeri Malang, Indonesia

Impacts of dams on surrounding groundwater levels

H Sulistiyono[1], D S Agustawijaya[1] and B W R Wardani[1]

[1]Faculty of Engineering, University of Mataram, Indonesia. 83125

h.sulistiyono@unram.ac.id

Abstract. The increasing number of dams built to address surface water problems has attracted the awareness of engineers about the impact of water in dams on surrounding groundwater levels. Some of the benefits of a dam include reducing peak flooding and storing excess water during the rainy season for use in the dry season. However, it is necessary to be aware that an increase in surface water in one place can have a negative impact on the surrounding environment. This paper describes the relationship between the water level in the dam and changes in the surrounding groundwater level. This study did not discuss the type of soil layer. The analysis was carried out by observing the level of the surrounding wells. Following Darcy's law of homogeneous isotropic steady-state flow, as well as the principle of the double-well pumping test, the well water level is used as a groundwater representation level. This research was conducted in the sub-districts around the Pandan Duri Dam. The results showed that water levels inside the dam had a significant impact on the surrounding groundwater levels. This is evidenced in this research that water levels of wells in Sakra District significantly changes following the change in water levels inside the Pandan Duri Dam.

1. Introduction

Dams are made for various purposes, including to overcome the problem of water shortages in the dry season and water excesses in the rainy season [1], [2], [3]. [4]. Besides providing advantages, the construction of dams can also cause various disadvantages. Advantages of dams are to reduce flooding, provide water for irrigation, supply household drinking water, provide water conservation, and facilitate water recreation. However, disadvantages of dams can include environmental damages, changes in sediment transport along river basins, disruptions of fish migration routes in riverside and unexpected rising groundwater levels [5].

Water inside a dam can cause the surrounding groundwater levels to rise [6]. The rising groundwater level can cause both positive and negative impacts on the environment. Positive impacts include ease of extracting water from the ground and increasing soil fertility. Negative impacts include decreasing soil stability, easily damaged foundations of buildings and roads, and an increasing risk of liquefaction [7], [8]..

This research aims to obtain the impact of water level inside a dam on the surrounding groundwater levels.

2. Literature review

Ardiansyah [8] conducted research on the Effect of Reservoir Water Fluctuations on Seepage Discharge Using the SEEP / W Model. The results of the model show the flow pattern reasonably follows the principle of the Casagrande method.

Hatmoko and Rauf [9]conducted research on Reservoir Water Level as an Indicator of Drought. Case Studies on Kedungombo Reservoir and Cacaban Reservoir. Results concluded that the reservoir water level can be used as an indicator of drought, is effective in early detection of drought disasters, and has a similar pattern to meteorological drought.

Meisyara and Fathani [10], conducted research on The Effect of Surface Water Reservoir Fluctuations on Deformation and Stability of Soil Dam-Test Model in the Laboratory. Results showed that the deformation that occurs in the soil dam model is more dominantly influenced by the water level. The higher the water level, the greater the deformation. Steeper upstream dam slopes cause greater deformation than gentle slopes. Deformation also occurs easily when the water level rises rapidly.

3. Methodology

The research procedure includes four following strategies: Preparations, Data Collections, Analisis, and Results. The proposed procedure is shown in Figure 1.

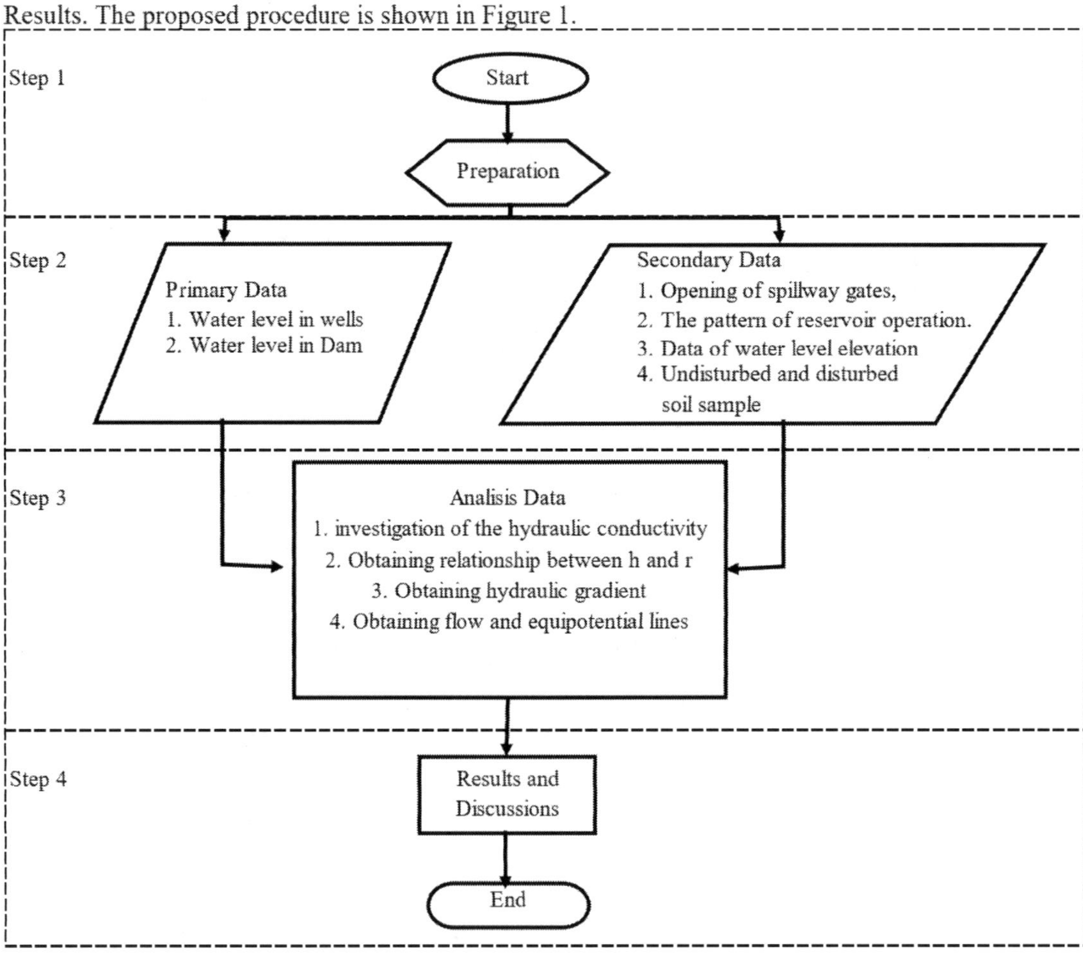

Figure 1 Proposed Procedure of the study

3.1. Preparations

3.1.1. Preliminary survey.

At this stage, observations were made starting from recording coordinates of the dam and wells, the physical condition and daily use of the wells, the administrative area, preparing well pumps and notebooks. Those wells are plotted in Figure 2 to Figure 7.

3.1.2. Research of literature.
The research of literature prepares references and theories needed to carry out research analysis. This research needs theory of groundwater flow patterns, information regarding the causes of groundwater fluctuation, and reasons why this research is important.

3.1.3. Summary of the problem.
At this stage the direction of the research must be ascertained so that the research can run well. In essence, this study will discuss whether the presence of water in a dam affects the surrounding groundwater level, provide a better understanding of flow patterns, and demonstrate the application of research to actual dams.

3.2. Data collections
Primary Data. Primary data in this research include levels of water inside the dam, also well water level data before and after daily uses. *Secondary Data.* Secondary data needed in this research are as follows: opening of spillway gates, the pattern of reservoir operation, data of water level elevation, and undisturbed and disturbed soil sample.

3.3. Analysis
The analysis begins with an investigation of the hydraulic conductivity in the field [11-14]. The strategy is through well pumping tests. The flow rate into the well which is assumed to be equal to the adjusted flow rate from well pumping is used to obtain the hydraulic conductivity in the equation below [4].

$$Q = -\,ki.A = -k\frac{dh}{dr}2\pi r \tag{1}$$

with: Q = flow rate into the well (cm^3/s), k = soil permeability coefficient (cm / s), i = slope of the water energy line, A = cross-sectional area of flow (cm^2), h = depth of water in the well (cm), r = distance of the well to control (cm), π = mathematical constant (= 3.14159).

The integration between r_1 and r_2 with the corresponding heights h_1 and h_2 is obtained through

$$k = \frac{2{,}3Q\log\frac{r_2}{r_1}}{\pi({h_2}^2 - {h_1}^2)} \tag{2}$$

with: Q = flow rate into the well (cm^3/s), k = permeability coefficient (cm / s), h = depth of water in the well (cm), r = distance of the well to control (cm).

Flow nets are graphical representations of the flow of water through the soil mass. The net flow is formed by a combination of flow lines and equipotential lines. These lines help define the seepage graphically. The flow line is the path through which water seeps into the ground from upstream to downstream. Flow lines can be drawn at any point where water starts to seep. Each flow line has the same k value. A graph depicting flow lines and equipotentials is called flownets [15], [16]. The flownet is used to calculate the amount of seepage that may occur. The flow line shows the direction of groundwater flow from the red zone which represents the area that has a higher potential pressure to the blue zone which represents the area that has a lower potential pressure. The flow rate can be calculated based on the net flow line using the Darcy equation:

$$q = Aki \tag{3}$$

With: q = flow rate (cm^3/s), A = cross-sectional area of flow (cm^2), k = permeability coefficient (cm / s), i = hydraulic gradient.

If the flow reviewed is only on one surface of the soil mass which is represented as flow between two flow lines in the flow net, then the equation (3) becomes:

$$q = k\frac{\Delta h}{l}b \tag{4}$$

With: q = flow rate on one surface of the soil mass (cm^3/s), k = permeability coefficient (cm / s), Δh= loss of potential energy between 2 successive equipotential lines, b = the distance between the two flow lines (cm), l = the distance between the two equipotential lines (cm).

4. Application

To demonstrate this research procedure, this research was conducted in several villages downstream of the Pandan Duri Dam in Sakra District, East Lombok Regency, West Nusa Tenggara Province as shown in Figure 2. Dam technical data that is important for this research include: inundation area: 315.70 Ha, storage volume: 27.20 million m^3, spillway width: 37.5 m, irrigation service area: 5168 Ha, peak elevation: + 284.00 m, riverbed elevation: + 246.00 m, normal MA elevation: + 281.50 m, MA flood elevation: + 282.75 m.

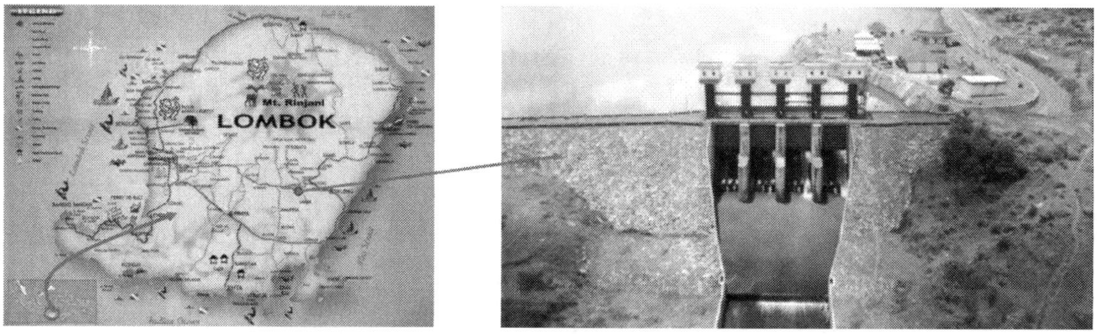

Figure 2 Study Location

4.1. *Data collections*

4.1.1. *Primary data.*

The survey was conducted in two periods: one is during a dry season and the other one is during a rainy season. The first period is carried out on 9 September 2019 during the dry season, and the second period is on 10 January 2020 during the rainy season. Each period measurement of the well water level elevation is carried out at 45 well points from 6 villages in the Sakra District area. The location and number of observations well points are shown in Figure 3 to Figure 8. The observation results are shown in Table 1 to Table 6.

Figure 3 shows locations of observation well in the Kelaga Dirik Village. Kelaga Dirik is the closest village to the dam. This village is about 80 to 200 meters downstream of the dam. The ten wells observed in this village are shown with yellow marks in Figure 3. Data from the observation wells in Kelaga Dirik Village are shown in Table 1. Table 1 shows the depth data for 10 wells in Kelaga Dirik Village taken during the dry season and the rainy season. From these data, it is known that the depth of groundwater is not always deeper during the rainy season. It was also noted that the closer the well to the dam does not necessarily have a higher groundwater level than the well further away from the dam. This is due to the variability of water use. Therefore, the calculation of the average is preferred in this analysis. In general, the condition of the wells in Kelaga Derik as follows: The average difference in groundwater level between the rainy season and the dry season is 0.22 m. The average depth of the groundwater level from the surface is 4.67 m. The average distance to the dam is 275 m. Figure 4 shows locations of seven observed wells in the Penujak Bongkot Village. They are shown with red marks in Figure 4. Penujak Bongkot is the second closest village to the dam. This village is about 420 to 740 meters downstream of the dam and located after Kelaga Dirik Village. Table 2 explains that the average difference in groundwater level between the rainy season and the dry season is 0.55 m. The average depth of the groundwater level from the surface is 3.97 m. The average distance to the dam is 281 m. At the same distance from Penujak Bongkot Village to the dam, to the right of Penujak Bongkot Village is Buwuh

Village as shown in Figure 5. Figure 5 shows locations of six observed wells with green marks in the Buwuh Village. Table 3 explains that the average difference in groundwater level between the rainy season and the dry season is 0.45 m. The average depth of the groundwater level from the surface is 3.43 m. The average distance to the dam is 240 m. Figure 6 shows five observation wells located in Dengkur Village. They are marked in pink in Figure 6. The village is located 1,800 to 2,100 meters downstream of the dam. Table 4 shows that the average difference in groundwater level between the rainy season and the dry season is 0.35 m. The average depth of the groundwater level from the surface is 3.32 m. The average distance to the dam is 1879.2 m. Figure 7 shows ten observation wells located in West Suangi Village. They are marked in white in Figure 7. The village is located 1,200 to 1,600 meters downstream of the dam. Table 5 shows that the average difference in groundwater level between the rainy season and the dry season is 0.28 m. The average depth of the groundwater level from the surface is 3.05 m. The average distance to the dam is 1399.3 m. Figure 8 shows seven observation wells located in Palung Dasan Village. They are marked in orange in Figure 8. The village is located 1,600 to 1,800 meters downstream of the dam. Table 6 shows that the average difference in groundwater level between the rainy season and the dry season is 0.41 m. The average depth of the groundwater level from the surface is 2.95 m. The average distance to the dam is 1720.9 m.

Table 1. Observations of 10 wells in Kelaga Dirik Village on Dry Season and Rainy Season

No of Well	Well Depth		ΔZ (m)	Distance to Dam (m)	Depth of Groundwater (m)
	Dry Season	Rainy Season			
1	9.6	9.27	-0.33	83.3	5.73
2	9.35	9.3	-0.05	105	4.7
3	6.2	6.5	0.3	106	3.5
4	9.2	9.4	0.2	109	5.6
5	9.33	9.55	0.22	110	4.45
6	3.8	3.9	0.1	114	6.1
7	6.43	7.1	0.67	144	6.9
8	3.26	3.58	0.32	148	4.42
9	7.35	7.88	0.53	169	2.12
10	6.53	6.78	0.25	212	3.22
Ave.	7.11	7.33	0.22	130	4.67

Table 2. Observations of 7 wells in Penujak Bongkot Village on Dry Season and Rainy Season

No of Well	Well Depth		ΔZ (m)	Distance to Dam (m)	Depth of Groundwater (m)
	Dry Season	Rainy Season			
1	9.20	9.82	0.62	575	2.10
2	23.25	23.88	0.63	657	5.18
3	7.33	7.90	0.57	687	2.01
4	17.32	17.90	0.58	722	3.24
5	16.27	16.76	0.49	732	2.10
6	24.23	24.99	0.76	738	1.12
7	15.78	15.99	0.21	749	12.01
Ave.	16.20	16.75	0.55	714	3.97

IOP Conf. Series: Earth and Environmental Science **847** (2021) 012001 doi:10.1088/1755-1315/847/1/012001

Table 3. Observations of 6 wells in Buwuh Village on Dry Season and Rainy Season

No of Well	Well Depth		ΔZ (m)	Distance to Dam (m)	Depth of Groundwater (m)
	Dry Season	Rainy Season			
1	5.12	5.35	0.23	1105	2.65
2	5.25	5.98	0.73	1099	2.02
3	6.40	6.78	0.38	1158	1.22
4	9.40	9.59	0.19	1178	1.41
5	8.23	8.80	0.57	1183	3.20
6	10.30	10.90	0.60	1235	4.10
Ave.	7.45	7.90	0.45	1159.7	2.43

Table 4. Observations of 5 wells in Dengkur Village on Dry Season and Rainy Season

No of Well	Well Depth		ΔZ (m)	Distance to Dam (m)	Depth of Groundwater (m)
	Dry Season	Rainy Season			
1	7.50	7.80	0.30	1863	2.50
2	9.55	9.95	0.40	1867	4.45
3	11.23	11.57	0.34	1884	3.77
4	11.44	11.84	0.40	1891	3.56
5	15.66	15.96	0.30	1891	2.34
Ave.	11.08	11.42	0.35	1879.2	3.32

Table 5. Observations of 10 wells in West Suangi Village on Dry Season and Rainy Season

No of Well	Well Depth		ΔZ (m)	Distance to Dam (m)	Depth of Groundwater (m)
	Dry Season	Rainy Season			
1	3.80	3.90	0.10	1322	3.10
2	4.78	4.98	0.20	1353	3.22
3	5.57	5.79	0.22	1354	2.43
4	7.65	7.84	0.19	1396	1.35
5	9.29	9.57	0.28	1404	5.76
6	12.85	12.98	0.13	1417	2.15
7	15.90	15.90	0.00	1419	2.10
8	16.25	16.87	0.62	1430	3.75
9	17.20	17.76	0.56	1444	2.80
10	18.20	18.67	0.47	1454	3.80
Ave.	11.15	11.43	0.28	1399.3	3.05

Table 6. Observations of 7 wells in Palung Dasan Village on Dry Season and Rainy Season

No of Well	Well Depth		ΔZ (m)	Distance to Dam (m)	Depth of Groundwater (m)
	Dry Season	Rainy Season			
1	4.25	4.50	0.25	1743	3.75
2	13.20	13.59	0.39	1725	1.80
3	13.26	13.78	0.42	1759	3.64
4	14.32	14.87	0.55	1665	2.68
5	15.25	15.83	0.58	1691	4.75
6	20.22	20.67	0.45	1722	2.78
7	20.75	20.98	0.23	1741	1.25
Ave.	14.46	14.89	0.41	1720.9	2.95

Figure 3. Map of Well Observation Location in Kelaga Dirik Village

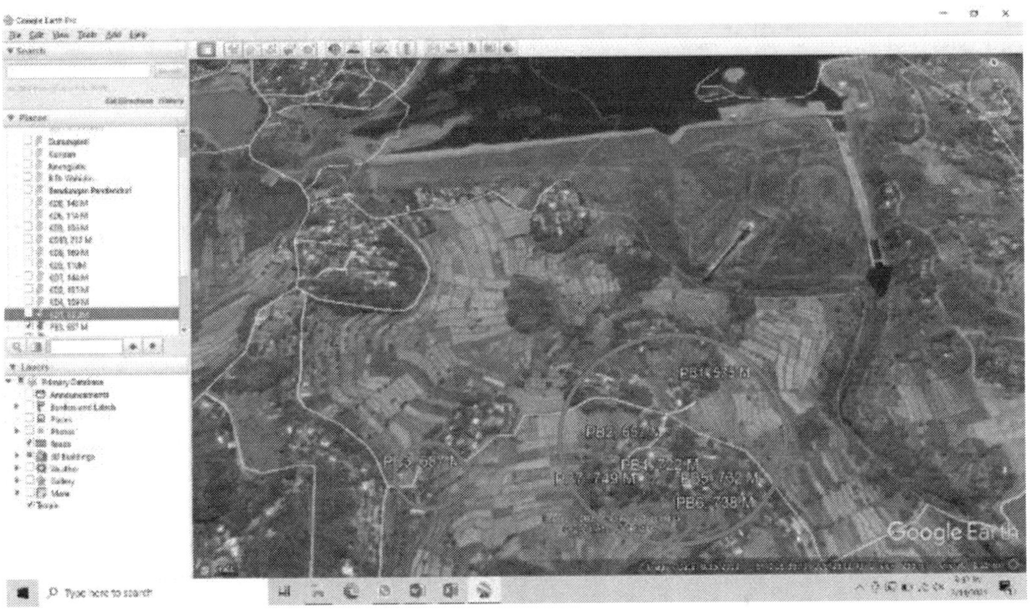

Figure 4. Map of Well Observation Locations in Penujak Bongkot Village

Figure 5. Map of Well Observation Location in Buwuh Village

Figure 6. Map of Well Observation Locations in Dengkur Village

Figure 7. Map of Well Observation Location in West Suangi Village

Figure 8. Map of Well Observation Locations in Palung Dasan Village

4.1.2. Secondary data.
Secondary data required is data in accordance with periods which include the following data: (a) Height Opening of the Pandan Duri Dam spillway gates from September 2019 to January 2020; (b) The pattern of reservoir operation; (c) Data of water level elevation of the reservoir inundation at Pandan Duri Dam; and (d) Undisturbed and disturbed soil sample

4.2. Analysis of groundwater level
To plot the groundwater level. The main reasons for drawing groundwater level contours are to visualize actual or theoretical groundwater levels, and to see groundwater distribution patterns or to find

anomalies. Figure 9 shows the simulated groundwater surface contour image. It can be seen that the distribution of groundwater level contours in the red and blue areas in Figure 9 is denser than the contour distribution of the groundwater level in the yellow area. This indicates that the yellow area is more permeable. Next is to relate the groundwater level to the elevation of surface water in the reservoir and to compare the groundwater level to the results of flow nets due to seepage as explained in the next section.

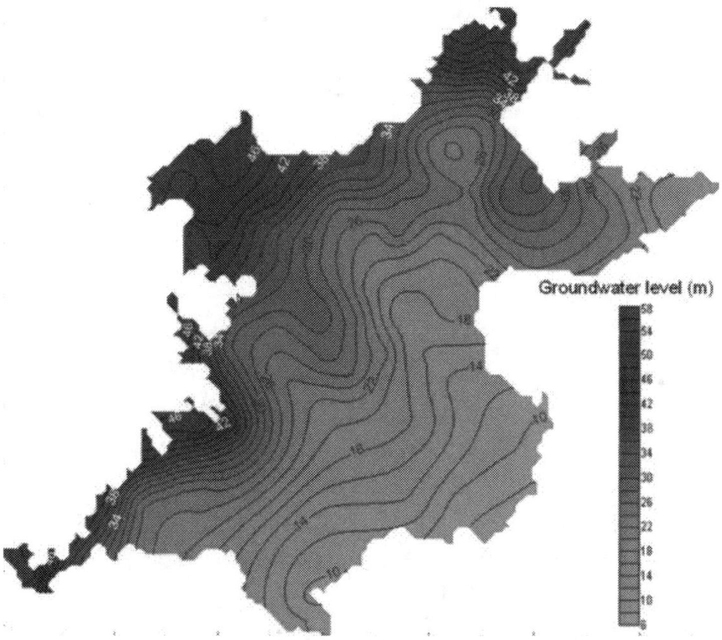

Figure 9 Groundwater level Map

4.3. Analysis of flow nets

From the comparison between equipotential lines due to seepage and the relationship between changes in reservoir water level and well water level, it was found that water inside the reservoir significantly affect to the surrounding groundwater levels. The results are presented in Figure 10. Figure 10 shows a significant relationship between the water level in the reservoir and the surrounding groundwater levels. It was found also in the simulation that the fluctuation of water level inside of the Pandan Duri Dam has a significant effect on the surrounding groundwater level.

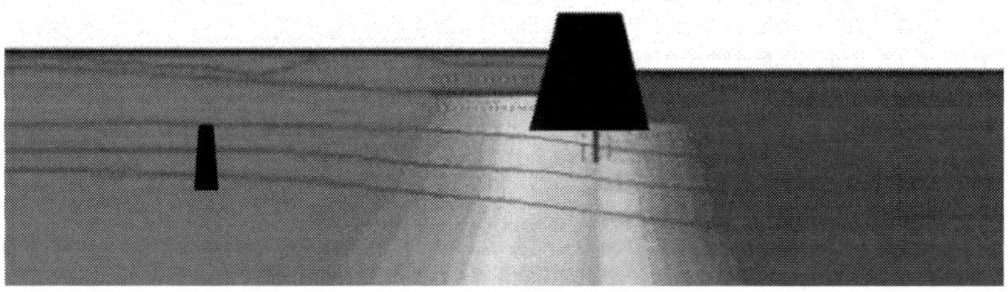

Figure 10 Flow Nets

5. Conclusion

Based on the results of the analysis, the following conclusions can be drawn:

a. This research was successful in proving that the presence of water in the reservoir of a dam affects the groundwater level of the surrounding area

b. This research proves that the groundwater flow pattern generally moves in and out of the groundwater filling system and is related to the fluctuation of the reservoir water level. This is especially evident during the rainy and dry seasons.
c. This research procedure was successfully demonstrated in the area around the Pandan Duri dam in Sakra District. East Lombok Regency. From the results of this research application. It is known that the water level fluctuation of Pandan Duri Dam has a significant effect on the groundwater level in the villages downstream of the dam.

References

[1] Suyono S and Takeda K 1977. Bendungan Type Urugan. PT. Pradnya Paramita. Jakarta.
[2] Hodge A 2000 "Reservoirs and Dams". Handbook of Ancient Water Technology. Technology and Change in History. 2. Leiden: Brill. pp. 331–339. ISBN 978-90-04-11123-3
[3] James P and Chanson H 2002 "Historical Development of Arch Dams. From Roman Arch Dams to Modern Concrete Designs". *Australian Civil Engineering Transactions. CE43: 39–56.*
[4] Ren K 2019 "A nature-based reservoir optimization model for resolving the conflict in human water demand and riverine ecosystem protection". Journal of Cleaner Production. 231: 406–418. doi:10.1016/j.jclepro.2019.05.221ICOLD-CIGB, 2017. Buletin 154. Dam Safety Management: Operational Phase of the Dam Life Cycle.
[5] Almeida R 2018 "A case study on environmental sustainability: A study of the trophic changes in fish species as a result of the damming of rivers through clustering analysis". Computers & Industrial Engineering. 135: 1239–1252. doi:10.1016/j.cie.2018.09.032
[6] Killian C D, Asquith W H, Barlow J R B, Bent G C, Kress W, Barlow P M, and Schmitz D W. 2019. Characterizing groundwater/surface-water interaction using hydrograph-separation techniques and groundwater-level data throughout the Mississippi Delta, USA. https://doi.org/10.1007/s10040-019-01981-6
[7] USGS. "About Liquefaction". Archived from the original on 2013-04-12
[8] Ardiansyah R, Sobriyah S, and Wahyudi A 2014 Pengaruh Fluktuasi Muka Air Waduk Terhadap Debit Rembesan Menggunakan Model SEEP/W (Effect of Reservoir Water Fluctuations on Seepage Discharge Using the SEEP / W Model). Matriks Teknik Sipil. eISSN: 2723-4223. https://jurnal.uns.ac.id/ matriks/article/view/37417
[9] Hatmoko W and Rauf A 2012 Tinggi Muka Air Waduk sebagai Indikator Kekeringan. Studi Kasus pada Waduk Kedungombo dan Waduk Cacaban (Reservoir Water Level as an Indicator of Drought. Case Studies on Kedungombo Reservoir and Cacaban Reservoir). Conference: Seminar Bendungan Besar 2012. Volume: 1.
[10] Meisyara N and Fathani T 2012 Pengaruh Fluktuasi Muka Air Reservoir Terhadap Deformasi Dan Stabilitas Bendungan Tanah-Uji Model Di Laboratorium (Effect of Surface Water Reservoir Fluctuations on Deformation and Stability of Soil Dam-Test Model in the Laboratory). Tesis. S2 Universitas Gadjah Mada.
[11] Darcy H 1856 Les fontaines publiques de la ville de Dijon. Paris: Dalmont
[12] Whitaker S 1986 "Flow in porous media I: A theoretical derivation of Darcy's law". Transport in Porous Media. 1: 3–25. doi:10.1007/BF01036523. S2CID 121904058
[13] Boersma L 1965 Field Measurement of Hydraulic Conductivity Below a Water Table. https://doi.org/10.2134/agronmonogr9.1.c14
[14] Fryar A and Mukherjee A 2021 Groundwater Hydrogeology. Editor(s): David Alderton, Scott A. Elias, Encyclopedia of Geology (Second Edition), Elsevier, Academic Press, 2021, Pages 399-407, ISBN 9780081029091, https://doi.org/10.1016/B978-0-12-409548-9.12115-3.
[15] Cedergren H H R 1977 Seepage, Drainage, and Flow Nets, Wiley. ISBN 0-471-14179-8
[16] Knappett J and Craig R F 2012 Craig's Soil Mechanics 8th edition, Spon Press. ISBN 978-0-415-56126-6

Malay heritage: rehabilitation of Uma Kabuong Limo house

M A A Muhd[1] and K Indra[1]

[1]Department of Architecture, Riau University, Jl. HR. Soebrantas Km 12, Pekanbaru, Indonesia

muhd.arief@lecturer.unri.ac.id

Abstract. Uma Kabuong Limo is an 80-year-old traditional Malay house located in an old village of Pulau Belimbing, Riau Province. This building represents Malay architecture. Currently, the building is experiencing damage to its structural and architectural components. Several factors cause building damage such as building age, weather, usage, and flooding. This damage can lead to collapse and loss of the building as a whole. This study aims to identify the characteristics and see the damage to the building elements of Uma Kabuong Limo. This research method is descriptive qualitative with synchronic reading analysis technique. The research stage was to collect data through field observations and interviews. The next stage is making the model through Building Information Modelling to see the comparison of building damage. The results showed that the architectural knowledge was manifested in construction techniques, building materials, and carvings. Damage to buildings can be seen in structural and architectural elements. The level of damage is in the category of heavy damage, so it is necessary to immediately carry out building rehabilitation. The results of this research will be used further in the stage of preparing the document for the rehabilitation model of the Uma Kabuong Limo.

1. Introduction

Buildings can reflect the cultural roots of a particular group resulting from cultural and environmental processes. This process comes from the management of life which has meaning and symbols in society in fulfilling demands. Old buildings are architectural assets that need learning and maintenance. One building that cannot be separated from life is a house. A house is a place for the interaction of residents with the surrounding to protect and fulfilling their living demands. Primitive houses lasted a very long time based on high-level traditions with a depth of local traditions [1]. The vernacular architecture includes residential houses and other buildings related to the context of the environment and local/local resources, individuals, or local communities [2]. In contrast to Uma Lontiok which functions as a traditional house, the Malay house Uma Kabuong Limo operates as a resident's house. Malay house Uma Kabuong Limo is a term in the local language as design for a house. Uma is home, Kabuong is Lorong, and Limo is five. So, the Malay house of Uma Kabuong Limo is a house with 5 alleys (Lorong). The Malay house of Uma Kabuong Limo is the original building of the Malay community in Kampar Regency, Riau Province. The Malay community in Kampar district is categorized as the mainland Malay community [3]. This construction is a hereditary building that has been inherited for a long time, the average age is 80-100 years. Professional craftsmen or architects but rather the ability of the community to jointly use local techniques and materials did not design this building. In the development process, it adapts to the climate, traditions, economy, and the environment. They scattered these buildings in groups to form villages.

Malay buildings have building constituent components comprising stairs, foundations, pillars, floors, walls, doors, windows, and roofs [4]. The Malay home building of Uma Kabuong Limo develops with excellent techniques and awareness of meaning through carvings, spatial orientation, and materials. This seems to be proportional to the opinion of who says that vernacular buildings are rough [5]. The current era of globalization has transformed effect on many aspects such as economy, social and building. Nowadays, there is a change in the transformation of houses in the villages of Kampar Regency. The community has switched to not using the Malay house of Uma Kabuong Limo as a place to live, but buildings with new shapes, materials, and techniques. So, that many of Uma Kabuong Limo's Malay houses were damaged. Some of the reasons for the damage to the building of the Malay house in Uma Kabuong Limo, such as being unoccupied, the old building age, due to weather, and flooding.

This research is an identification of the damage to the building components of the Malay household Uma Kabuong Limo. The discussion will be carried out by grouping building materials based on structural aspects and architectural details. The results of this study can be used as a formulation of action for the restoration of the Malay house building in Uma Kabuong Limo. The government of Kampar Regency can take advantage of the proceeds of the research in preparing a program to maintain the Malay house building in Uma Kabuong Limo, which is in the tourist village of Pulau Belimbing, Kuok District.

2. Literature review

Settlement is defined as an individual-made formation of a natural process by providing support to live in and organize human life as individuals or groups [6]. Meanwhile, that there is socialization between individuals by introducing enforced norms and customs [7]. As an old village, Pulau Belimbing has a hierarchy of space through inter-ethnic levels and internal spaces of the tribe. Furthermore, stated that it equipped tribal settlements with supporting facilities and houses. Apart from functioning as a residence, a house is also a symbol and a reflection of the values of its owner [8]. Kampung Pulau Belimbing is an area with the category of Malay mainland Riau, so the building style includes Malay architecture [9].

As a construction that has attributes, it can be seen based on vernacular architecture. Vernacular architecture applies cultural and environmental elements that are manifested in architectural physical forms such as floor plans, structures, details, ornaments, etc. [10]. Vernacular's architecture in all parts of the world has a similar type and puts forward specific local themes. The Malay house symbolizes the owner's social status by providing better carvings and material quality [11]. The roof of traditional Malay houses not only reflects the origin of the residents but also their social and financial position. Malay house has been three physical components in the way of construction: that the house is in the form of a stage, gable shape, and roof finishing with gable-finials [12]. The Typology of traditional Malay houses is a home on stilts or a classed house and has high poles [13].

The Malay house of Uma Kabuong Limo is an old house with an average age of 80-100 years and has architectural characteristics so that this factor can be used as a cultural heritage building. Therefore, the buildings must be carefully based on the actions of cultural heritage buildings. The Indonesian government has regulated reserve cultural buildings in Law number 11 of 2010 concerning cultural heritage. Cultural heritage criteria such as being 50 years or more, representing a style period of at least 50 years, having special and scientific significance, and strengthening the national personality. Cultural heritage buildings need conservation, namely the entire process of maintaining a place to maintain its cultural significance [14]. The goal of conservation is to care for places of cultural heritage value [15]. There are seven levels of conservation, prevention, preservation, consolidation, restoration, rehabilitation, reproduction, and reconstruction. The action chosen in the Uma Kabuong Limo Malay house is restoration, restoring, and maintaining the original concept or finding the initial shape of the object. Restoration and re-integration of details and drawings are carried out based on archaeological evidence, original design, and authentic evidence.

3. Methodology

The research method is a descriptive qualitative analysis that involves theory, hypothesis, and subject determination. It is supported by data collection, processing, and analysis [16]. The research target used was the Malay house of Uma Kabuong Limo belonging to the Nurdin Abdullah family in Pulau Belimbing Village, Kuok District, Kampar Regency, Riau. Some of the reasons for choosing the building as an object, such as the construction is 80 years old, the house has all the components of a Malay building; the house is in an old village; the house is owned by Shiompu (traditional elders), the house has been registered by the Riau Center for Settlement and Building Development Information Ministry of Public Works and Housing (see figure number 1). This research has 4 steps. First, collecting data is doing by field observations, interviews, and secondary data published data collection out. The collection was carried out using the synchronic reading technique, which is to find synchronization of empirical facts found in the field [17]. Secondly, creating building simulation model through Building Information Modelling (Archicad Software). Third, the study carried out is to see the correlation between the conditions that should be simulated through the engineering model of Building Information Modeling with the phenomena/data obtained. Fourth, result of the research through comparison of the damage to house components analysis with literature. It based the percentage of damage to buildings on Permen PU No. 24 of 2018, while the calculation of damage is through comparison of modeling with facts found in the field. The research variable is characteristic of building and component building condition.

Figure 1. Location and Object Map of Uma Kabuong Limo Malay house.

4. Results and discussions

4.1. The characteristic of Uma Kabuong Limo

The Malay house of Uma Kabuong Limo is found on the island of central Sumatra, with the largest distribution center in Kampar Regency. It classified environmental characteristics as tropical climates. The characteristics of a tropical climate such as high humidity and rainfall, non-extreme weather changes, dry and rainy seasons, and so on. The building has a tropical character considering the orientation of the building, heat protection (sun shading) with lots of openings, and bright-colored materials [18]. The Malay house building in Uma Kabuong Limo responds to the tropical climate by applying a sloping roof, an overstock, a vent grid, living windows, and wood materials. The color used is the original wood color, which is dominated by light brown.

Figure 2. Uma Kabuong Limo Malay House

Figure 3. Malay House Plan of Uma Kabuong Limo

The Malay house building of Uma Kabuong Limo has different functions. 1st Kabuong called Serambi functioned as a room having 5 passageways (Lorong). Each hallway has guests. 2nd Kabuong is named Biliok which serves as the central area. The main bedroom is for the parents. The bedroom is outside connected to Tompek Kojo or the work area. 3rd Kabuong is called Biliok Pajie which functions as a child's bedroom. 4th Kabuong functions as a circulation link between functions. 5th Kabuong is a Dapu that functions as a kitchen. This function illustrates the zoning in the area. The semi-public zone is located in 1st Kabuong, where there is an interaction between house owners and guests. Guests who may enter are recognized and important guests. The private zone is in 2ns Kabuong and 3rd Kabuong. Children may not enter Biliok / parents' room. Meanwhile, parents may enter the child's Biliok (room) limitedly. The semi-private zone is in Dapu (kitchen) or 5th Kabuong (see figure number 3). This space may be used by all homeowners for cooking and eating activities. The public zone is located in the floor's basement. This building has a stage structure so that the lower space serves as a public space. Activities in this room such as interacting with neighbors, welcoming unfamiliar guests, and wood storage space.

Malay house on stilts. The average building height of 1.5 Malay houses Uma Kabuong Limo is a construction typology of 2 m. On the front and back, on the left side of the building, there are stairs connected to the door. The number of steps is an odd number such as 5, 7, 9, and so on. The structural components of the Malay house Uma Kabuong Limo are divided from the lower structure in contact with the ground, the middle structure as a building frame, and the roof structure as a support for the roof and retaining storage space (Salang). The structure material is wood except for foundations. The stiffener of structural components did not use nails, but used a connection as knots, holes, and pegs. However, using nails in the roof structure. Uma Kabuong Limo is seen as a traditional Malay architecture on the aspects of the roof, materials, ornaments, this is following literature number 11.

Table 1. The Structure Components of Uma Kabuong Limo

Components	Description	Figure
The lower structure uses a stone foundation	Supporting building loads and distributing loads to the ground	
The middle structure uses columns and beams. Material using wood	The building frame uses a beam and column bonding system. There are 2 types of beams, bonding beams between columns and retaining beams for wooden floors.	
The upper structure is a roof frame that uses wood material	The roof frame functions as a holder for the roof cover. The form of the roof that is often encountered is a saddle.	

Architectural component materials use wood materials. Architectural components are found on walls, floors, ceilings, roof coverings, stairs, doors, windows, and carvings. The Malay house, the Malay house Uma Kabuong Limo, has carvings with various motifs such as Bees depending on the roof trim and Gasiong on the stairs. Carvings are a symbol of the homeowner's social status, so the more carvings, the higher the owner's social status.

Table 2. Architectural components of Uma Kabuong Limo

Components	Description	Figure
Walls	The wall as a building envelope uses wood planks. There are 2 types of patterns for the arrangement of the planks. 1. The diagonal pattern is used on full walls with no windows 2. The vertical pattern is used on the wall under the window.	
Ceilings	The ceiling uses wood planks. The pattern of the arrangement of the boards is in the direction of the width of the building.	
Floor	The floor uses wood planks. Arrangement pattern along the length of the building. The floor is installed above the floor beams, which are arranged in the direction of the width of the building.	
Roof cover	The roof cover uses zinc plate.	
Door and window	Doors and windows use wooden frames and are closed using wooden planks. Doors and windows have vents using a grid (jalousie).	
Stairs	The stair uses a wooden structure and a cover from wooden planks. The lower floor uses a pile of stones or cast cement. On the outside there are dividers and handles.	

4.2. Damage to Uma Kabuong Limo's Malay house

Damage to buildings is visible on all sides of the building. The heaviest damage occurred in Lorong 5 (kitchen) was damaged 100% or collapsed so that all components could not be used. Damage to buildings was found in structural and architectural components. The structural components of the damage were dominated by beams 57.14% and 37.5% floor beams, while columns were 30%. From the remaining damage to the columns and beams, Uma Kabuong Limo is a traditional Malay house following literature number 12 and 13 that a traditional Malay house is a house on stilts and has high pillars. Architectural damage is the biggest damage. The damage was to doors and windows 76.47%, roof coverings 60.10%, and floors 55.79%. The windows and doors of Uma Kabuong Limo have detailed arrangements of fields and carved ornaments so that it is in accordance literature number 10 that cultural elements are manifested in physical form in the form of ornaments and details. The damage can be seen from weathering due to wet (rain) and fungi. Meanwhile, the roof of the building experiences corrosion / rust. The percentage of total damage to all components of the Uma Kabuong Limo building is 55.14%. Based on the comparison between the Minister of Public Works Regulation No. 24 of 2018-part 3 page 17 with the fact that the building was mostly structural and non-structural, it required special maintenance for the Uma Kabuong Limo building.

Figure 4. The original look of Uma Labuong Limo

Figure 5. Computer model of Uma Kabuong Limo using Building Information Modelling (Archicad 24)

Table 3. Analysis of damage to the elements of the Malay house of Uma Kabuong Limo

Num.	Building components	Amount	Damage amount	Percentage (%)
1	Foundation	20 pieces	3 pieces	15
2	Columns	20 pieces	6 pieces	30
3	Floor beams	16 pieces	6 pieces	37,5
4	Wall beams	14 pieces	8 pieces	57,14
5	Roof truss	24 pieces	10 pieces	41,66
6	Wall	126,5 m^2	90,75 m^2	71,73
7	Ceilings	72 m^2	72 m^2	100
8	Floor	72 m^2	40,17 m^2	55,79
9	Roof cover	86,4 m^2	51,93 m^2	60,10
10	Door and window	34 pieces	26 pieces	76,47
11	Stairs	2 pieces	2 pieces	100
12	Curving	50,6 m	8,3 pieces	16,40

5. Conclusion

The Malay house of Uma Kabuong Limo is the original house of the mainland Malay people who are scattered in Kampar Regency. Unlike the Lontiok Malay house as a traditional house used by traditional elders and customary events, the Malay house Uma Kabuong Limo Malay house is a resident's house, so it has a unique form and construction structure. This building is said to be the Malay House of Uma Kabuong Limo because it has 5 volume sides with their respective functions. Materials use wood and stone. The building has spatial planning rules by dividing the zone into public, semi-public, private, and private zones. For the people of Pulau Belimbing, a house is not just a place to live but a symbol of social status so that it shows unique qualities and is equipped with a variety of carvings. This building is suitable for data collection and rehabilitation because it has architectural characteristics and knowledge, especially Malay. Changes in the times caused people to leave the Malay house. The Malay house of Uma Kabuong Limo so that many houses were found damaged or even destroyed. Based on the examination, it was found that there were defects in all aspects, structural components, and architectural components. Overall, the building suffered damage in all components so that it could be categorized as heavy damage.

Acknowledgments

The author would like to thank the Institute for Research and Community Service, University for the research funding with contract number 590 / UM.19.5.1.3 / PT.01.03 / 2021. This research fund is used as operational funding and model making of the Malay House Uma Kabuong Limo, so that the research runs well.

References

[1] Rapoport A 1982 The meaning of built environment Sage Publications
[2] Brunskill R 1993 The traditional buildings of Cumbria in B Farmer and H Louw Companion to contemporary architecttural thought London dan New York: Routledge Pages 78-81
[3] Ernawati E 2009 Etnobotani masyarakat suku Melayu Daratan : studi kasus di Desa Aur Kuning kecamatan Kampar Kiri Hulu Kabupaten Kampar Provinsi Riau Institut Pertanian Bogor
[4] Zairin Z, Cinthia J M, and Nurul I A 2020 Architectural identification of traditional malay houses in Sumatra island : comparative study on the architectural components of Shapes Arsir: Jurnal Arsitektur 4 (2) pages 95-99
[5] Masner M 1993 Is there a modern vernacular? in B Farmer and H Louw Companion to contemporary architectural thought London dan New York: Routledge Pages 198-201
[6] Yunus H S 1987 Geografi permukiman dan beberapa permasalahan permukiman di Indonesia Fakulti Geografi Universitas Gajah Mada
[7] Budiharjo E 1998 Sejumlah masalah permukiman kota Penerbit Alumni
[8] Husaini M A A 2016 Morphological Characteristic of Pulau Belimbing Village in Riau Province, Indonesia International Journal of Education and Research (4) 483-492
[9] Astar H 2013 Evariasi vokal bahasa Melayu di Riau Daratan dan Kepulauan Gramatika : Jurnal Ilmiah Kebahasaan dan Kesastraan 1 (2); 93-102 doi:10.31813/gramatika/1.2.2013.31.93—102
[10] Sumalyo Y 1993 Arsitektur kolonial Belanda di Indonesia Gajah Mada University Press
[11] Faisal G 2019 Arsitektur Melayu : rumah Melayu Lontiok Suku Majo Kampar Langkau Betang: Jurnal Arsitektur 6 (1) DOI: 10.26418/lantang.v6i1.31007.
[12] Firzal Y 2015 Reconstructing socio cultural identity: Malay culture and architecture in Pekanbaru, Indonesia PhD dissertation Newcastle University
[13] Mudra M A 2004 Rumah Melayu: memangku Adat menjemput zaman Adi Cita Karya Nusa
[14] Burra Charter 1999 Australia
[15] ICOMOS 2000 Burra Charter: The Australia ICOMOS charter for places of cultural significance Australia
[16] Creswell J W 1997 Qualitative inquiry and research design, choosing among five traditions

London and New Delhi Sage Publications
[17] Darjosanjoto E T S 2006 Penelitian arsitektur di bidang perumahan dan permukiman ITS Press Pages 86-87
[18] Karyono T H 2016 Arsitektur tropis: bentuk, teknologi, kenyamanan, penggunaan energi. Penerbit Erlangga

What causes Ngancar River in Wiroko Temon sub-watershed vulnerable to flooding?

R M S Prastica[1] and A J Fanani[1]

[1]Civil Engineering Department, Universitas Gadjah Mada, Yogyakarta, Indonesia

rian.mantasa.s.p@ugm.ac.id

Abstract. The vulnerability to flooding becomes a concern in every region to optimize the mitigation strategy, hence the disaster impact can be reduced from the environment. What factors become the main triggering issue toward water-related disasters? Every region has a different answer. Can a river that does not have historical flood events become vulnerable to flood today? This study investigates the flooding factors in Ngancar River, Wiroko Temon sub-watershed, Wonogiri. The research employs hydrological and hydraulics analysis. The results reveal that Ngancar River experienced land-use changes over time which lead to the dynamic of the discharge return period being bigger. As a result, river capacity today cannot accommodate the flooding volume. The research concludes that the hydraulics condition needs to be improved for flood mitigation. Further hydrological analysis is suggested to be conducted.

1. Introduction

Flooding becomes threat worldwide, hence countries are struggling to minimize the impact of the disaster to the environment through various ways [1]. In structural path, government construct dam, modify river canal, and increase the water infrastructure resilience to fight the flooding [2]. On the other side, non-structural methods present top decision strategies from various stakeholders to decrease flooding impact to the environment [3].

How can we know that a region is vulnerable to flooding? What factors can trigger that phenomenon? This research identifies a phenomenon of flooding in Ngancar River, Wiroko Temon sub-watershed, Wonogiri region. According to the historical events, hydraulics factors in the study area do not affect the flooding happened. Meanwhile, several disaster events in the past years reveal that rivers cannot accommodate the flood volume. Does Ngancar River in Wiroko Temon sub-watershed vulnerable to flooding due to hydraulics factors? This research is a case study to investigate the hydraulics condition [4] of Ngancar River towards the flooding characteristics in the study area and expected to conclude whether river capacity becomes the main factor of the disaster.

2. Methods

This research employs hydrological analysis to obtain the return period discharge in the study area [5]. Nakayasu hydrograph synthetic unit is used to analyse the flooding in return periods presented [6]–[8]. This data becomes the basis to simulate the flooding through HEC-RAS modelling approach as hydraulics analysis supporting tool [9]. This software can visualise the river water profile and identify the flooding volume in several river segments [10]. The actual existing hydraulics of river condition is obtained from Google earth features to be coupled with HEC-RAS features. As a scenario of flooding mitigation in the future time, new design of river is presented even there is still no evidence that flooding volume exceeds the river capacity.

Initially, sub-watershed is identified as depicted in Figure 1. Three rainfall stations are presented as hydrological analysis, depicted in Figure 2 (a) and affect two main regions of sub-watershed, as depicted in Figure 2 (b). Firstly, the research will employ Thiessen method, as shown in Figure 2 (a). Meanwhile, there are only two affected areas in the study area, hence, the research uses average rainfall method to analyse the rainfall data [11], as observed in Figure 2 (b).

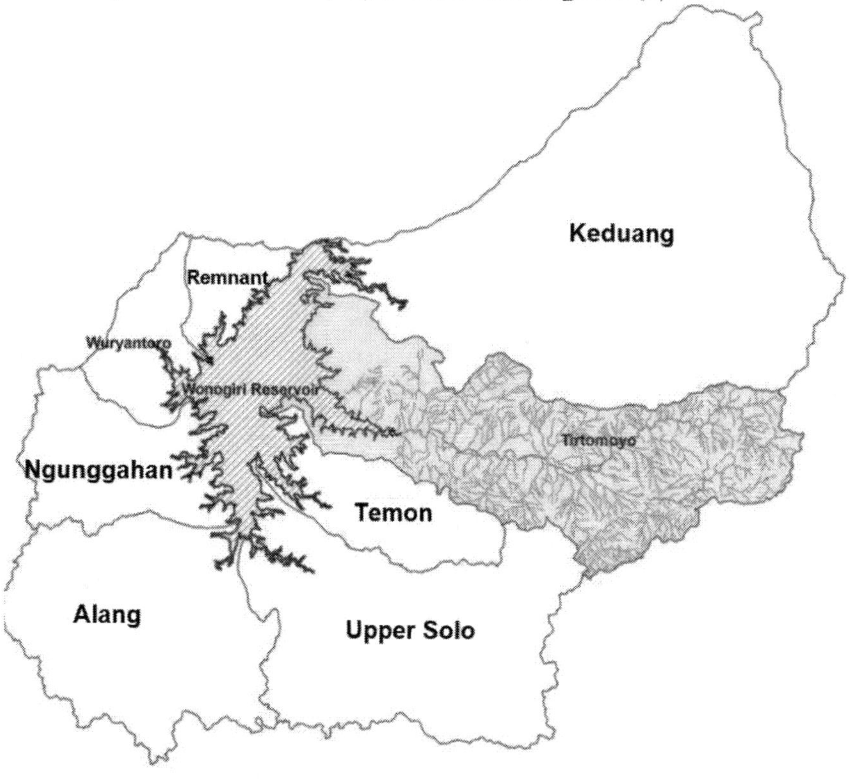

Figure 1. Ngancar River in Wiroko Temon sub-watershed [12]

3. Results and discussion

The results in this research reveal the hydrological analysis condition and hydraulics condition of Ngancar River supported by HEC-RAS software [13].

3.1. Flood analysis

The study of natural resources cannot be separated from hydrological analysis. With the hydrological analysis, we get the rainfall data as the basis for searching for flood discharge data. Flood discharge data analysis can be done with several methods, one of the methods is the Nakayasu hydrograph synthetic unit (HSU) method. Before getting flood discharge data, the research employed rain data using the Normal method, Log Normal method, Log Pearson Type III method, and Gumbel method [14]. The final result of discharge return period from Nakayasu HSU is shown in Table 1 and Figure 3.

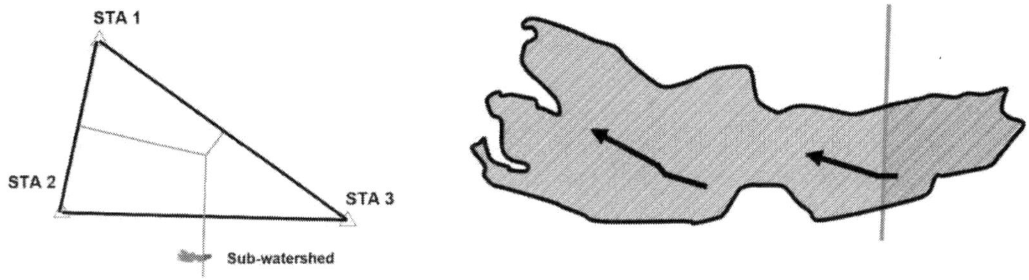

(a) (b)

Figure 2. (a) First condition: Thiessen method analysis plan and (b) Existing condition: areas affected by rainfall stations will be analysed by average rainfall method

Table 1. Nakayasu hydrograph synthetic unit

Return period discharge	Q_{max} (m^3/s)
2 years return period	34.17
5 years return period	44.29
10 years return period	54.29
25 years return period	73.77
50 years return period	94.27
100 years return period	126.65
500 years return period	327.97
1000 years return period	701.76

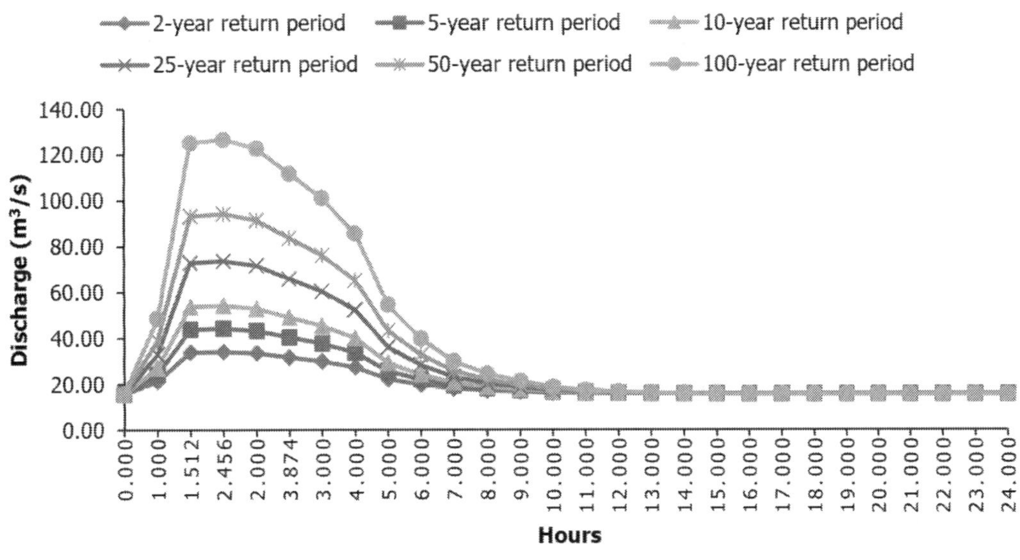

Figure 3. Flood hydrograph from Nakayasu HSU

The rainfall analysis used in this study was expected to obtain the calculation of the flood discharge at the 2, 5, 10, 25, 50, 100, 500, and 1000 years return periods. With this flood discharge calculation, it is expected that it will become the initial data for flood prevention in the future.

3.2. Validation of data analysis

To validate the discharge analysis result, a previous analysis from other research is presented in this research. Table 2 provides the results from the research.

Previous research by Mulyanto [12] observes the flood discharge in the Tirtomoyo sub-watershed, meanwhile Wiroko Temon is part of its sub-watershed. Hence, the discharge happened in the study area is smaller than research result by Mulyanto [12]. As initial hypothesis, the result can be validated but need more depth investigation in the future research about this. A result provided by Pediano et al. [15] by using other methods, Muskingum-Cunge and Hydrocad methods reveal that the data cannot be compared. Comparison should meet the same methods to obtain better validity. Hence, further analysis will not need any supporting data from different approaches.

In the Wiroko Temon sub-watershed there are land uses that change over time which can trigger the dynamic change of water resources. Therefore, the discharge analysis is needed using the land cover coefficient data of Wiroko Temon sub-watershed. To complete this step, the data needed in this research are cover area percentage, land use, coefficient of land use [16].

Table 2. Data validation analysis

Return period discharge	**Q_{max} (m^3/s)**	**Q_{max} (m^3/s)**	**Q_{max} (m^3/s)**
	Mulyanto, et al. [12] with Nakayasu method	**Pediano, et al. [15] with Muskingum-Cunge and Hydrocad methods**	**This research with Nakayasu method**
2 years return period	197.42	0.5356	34.17
5 years return period	254.30	3.1488	44.29
10 years return period	287.76	6.6097	54.29
25 years return period	N/A	14.4799	73.77
50 years return period	312.87	23.1901	94.27
100 years return period	352.66	33.9149	126.65
500 years return period	N/A	N/A	327.97
1000 years return period	450.99	84.0819	701.76

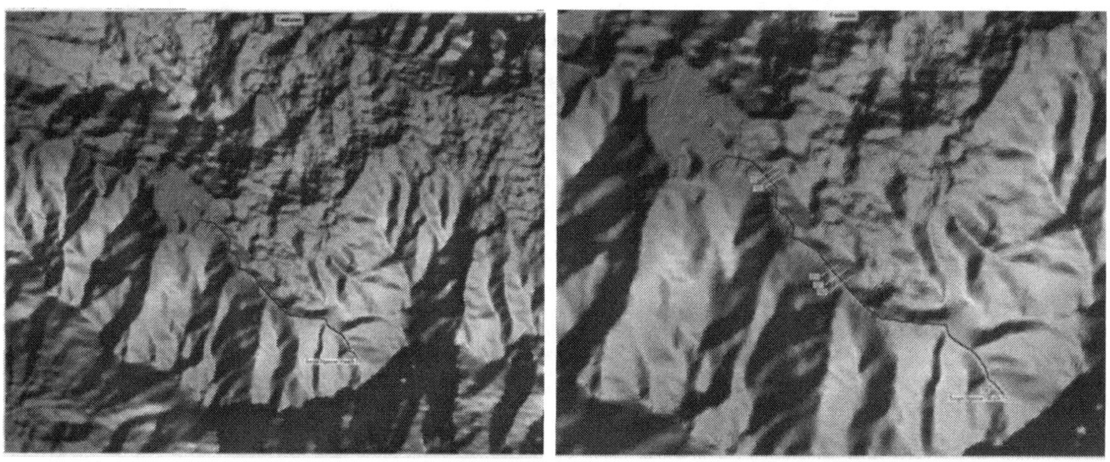

(a) (b)

Figure 4. Contour inserted to the HEC-RAS features (a) by drawing a river and (b) put the river sections observed in the study

The data above is obtained from the product of Water Resources Department of the Ministry of Public Works and Housing, Indonesia. The calculation of runoff coefficient analysis can be done by multiplying the percentage of land use with the coefficient of land class (C) and will obtain a weighted C value. The weighted coefficients for the return periods of 2 years, 5 years, 10 years, 25 years, 50 years, and 100 years are 0.449, 0.483, 0.505, 0.547, 0.579, and 0.643, respectively.

3.3. HEC-RAS analysis

The hydraulics analysis in this research concern on one of the rivers in the sub-watershed, Ngancar River. The channel modelling of the Ngancar River uses the help of the HEC-RAS application on both two scenarios, the existing condition channel and the channel that has been redesigned [17]by using the previously obtained discharge data. This modeling is used to obtain prediction of flood discharge at the return period of 2, 5, 10, 25, 50, 100, 500, and 1000 years. The following stages are involved to model the river by obtaining a HEC-RAS modelling approach. Initially, insert the map downloaded from national digital elevation model (or usually known as DEMNAS) data into the HEC-RAS application and create a river flow [18] according to the existing contour on the map, as depicted in Figure 4. Then, determine cross-sections with a number of three segments each on the upstream, middle stream, and downstream. Next, enter the value of the Nakayasu method result calculation in the unsteady flow program on the 2, 5,10, 25, 50, 100, 500, and 1000 year return periods. To get the results of the unsteady flow analysis, run the data that has been entered. Finally, the research is getting run results from unsteady flow.

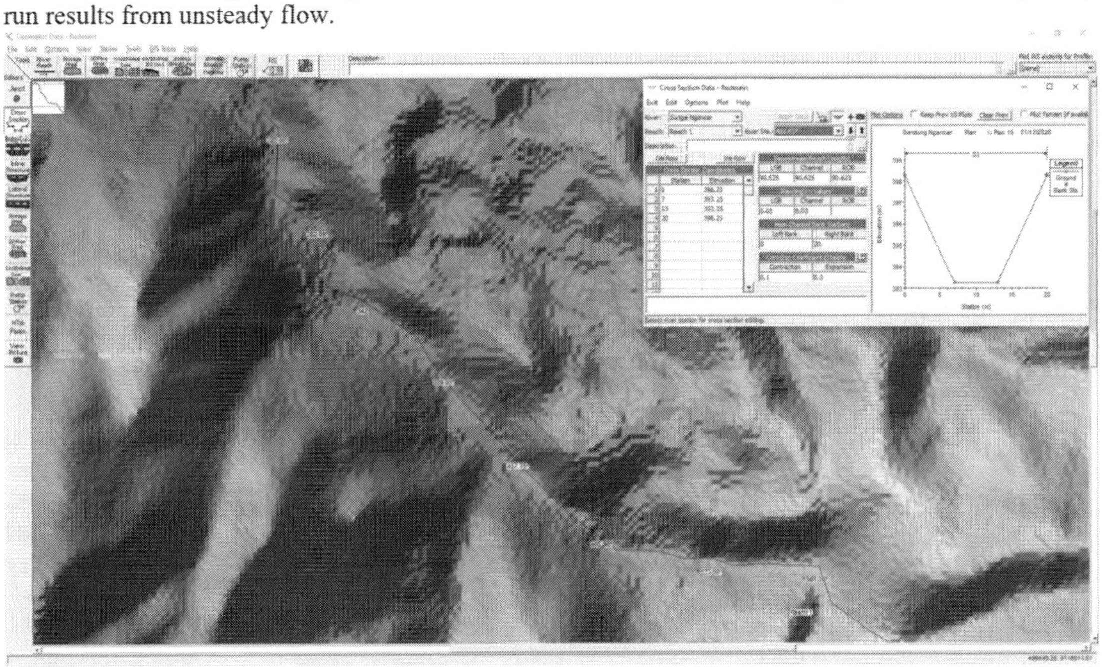

Figure 5. Normalization scenario for flood mitigation

Due to the results of the 100 years return period analysis, several points experienced flooding or water overflow in several places, as depicted in Figure 6. The next step is remodelling the river to avoid flooding. How to model this scenario? First, make a design as planned. Input all data of discharge return period with the condition of unsteady flow to run the flooding, as depicted in Figure 5. Make sure the analysis of the new design can accommodate the flooding volume according to the hydrological analysis.

As the presented result, Ngancar River cannot accommodate flooding for 100-year return period but can load flood volume below it. The hydraulics condition factor reveal that flooding events in the study area are correlated. It means, the hypothesis that hydraulics factors play an important role in causing flooding is validated through the analysis.

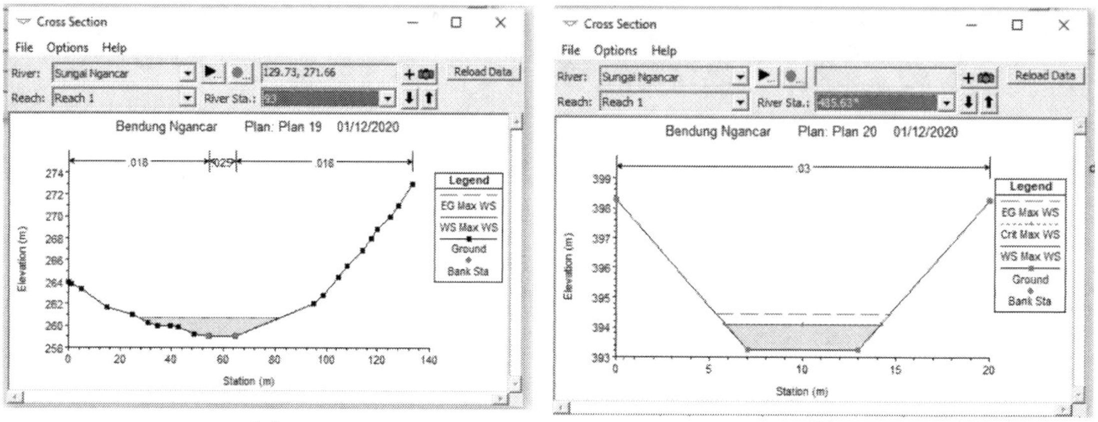

(a) (b)

Figure 6. HEC-RAS analysis result from (a) upper stream part of the existing river condition, (b) re-design the upper stream part

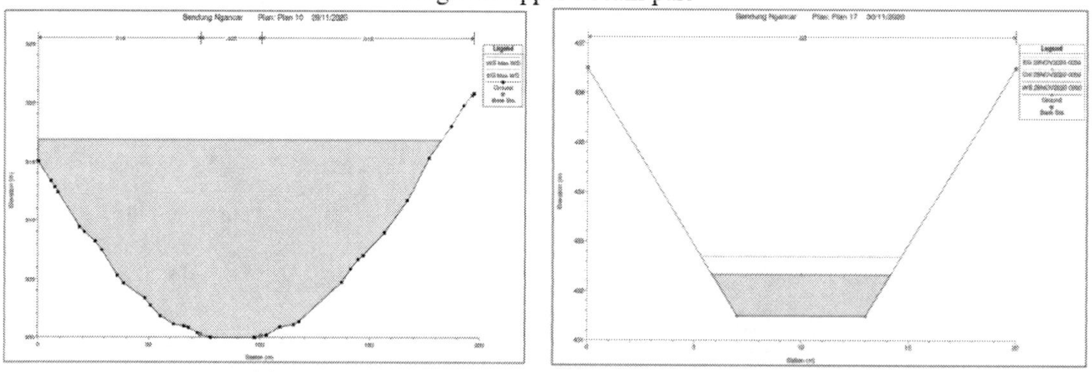

(c) (d)

Figure 6. HEC-RAS analysis result from (c) middle stream part of the river existing condition which cannot load the flooding, and (d) the normalization strategies to tackle flooding result

4. Conclusion and Recommendation

To conclude, the research hypothesis deals with the research result that flooding events in Ngancar River is triggered by river capacity which cannot accommodate the hydrological condition in the study area. For further investigation, analysis regarding discharge validation needs to be improved. Moreover, several mitigation scenarios to prevent flooding can be presented to provide better decision for mitigation strategies to fight flooding.

Acknowledgments

This research is fully supported by Universitas Gadjah Mada, Research faculty grant, 134/UN1.SV/K/2021. The authors fully acknowledge the Department of Civil Engineering and Hydraulics and Water Infrastructure Laoratory, Universitas Gadjah Mada.

References

[1] Nkwunonwo U C, Whitworth M and Baily B 2020 A review of the current status of flood modelling for urban flood risk management in the developing countries **7** 00269.

[2] Bechtol V and Laurian L 2005 Restoring straightened rivers for sustainable flood mitigation **14** 6–19.

[3] Chowdhooree I and Islam I 2018 Factors and actors for enhancing community flood resilience: An experience from a river-side settlement in Bangladesh **9** 153–169.

IOP Conf. Series: Earth and Environmental Science **847** (2021) 012003 doi:10.1088/1755-1315/847/1/012003

[4] Saputra A A, Saputro D D and Prastica R M S 2021 Initial investigation of hydraulics engineering for flood mitigation: A case study in Krueng Aceh Sub-watershed **622**.
[5] Prastica R M S, Apriatresnayanto R and Marthanty D R 2019 Structural and green infrastructure mitigation alternatives prevent Ciliwung River from water-related landslide **9** 1825–1832.
[6] Kang M S, Koo J H, Chun J A, Her Y G, Park S W, and Yoo K 2009 Design of drainage culverts considering critical storm duration **104** 425–434.
[7] Prastica R M S, Maitri C, Hermawan A, Nugroho P C, Sutjiningsih D and Anggraheni E, 2018 Estimating design flood and HEC-RAS modelling approach for flood analysis in Bojonegoro city **316**.
[8] Kuncoro M A, Winarto S and Purnomo Y C S 2018 Studi Penanggulangan Banjir di Kali Batan Kabupaten Kediri **1**.
[9] J. Li *et al* 2020 Digital Elevation Models for topographic characterisation and flood flow modelling along low-gradient, terminal dryland rivers: A comparison of spaceborne datasets for the Río Colorado, Bolivia **591** 125617.
[10] Geravand F, Hosseini S M and Ataie-Ashtiani B 2020 Influence of river cross-section data resolution on flood inundation modeling: Case study of Kashkan river basin in western Iran **584** 124743.
[11] Chow V T, Maidment D R, Mays L W and Ven Te Chow L W M, Maidment D R 1998 Applied Hydrology1–294.
[12] Mulyanto J 2012 Analisis Banjir Tahunan Sungai Tirtomoyo Wonogiri.
[13] Quirogaa V M, Kurea S, Udoa K and Manoa A 2016 Application of 2D numerical simulation for the analysis of the February 2014 Bolivian Amazonia flood: Application of the new HEC-RAS version 5 **3** 25–33.
[14] Syaifurahman A, Nurzain D A and Prastica R M S Prastica 2021 River engineering for flooding characteristics in the upperstream part of Bengawan Solo Sub-watershed **622**.
[15] Pediano P D, Hadiani R and Suyanto 2014 Penelusuran Banjir Di Das Temon Dengan Metode Muskingum-Cunge Menggunakan Hydrocad Matriks Tek. Sipil **2** 718–726.
[16] Zope P E, Eldho T I and Jothiprakash V 2016 Impacts of land use-land cover change and urbanization on flooding: A case study of Oshiwara River Basin in Mumbai, India **145** 142–154.
[17] Lin E, Shaad K and Girot C 2016 Developing river rehabilitation scenarios by integrating landscape and hydrodynamic modeling for the Ciliwung River in Jakarta, Indonesia **20** 180–198.
[18] Hameed H M 2017 Estimating the effect of urban growth on annual runoff volume using GIS in the Erbil Sub-Basin of the Kurdistan Region of Iraq **4**

Effectiveness of vibratory added mixing concrete with heating-grinding recycled coarse aggregate

N N Kencanawati[1], H Hariyadi[1], A Akmaluddin[1], I D M A Karyawan[1], F Mahmud[1] and P N Saputro[1]

[1]Civil Engineering Department, Mataram University, Jl. Majapahit 62 Mataram 83125 Indonesia

nkencanawati@unram.ac.id

Abstract. Concrete is a building material that is widely used in the construction industry. With the increasing need for concrete, the demand for constituent materials will increase. To reduce dependence on natural materials, the recycling process of the concrete waste has been encouraged recently. This study applied a heating and grinding method of waste concrete to obtain higher quality recycled coarse aggregate. The addition of vibrations in the mixing process is aimed to enhance the recycled concrete compressive strength. To produce recycled coarse aggregate, the concrete waste was heated to a temperature of 100°C for 24 hours followed by a grinding process using a Los Angeles machine for 600 cycles. Additional vibrations for 30, 60, and 90 were applied in a concrete mixer during the mixing process. Conventional concrete with normal aggregate and no additional vibration during mixing was also provided for comparison. According to the experimental results, the quality of natural coarse aggregate and the recycled one using heating and grinding methods is not much different. Compressive strength of recycled aggregate concrete with vibrations of 30, 60, and 90 seconds increases 21.1%, 22.1%, and 22.6% sequentially correlated to normal concrete. Recycled concrete using heating and grinding methods with the addition of vibration during the mixing process is effective to be used as an alternative to normal coarse aggregate.

1. Introduction

Concrete is the most commonly used construction material. This is because concrete has advantages over other materials. From an economic point of view, the price of concrete is relatively cheap because the ingredient comes from local materials. In addition, concrete has high compressive strength, can be molded into various shapes, and has good resistance to weather and the environment. These advantages make concrete the main choice as a construction material [1–5].

In construction, there are also many concrete wastes resulting from testing and demolition of buildings. A huge amount of construction demolition scrap creates environmental problems for instance the landfill problem due to a large amount of waste generated from the old concrete structures. Another problem that arises is the depletion of natural aggregates because they are used continuously in construction and since the formation of natural aggregates takes a very long time. Thus an alternative is needed to reduce the exploitation of natural aggregates [6–9].

To balance between landfill problem and depletion of natural aggregates, it is in demand to handle recycled aggregate of concrete ruin. The traditional method of recovering recycled aggregate from trash concrete does not reveal any notable quality to be utilized in the assembling of new concrete. The recycled aggregate concrete possesses lower mechanical properties than ordinary concrete. The bonding mortar on coarse aggregates leads to lower quality recycled aggregates [5, 10-11].

The pulsed power method has been used in Japan to reduce the mortar content in aggregates. High stress power is applied to the concrete waste and causes the release of mortar from the aggregate surface. Concrete made from recycled aggregates has a compressive strength that equals concrete made by natural aggregates. However, the pulsed power application is still difficult to be utilized in Indonesia due to huge electricity consumed. Therefore, the other suitable method shall be initiated to produce higher recycled aggregate quality [12–14].

Research conducted by [15-16] concluded that the characteristic of concrete with recycled coarse aggregate produced by the heating-grinding method showed an 8% difference in strength compared to normal aggregate concrete. Thus, recovered coarse aggregate generated by thermal-mechanical methods can be used as an option to normal coarse aggregate. However, efforts to enhance the compressive strength of concrete using this recycled aggregate are continuing ahead. Considering research conducted by [17] stated that the concrete compressive strength can be increased by providing vibration for 1 minute in the middle of the mixing process. Therefore, the vibratory added mixing is adopted in this research to be more effective to promote the characteristic of heating-grinding recycled coarse aggregate concrete. Thus, it is expected to be potential as an alternative material for sustainable concrete shortly.

2. Method and materials

2.1. *Recycled aggregate production*

The method in the concrete recycling process to extract the recycled aggregate began with collecting the concrete waste and cut it into smaller sizes (20-30 cm). The heating process was conducted by placing the concrete waste in the oven at a heat of 100° C for one day. This process was intended to achieve evaporation of water contained in the mortar pores so that it caused pressure and made it easier to release the mortar from the aggregate during the grinding method.

Grinding was intended for separating coarse aggregate from the existing mortar. This method was carried out with a machine that rotates 600 cycles. The grinding process was assisted by 11 steel balls inside the machine during rotation to rub the old mortar on the aggregate surfaces [15-16].

2.2. *Concrete mixture proportion*

Two types of concrete were made, the first was the normal concrete prepared from fresh natural coarse aggregate; meanwhile, the second was a concrete production of recycled coarse aggregate. Fine aggregate for both concrete was the same, which was natural sand having a maximum diameter of 4.75 mm. Either natural coarse aggregate or recycled coarse aggregate were having a maximum diameter of 20 mm. The density of fine aggregate and natural aggregate were 2.8 and 2.76 each.

Water cement ratio was held equal among the concrete mixture, which was 0.5, and slump value was designed as 60 mm. The mixture proportion of the concrete contains fresh coarse aggregate and recycled coarse aggregate is presented in Table 1.

Table 1. Concrete mixture proportion.

Concrete type	Weight (kg/m^3)			
	Water	Cement	Fine aggregate	Coarse Aggregate
Normal Concrete	225	450	816	957
Recycled Concrete	225	450	776	911

Based on the Table 1, there was a change in the need for normal coarse aggregate and recycled coarse aggregate. This is due to the change in density. Clearly, the density affects the composition of the concrete mixture where the density of normal coarse aggregate was 2.76, meanwhile, the density of recycled coarse aggregate was 2.57.

2.3. Concrete mixing and specimens

Generally, the mixing process of normal coarse aggregate concrete were according to standardized mixing process from national code for making concrete specimen [17]. However, there was a modification of mixing stage during making recycled coarse aggregate concrete. The process of giving vibrations was conducted during the concrete mix process with a stick vibrator concrete. This technique of adding vibrations was carried out inside the concrete mixer machine. The vibration was applied at two different points inside the concrete mixer to ensure the vibration effect was uniformly distributed on the fresh concrete. Vibrations were added during the mixing process for 30, 60, and 90 seconds. The variation of the specimen is given in Table 2 meanwhile the method of adding vibration during mixing is illustrated in Figure 1. Cylinder concrete standardized according to the national code for compressive strength measurement was used as specimen. The sizes were 150 mm in diameter and 300 mm in height. The curing time of concrete took place for 28 days to ensure that the hydration process was complete.

Table 2. Variety of vibratory added mixing specimen

Concrete type	Vibratory added mixing (s)
Normal coarse aggregate concrete (NCA)	0
1st Recycled coarse aggregate concrete (RCA-0)	0
2nd Recycled coarse aggregate concrete (RCA-30)	30
3rd Recycled coarse aggregate concrete (RCA-60)	60
4th Recycled coarse aggregate concrete (RCA-90)	90

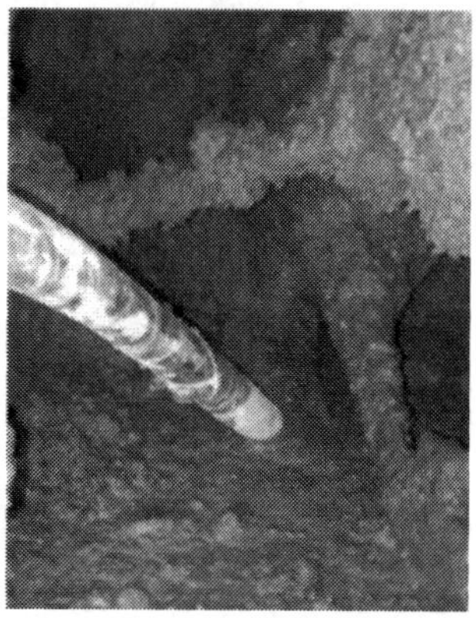

Figure 1. Method of vibratory added mixing

2.4. Concrete testing

After curing time was completed, the specimens were loaded under compression using compression testing machine until failure. The concrete testing was according to National Standard for compression testing of concrete using cylinder specimen [18]. Prior to the testing, the top surface of the specimens was flattened using sulfuric paste, to achieve a good load transfer from the machine to the concrete. Figure 2 shows the concrete testing.

Figure 2. Concrete testing

3. Result and discussions

3.1. Recycled aggregate physical properties

Table 3 present the characteristics of physics from recycled coarse aggregate compared to the requirement given in the national code for the qualities of coarse aggregate for concrete material making. Meanwhile, Figure 3 illustrates the comparison of physical properties between the normal and recycled coarse aggregate. In general, there is a difference between normal coarse aggregate and recycled coarse aggregate physical properties. The decrease occurred in density and water absorption of recycled coarse aggregate due to adhered old mortar in the surfaces of recycled coarse aggregate. In addition to this reason, the fineness modulus increases. However, the overall physical properties of recycled coarse aggregate still satisfy the aggregate quality standard [19]. In addition, the sieve analysis of the coarse recycled aggregate meets the requirement from the code as shown in Figure 4; even though the result approaches the lower limit the most.

Table 3. Physical properties of coarse recycled aggregate compared the specification from the code

Physical properties	Recycled coarse aggregate	Code specification
SSD density (kg/m^3)	2.57	2.4 – 2.9
Bulk density (kg/m^3)	2.48	2.4 – 2.9
Water absorption (%)	3.32	< 5%
Fineness modulus	6.72	6.0 – 7.1

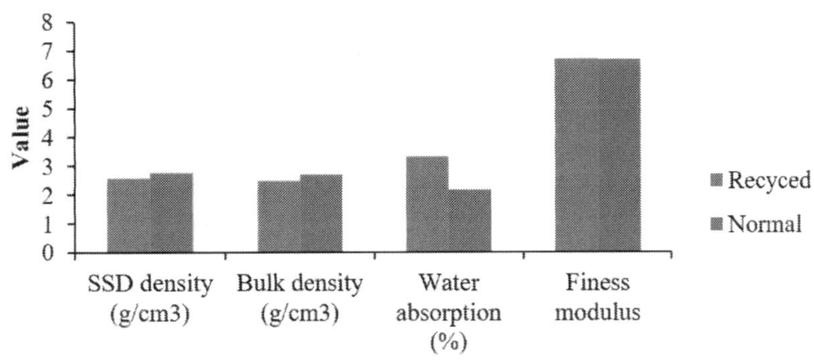

Figure 3. Physical properties of recycled and normal coarse aggregate

Considering the physical properties of conventional recycled aggregate as described in [10], the values obtained in this study show better quality because the surfaces of conventional recycled aggregate are still covered by old mortar entirely. Meanwhile, in this study due to a comprehensive treatment using heating and grinding process; thus, the attached existing mortar can be dismantled effectively. The visualization of recycled coarse aggregate in this study is presented in Figure 5.

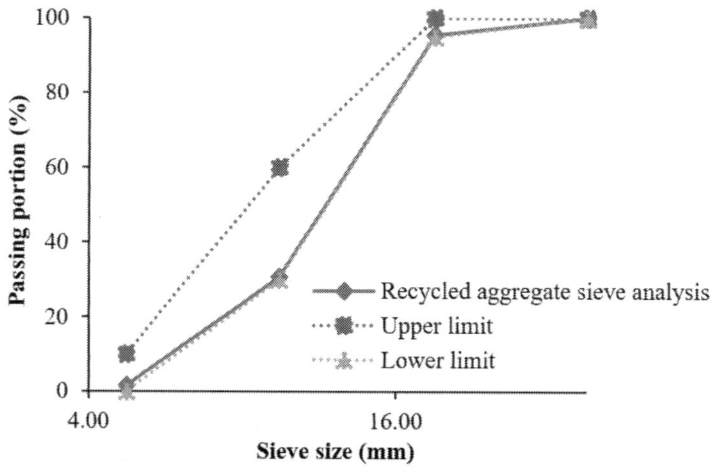

Figure 4. Sieve analysis of recycled coarse aggregate

Figure 5. Visualization of recycled coarse aggregate

3.2. Fresh concrete properties

Concrete with normal coarse aggregate shows the highest slump value. In general, the slump value of concrete with recycled coarse aggregate shows a smaller slump than the slump of normal concrete. This is due to the high water absorption of recycled coarse aggregate, even the SSD aggregate condition was maintained during mixing.

The effect of the vibratory added mixing further exacerbates the decrease in the slump value. Vibration for 90 seconds almost causes the concrete without a slump, which is around 1.17 cm. The slump of recycled concrete with a vibration of 30 and 60 second are in the range of 3-4 cm. The heat released by the vibrator is also assumed to cause a large reduction in the slump value. Concrete slump is presented in Figure 6 and 7.

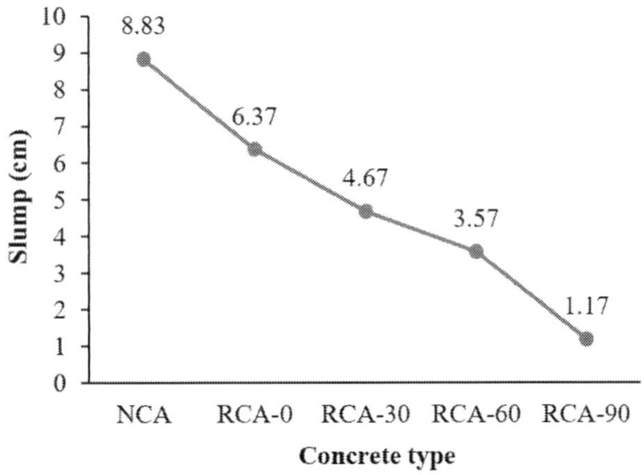

Figure 6. Concrete fresh properties

Figure 7. Slump testing

3.3. Compressive strength

Based on Figure 8, it can be seen that the application of recycled coarse aggregate in the concrete mixture can reduce the strength of concrete by 2.5% seen from the compressive strength of concrete with a normal coarse aggregate of 29.78 MPa and concrete from the recycled coarse aggregate of 29.04 MPa.

The reduction in the compressive strength of the concrete occurs because the mortar is still attached to the recycled coarse aggregate which decreases the quality of the recycled coarse aggregate. However, the use of recycled coarse aggregate with the addition of vibrations of 30, 60, and 90 seconds can improve the strength of concrete by 21.1%, 22.1%, and 22.6% compared to the compressive strength of concrete with a normal coarse aggregate of 29.78 MPa. The compressive strength of recycled coarse aggregate with the addition of vibrations of 30, 60, and 90 seconds are 36.05 MPa, 36.35 MPa, and 36.51 MP respectively. The longest vibration effectively achieves the highest compressive strength. The addition of vibration improves the bonding between recycled aggregate and new cement paste. In addition, the excess pores which exist in the old mortar of recycled aggregate can be filled by the new mortar. As the result, the concrete becomes denser and eventually, the compressive strength increases. More than one stage mixing has been reported by [5]. This result is in line with the result of this study that adding some processes during mixing enables to improve the performance of the recycled concrete.

Even the recycled aggregate concrete compressive strength is still lower than that of normal aggregate concrete; the vibratory added mixing can improve the recycled concrete effectively by around 20%. Among the addition of vibrations of 30, 60, and 90 seconds, each does not contribute many different results significantly. The compressive strength is only different 0.5 - 1.3% among the existing vibratory methods. Contrary to the lowest workability level based on the slump test, the 90 seconds vibratory results the largest compressive strength. An addition of either mixing water or superplasticizer is recommended to maintain the good workability level.

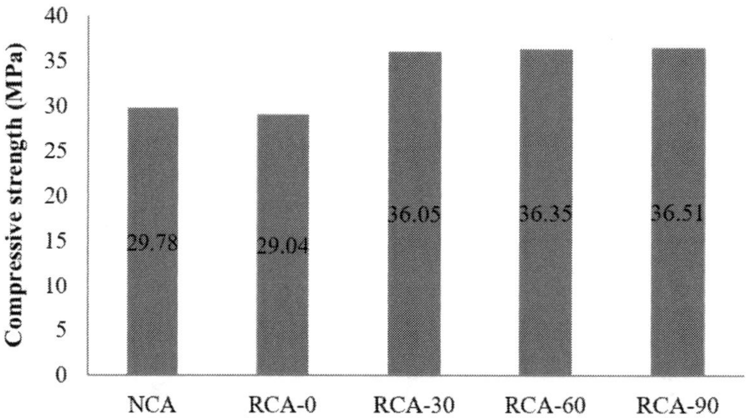

Figure 8. Concrete compressive strength

4. Conclusion

The physical properties of recycled coarse aggregate meet the aggregate quality standard based on the national code. The slump value of concrete with recycled coarse aggregate shows a lower slump than the slump of normal concrete. The decrease is seen significantly due to vibratory added mixing concrete. The heat released by the vibrator is assumed to cause a large reduction in the slump value. The use of recycled coarse aggregate in the concrete mixture reduces the strength of concrete by 2.5% compared to normal coarse aggregate concrete. However, the application of vibratory during mixing enables to improve the recycled concrete compressive strength effectively by around 20% compared to the compressive strength of normal concrete. Furthermore, an addition of either mixing water or superplasticizer is recommended to maintain the good workability level of recycled concrete made with vibratory added mixing.

References

[1] Tjokrodimuljo K 2007 Concrete Technology (in Indonesian).
[2] Mindess S 2019 1 - Sustainability of concrete 3–17.
[3] Mehta P K and Monteiro P J M 2014 Concrete: microstructure, properties, and materials.
[4] Woodson R D 2009 Concrete structures: protection, repair and rehabilitation.
[5] Chen W *et al.* 2019 Adopting recycled aggregates as sustainable construction materials: A review of the scientific literature **218** 483–496
[6] Behera M, Bhattacharyya S K, Minocha A K, Deoliya R and Maiti S Recycled 2014 aggregate from C&D waste & its use in concrete – A breakthrough towards sustainability in construction sector: A review **68** 501–516
[7] Kabirifar K, Mojtahedi M, Wang C and Tam V W Y 2020 Construction and demolition waste management contributing factors coupled with reduce, reuse, and recycle strategies for effective waste management: A review **263** 121265
[8] Pacheco-Torgal F, "1 - Introduction to advances in construction and demolition waste," in Advances in Construction and Demolition Waste Recycling Pacheco-Torgal F, Ding Y, Colangelo F, Tuladhar R and Koutamanis A 2020 1–10.
[9] Villoria-Sáez P, Porras-Amores C, and del Río Merino M "2 - Estimation of construction and demolition waste," in Advances in Construction and Demolition Waste Recycling, Pacheco-

Torgal F, Ding Y, Colangelo F, Tuladhar R and Koutamanis 2020 13–30.

[10] Xiao J, Li J and Zhang C 2005 Mechanical properties of recycled aggregate concrete under uniaxial loading **35** 1187–1194.

[11] Kenai S 2018 "3 - Recycled aggregates," in *Waste and Supplementary Cementitious Materials in Concrete*, Siddique R and Cachim P 79–120.

[12] Kencanawati N N, Iizasa S and Shigeishi M 2013 Fracture process and reliability of concrete made from high grade recycled aggregate using acoustic emission technique under compression **46** 1441–1448

[13] Narahara S *et al.* 2007 Evaluation of concrete made from recycled coarse aggregates by pulsed power discharge **1** 748–751

[14] Bluhm H 2006 Pulsed Power Systems.

[15] Kencanawati N N, Akmaluddin A, Merdana I N, Nuraida N, Hadi I R and Shigeishi M 2017 Improving of Recycled Aggregate Quality by Thermal-mechanical-chemical Process **171** 640–644

[16] Kencanawati N N, Fajrin J, Anshari B, Akmaluddin A and Shigeishi M 2015 Evaluation of High Grade Recycled Coarse Aggregate Concrete Quality Using Non- Destructive Testing Technique **776** 53–58

[17] SNI 03-2834-2000 2000 Indonesia National Standard Code: Procedure for Making Normal Concrete

[18] SNI 1974:2011 2011 Cara uji kuat tekan beton dengan benda uji silinder

[19] SNI 03-1750-1990. 1990 Indonesian National Standard: Aggregate for Concrete, Quality and Testing Methods

Sustainable principles in designing Lombok's Beachside Area: a Gading Beach case study

G A P P Kamase[1], R S Saptaningtyas[1] and T Handayani[1]

[1]Architecture Department, Faculty of Engineering, University of Mataram, Jalan Majapahit no. 62, Mataram, Indonesia

giska_ayu@unram.ac.id

Abstract. Recognized as a global best halal tourism, Lombok in West Nusa Tenggara (NTB) is anticipating significant tourism growth. As a response to this, the local government are trying to accelerate the supporting facilities for tourism development, including the Gading Beach area. On contrary, however, buildings are known among the world's biggest energy consumption. Therefore, the design proposal of these new buildings in the Gading Beach area ought to consider a holistic sustainable approach to create buildings that will encourage environmental, social and economic benefits. Five key sustainable design principles were utilized for the development of the Gading Beach area. Those principles are including site and ecology, community and culture, health, materials and energy. The implementation of these design strategies in the Gading Beach area is expected to exhibit an effective application of a sustainable approach in beachside buildings.

1. Introduction

In 2016, Lombok in West Nusa Tenggara (NTB) was recognized as the best halal tourism globally [1]. In addition, Lombok also scored the highest point in Indonesia Muslim Travel Index (IMTI) 2019 [2], where access, communication, environment, and services (ACES) are used as the main assessment criteria. These awards indicate the enormous potential of Lombok tourism, as well as the high demand for supporting infrastructures in the coming years as a response to this development.

One of the local government's attempt to this matter was to improve several tourism spots across Lombok, including Gading Beach. The design must be developed to accommodate various tourism activities, as well as to provide a unique and attractive experiences. Buildings, on the other hand, are also known as the world's biggest carbon emission contributor [3,4,5]. More than 30% of global energy was allocated for buildings [6,7,8], from construction to demolition process, including the transportation of building materials, maintenance, and daily operation [3].

Designing a beachside area is also significantly distinct from another area due to its particular soil structure, possibilities to abrasion, climatic condition and humidity, including its typical windy condition. Accordingly, the design process of the Gading Beach area should be undertaken comprehensively by considering the site's characteristics and appropriate materials selection. The design should focus not only on the relationship between the user and the building but also on the interaction between the building and the environment. Therefore, the design proposal of these new buildings in the Gading Beach area will be using sustainable design principles as guidelines.

2. Literature review

2.1. The current issues
6.44 billion tons of $CO2$ were released into the atmosphere, data in 2002 shows [9]. Due to human activities, the current impact of global warming is inevitable, and it occurs at an accelerating rate. Buildings are among the biggest contributor to carbon emissions that lead to this condition [3,4,5] as they contribute to more than 30% of global carbon emissions [3,6,7,8]. As architects, therefore, it is essential to address energy use in the building by designing a low energy building while ensuring occupants' comfort and well-being. However, what does it mean? Is it a realistic approach? More will be discussed below.

2.2. Sustainable design is the future
Currently, there is no single established definition of a sustainable building [7]. In some literature, it is described as a healthy building that was designed to reduce negative environmental impacts [10,11]. However, some argue that sustainable design should not only take environmental aspects into consideration [12]. It, instead, should involve a more pragmatic, holistic and comprehensive approach that oriented for the future [13,14,15,]. To be successful, a sustainable approach should integrate energy efficiency, climate, renewable resources and the users into the design [4,7], while considering the entire building life cycle [7,13]. These principles are expected to improve occupants' comfort, health, well-being, and productivity [4,14] and leads to a lower operating cost in building [5]. This condition indicates that the sustainable design approach is having a prominent value growth for the upcoming years and beyond [16].

2.3. Sustainable design principles
Five key principles in sustainable design described as follows.

2.3.1. Site and ecology
To date, nature plays a crucial part in fulfilling basic human needs as well as enhancing their quality of life in several ways. Natural environments influence human's life at the psychological and physical level as they are able to provide shading, shelter, and cooling, while also purifying water and air that is essential to human beings [17]. During the design process, it is necessary to consider building shape, orientation and cross-ventilation to ensure that the buildings gain enormous benefits that the natural environments provided.

A building can reduce energy use for room conditioning by 10-20% only by organizing an effective shape and orientation [13]. Without compromising occupants' comfort, there is also potential to reduce mechanical ventilation by 70% by making effective use of a cross-ventilation system [13]. To minimize the negative impacts of the built structure on the environment, therefore, a design that responded to its site plays a significant role in a sustainable approach [7]. In addition, whenever possible, mixed-used development should be taken into consideration to achieve efficient land use while providing an attractive and more engaged development [17].

2.3.2. Community and culture
A sustainable building is not only assessed through environmental objectives, but also economic and social goals [8], which include community participation. Local knowledge from the community is crucial during the development process as it helps to define their needs, concerns, and current issues [17]. However, community participation should go beyond the design and completion process of the building. By giving them control to manage the facilities, they will develop a sense of ownership that results in a more responsible guardianship. Moreover, involving the community during the development process helps to establish a fairer society which often leads to a more productive and collaborative contribution [17].

2.3.3. Health

Occupants' well-being is a vital aspect of a sustainable approach. Designing a healthy building, therefore, is essential, not only to prevent negative physical and psychological impacts but also to provide an enjoyable space for occupants which can uplift their spirit. Temperature, air movement and air quality, humidity, noise, and lighting are among many parameters that can help to define comfort [17]. Occupants can be benefited from the connection between buildings and nature. For instance, a space with sufficient natural daylight encourages users to be more active and positive, while a darker space often leads to depression [17]. However, daylight amount in a building should be carefully designed since excessive natural daylight may lead to glare discomfort while damaging occupants' eye and skin [17].

2.3.4. Materials

Annually, building industries consume 3 billion tons of raw materials [18] which account for 40% of total worldwide use [18,19]. Materials play a robust role not only in shaping a building's buildability, aesthetics, and cost [17], but also to ensure occupants' health and well-being [13]. Accordingly, the material selection process is essential. Material resource, manufacturing process, energy, transport, longevity, and waste amount are the key aspects to consider when selecting materials for sustainability [4,17] as well as its life cycle [13]. Integrating all of these aspects into a holistic design approach can reduce negative environmental impacts. All things considered, there is an urge to use low-carbon or low-impact materials from the outset of the project [5,13].

2.3.5. Energy

To achieve an energy-efficient building, a three-stage design approach should be employed. The priority is to provide a low-energy design solution, followed by the utilisation of efficient appliances and the application of renewable energy [3,17]. These methods must be undertaken in this specific order, otherwise, it will be inefficient [3]. Passive measures are best for achieving energy efficiency in building, particularly in a tropical climate. These include building orientation, sufficient daylight, and the integration of natural ventilation [17]. Building orientation in respect to the sun, the wind, and the site characteristics help to provide adequate natural light into the building while reducing the undesirable solar gain [17]. Cross-ventilation, in addition, is also the most energy-efficient way of providing fresh air which can help to cool the building [17].

Another efficient method to reduce energy use is through the integration of landscape elements. Known as the best natural shading, trees can reduce light varying from 10-70% depending on its leaves condition [20]. It is also able to reduce electrical ventilation on building up to 50% when its shading and cooling feature are combined [21]. Integrating softscape elements into the design, not only can protect buildings from direct sunlight and help to control microclimate but can also help to reduce noise, improve privacy, enhance habitats for flora and improving aesthetic quality [17].

3. Methodology

This study aimed to find the most effective design for Gading Beach, Lombok using a sustainable approach. The methodology used in this design is including site visit and survey, discussion with the local community and stakeholders, literature studies, and comparative studies. Site visit and survey records the existing features of the site, including microclimate, topography, and existing landscape. By doing so, it helps to gain a comprehensive understanding regarding site condition, including its strengths, weaknesses, and possible opportunities for the design. All this site knowledge is complemented by a discussion with the local community and stakeholders since they are able to provide specific information regarding their concerns, issues, needs and priorities. Then, an assessment can be made of what kind of method is suitable for Gading Beach development.

All data and information related to the design case are collected and assessed during literature studies. The resources are including research papers, articles, and related websites. In addition, a study on the existing building design, particularly ones with the sustainable design approach, also gives an overview of the practical applications in the field. The combination of site survey, literature and comparative studies help to establish the most appropriate approach that can be integrated into the design.

4. Result and discussion

As suggested in the previous section, sustainability is a holistic approach that integrates many aspects into the design. It not only focuses on minimizing ecological footprints during a building's life cycle but also improving the user's psychological and physical well-being. To achieve these purposes, therefore, the design of the Gading Beach area ought to consider the local climate conditions, including five sustainable aspects below.

4.1. Local climate conditions

Located in the south area of Mataram, West Nusa Tenggara, the local climate in the Gading Beach area is similar to most tropical climate areas in Indonesia. Throughout the year, the climate generally has two seasons, the dry and rainy seasons, which each season requires specific design approach of a building. As a response to a high rainfall intensity, all buildings in the area were designed with a tilted roof, ranging from 10° – 15°. In addition, the temperature in the area ranges from 21° – 34° C with extensive sunlight all around the year.

Regarding sunlight intensity, particularly in tropical climate, west and east side are two orientations that best avoided. In the design of the Gading Beach, however, the west side is the main attraction of the area. It is where the beach, the sunset and Mount Agung are located hence this orientation offers the best view, beautiful scenery, and sunset for visitors. Therefore, site plan in figure 1 shows that most buildings are oriented to this side.

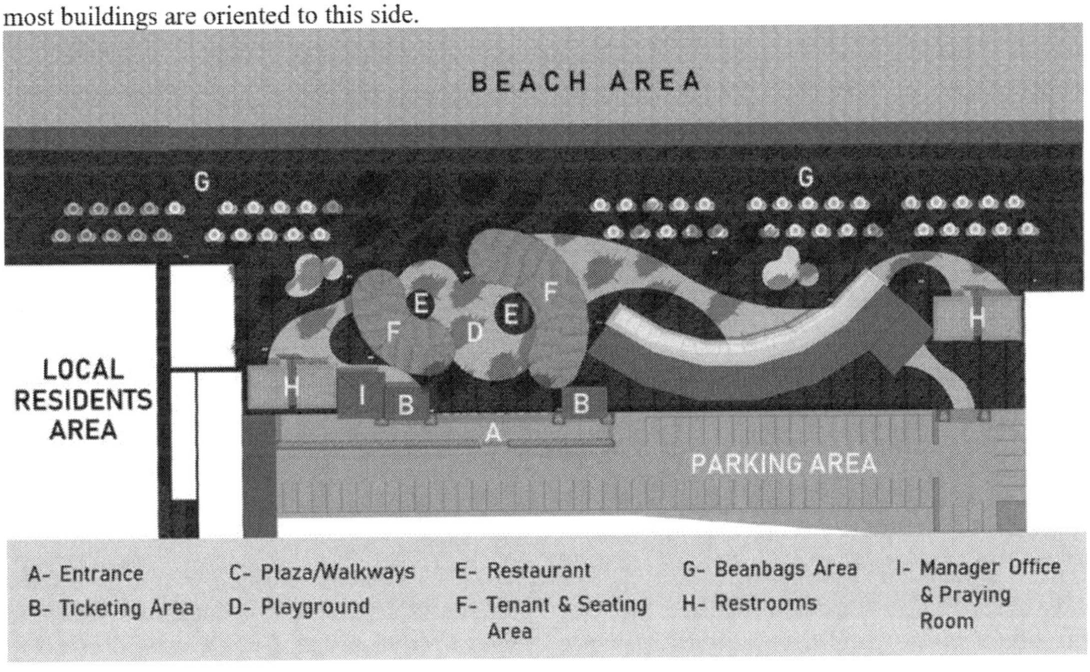

Figure 1. Building masses on the site that accommodate various purposes.

Despite the view, it has to offer, the building orientations to this side are susceptible to glare and visual discomfort for occupants. Therefore, to reduce glare and sunlight intensity, the main building was designed with a curved shape to avoid perpendicular orientation to the west side as shown in figure 2a. Figure 2a and 2b illustrates the two-meters wide shading that the main buildings have on all sides. Combined with the shades created from the adjacent landscapes, these aspects help to reduce the sunlight intensity significantly.

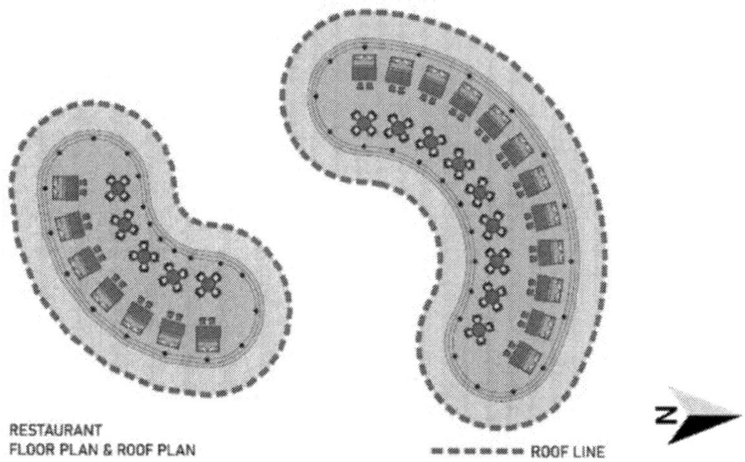

Figure 2a. Restaurant floor plan and roof plan. The floor plan is showing the curved-shape of the building while the roof plan is showing the two-meters wide shading that surrounding the building.

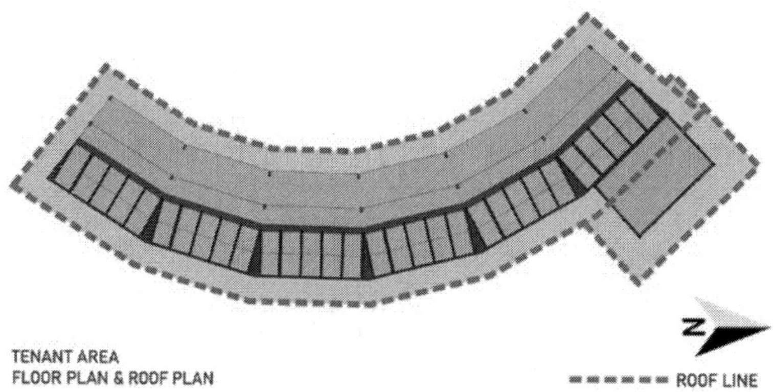

Figure 2b. Tenant and seating area floor plan and roof plan. The roof plan is showing the two-meters wide shading that surrounding the building.

4.2. Site and ecology

To establish a comfortable built environment, the mutual benefits between the development site and its natural environment needs to be well-considered. Figure 1 shows the relationship between site and the designated buildings on the site. Mixed-used development was used in the Gading Beach design to encourage efficient land use while generating an attractive development. The designated area was defined to accommodate various purposes, from refreshing area to culinary to water sports and playground. These various purposes were categorized into several buildings, where each building mass was located into different zoning on the site. This massing plan promotes improvement on pedestrians while conserving the existing trees on the site. Hence, the natural habitats for wildlife on the site can be protected. New vegetation was also introduced to enhance the existing landscape. Enough vegetation on the site will serve not only environmental benefits but also aesthetic, psychological and physical benefits.

IOP Conf. Series: Earth and Environmental Science **847** (2021) 012005 doi:10.1088/1755-1315/847/1/012005

4.3. Community and culture
Residents and local government involvement are one of the sources of input during the development of the Gading Beach area. Several informal discussions with their representatives were undertaken to gather information regarding their priorities, concern, and expectations. This involvement not only helps to increase local people awareness about the development plan but also give them a sense of ownership. In the future, after the completion of the buildings, the government are expected to collaborate with the residents to manage the whole area. When buildings become part of the community's, they have been proven to be more economically and socially sustainable because the community has gained a responsible guardianship to the area. As a result, the Gading Beach area can be a significant asset for West Nusa Tenggara, particularly Mataram, for many years to come.

4.4. Health
Designing for occupants' comfort and well-being is not a luxury. It, instead, a necessity to establish a healthy built environment. As shown in figure 3, the main idea of the Gading Beach design was to provide contact with nature by creating a vague distinction between indoor and outdoor spaces. This idea is translated into the buildings, where the use of massive walls is minimized. As a result, it allows each building mass to have sufficient access to natural daylight and ventilation throughout the day. This design also encourages air movement, improve the air quality, and help to achieve a comfortable temperature. All these parameters are essential in creating a healthy environment.

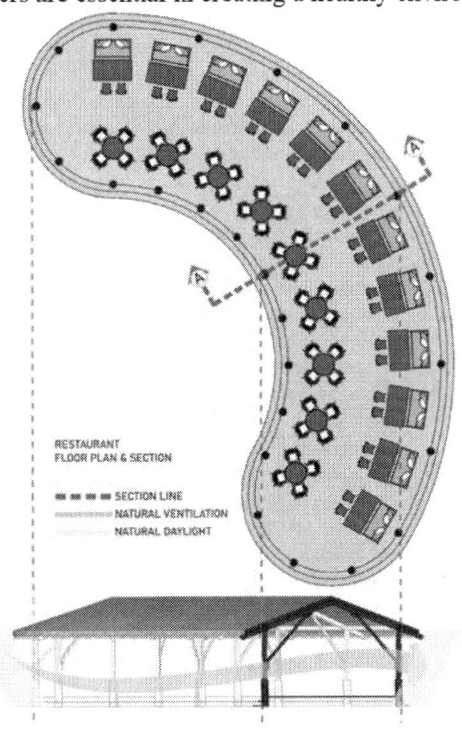

Figure 3. The floor plan and section of the restaurant.

Both images indicate the open design where no massive walls. This design promotes connection to nature and surrounding view while enhancing occupants' health and comfort by providing sufficient natural daylight and ventilation.

On other hand, the minimum use of massive walls also provides limitless visual access to the surrounding views. Combined with multiple options of seating area as illustrated in figure 4a and figure 4b, the visitors are expected to find it enjoyable and delightful as they are more connected to nature. Accordingly, it will uplift their spirit and benefitted them psychologically.

Figure 4a. Bean bag seating area as one of the seating options in the area

Figure 4b. Multiple options of seating area helps to create a more enjoyable spaces which help to improve visitors' happiness.

4.5. Materials

To comply with the initial target to maximize the use of natural resources, buildings in the Gading Beach area are mostly constructed using wood as indicated in figure 5 and figure 6. Wood is selected due to several reasons. First, it is renewable and natural. Second, it is locally available in the area, thus, the energy used for transporting materials to the site can be significantly reduced. Third, it can be easily disassembled at the end of the building's life which will increase the opportunity for reusability and recyclability. Accordingly, there will be a waste reduction. Forth, it is also durable and suitable for beachside microclimate. All these factors contribute to minimizing the embodied energy produced throughout the buildings' life cycle.

Figure 6 also illustrates that recycle materials were also utilized in several areas for finishing and interior materials, for instance, food court cladding and dining area tables, also restroom cladding, and children playground. There are many several benefits to doing this, such as energy reduction for resourcing and manufacturing process, waste reduction, and cost reduction. Moreover, each of these recycled materials has specific visual characteristics that can be aesthetically appealing.

Figure 5. Wood as the primary material for the buildings in the Gading Beach area.

Figure 6. Wood as primary material combined with the recycle wood for food court cladding.

4.6. Energy

As discussed above, the buildings in the Gading Beach area were designed to relate to nature as seen in figure 7. This approach is not only beneficial in creating a healthy and attractive environment, but it also helps to establish an energy-efficient building. Figure 8 illustrates that minimum use of massive walls provides sufficient daylight and promotes cross-ventilation that will passively cool and ventilate the building.

Figure 7. Massing design that allows contact with nature.

Figure 8. The minimal use of massive walls to create connection between spaces.

The bird eye view perspective in figure 9 shows that all buildings in the Gading Beach area were designed to be slim and break into different massing. This approach to ensure that each building has sufficient access to natural resource. Combined with the ability of vegetations on the site in providing natural shading, the energy used for electrical lightings and mechanical cooling can be significantly reduced while still ensure occupants' comfort.

Figure 9. Bird eye perspective of the Gading Beach area.

5. Conclusion

Buildings are one of the biggest contributors to global energy use [3,4,5,6,7,8] have potential lives spanning many years to come and affecting future generations. To support tourism, many beachside areas are being developed, particularly in Lombok. This development involves the construction of new buildings. Therefore, a comprehensive approach needs to be undertaken during the design process of the new building. Previous studies show that sustainable design is a holistic and comprehensive approach [13,14,15] that will not only environmentally beneficial [11,12] but will also enhance occupants' health, comfort, and well-being [4,10,14,17].

Site, community, health, materials, and energy are five key principles of sustainable design [17] that was implemented into the design of the Gading Beach area. Specifically, these aspects include climate, building shape and orientation, solar protection, renewable resources, and the users into the design process [4,7,13]. Therefore, this study is expected to illustrate a comprehensive application of sustainable principles in building, particularly in the beachside area.

This research, however, has many limitations and needs to be improved in the future. The sustainable design strategies described above have not been evaluated using the assessment tools/criteria. Therefore, an evaluation tool needs to be formulated to help to assess the success level of sustainable buildings in the future.

Acknowledgments
This research was supported by Mataram Research and Development Department (Balitbang). We thank Mataram Regional Development Planning, Research and Development (Bappeda), Mataram Department of Tourism, Mataram Department of Fisheries and Gading Beach local communities for their valuable inputs during the development of this work.

References
[1] World Halal Tourism Award 2016
[2] Indonesia Muslim Travel Index (IMTI) 2019
[3] Pelsmakers S 2015 *The Environmental Design Pocketbook* (London: RIBA Publishing) 2nd Edition
[4] Mehta G, Mehta A and Sharma B 2014 Selection of materials for green construction: A review *J. of Mech. and Civ. Eng.* **11** 80–83
[5] Mokal A, Shaikh A, Raundal S, Prajapati S and Phatak U 2015 Green Building Materials: A Way Towards Sustainable Construction *Inter. J. of Appl. or Innov. in Eng. and Manage.* **4** 244–249
[6] IPCC 2001 *Climate Change 2001 Synthesis Report: The Third Assessment* Report (Cambridge: Cambridge University Press)
[7] Kotkar A and Salunkhe H 2017 A review paper on green building research *Inter. J. of Adv. Res. in Sci. and Eng.* **6** 901–906
[8] Bonde M 2016 *Green Buildings: Exploring Performance and Thresholds* (Stockholm: KTH Royal Institute of Technology)
[9] Worldwatch Institute 2003 *State of the World: Progress Towards a Sustainable Society* (London: Earthscan) 20th Edition
[10] Kibert C 2008 *Sustainable Construction: Green Building Design and Delivery* (Hoboken: Wiley) 2nd Edition
[11] RICS 2009 *Sustainability and Commercial Property Valuation* (Reading: Columns Design Ltd) 2nd Edition
[12] Berardi U 2013 Clarifying the new interpretations of the concept od sustainable building *Sustain. Cities and Soc.* **8** 72–78
[13] Bauer M, Mosle P and Schwarz M 2010 *Green Building: Guidebook for Sustainable Architecture* (New York: Springer)
[14] Parashar A and Parashar R 2012 Construction of an eco-friendly building using green building approach *Inter. J. of Sci. and Eng. Res.* **3** 1–7
[15] Girarder H 1999 *Creating Sustainable Cities* (Totnes: Green Books)
[16] Lorenz D and Lutzkendorf T 2011 Sustainability and property valuation: Systematisation of existing approaches and recommendations for future action *J. Prop. Invest. Finance* **29** 644–676
[17] Sassi P 2006 *Strategies for Sustainable Architecture* (New York: Taylor & Francis)
[18] Hamner B 2007 Effects of green purchasing strategies on supplier behaviour *Green. the Supp. Chain* **2** 25–37
[19] Carter C 1996 Inter-organizational atecedentsand determinants of environmental purchasing: Initial Evidence from the Consumer Products Industry *Dec. Sci.* **29** 659–684
[20] Thomas R 1999 *Environmental Design* (London: E & FN Spon)
[21] Foster R 1994 *Landscaping That Saves Energy and Dollars* (Old Saybboork: Globe Pequot Press)

Effectiveness *Vetiveria zizanoides* plant in ability absorption of heavy metals in coal mining waste

A Andrawina[1], R Ernawati[1], T A Cahyadi[1], W S Bargawa[1], N A Amri[1], and I Ferdian[2]

[1]Department of Mining Engineering, UPN Veteran Yogyakarta, Yogyakarta, Indonesia
[2]Environmental Management Department, PT. Bukit Asam, Tbk, Tanjung Enim, Indonesia

andrawina17@gmail.com

Abstract. *Vetiveria zizanoides* plants have considerable potential for phytoremediation, because these plants show relatively high ability to accumulate and translocate heavy metals in wetlands and are able to tolerate acidic water pH. The results of the study showed that the *Vetiveria zizanoides* plant has a fairly good survival rate in acid mine water conditions and has a fairly high ability to absorb Fe, Mn and sulfur metals. This is evidenced by the treatment of *Vetiveria zizanoides* plants by planting 20 *Vetiveria zizanoides* stems into acid mine water media without the addition of organic substrates, seen on day 20 after treatment of 20 *Vetiveria zizanoides* stems there was only 1 plant that died, the rest still survive in acidic water conditions. In this treatment, the *Vetiveria zizanoides* plant was able to absorb Fe, Mn and sulfur with an efficiency increase of 93% for Fe metal, 40% for Mn metal and 43% sulfur for 20 days. However, to improve the quality of acid water in this treatment, it was not optimal, so treatment was given with a combination of *Vetiveria zizanoides* and organic substrates. Treatment 3 was the most effective in improving acid mine water quality, because it had an average effectiveness of 58% for increasing the pH of the air and 97% for decreasing the concentration of Fe and Mn metals. With the efficiency of increasing the metal concentration in plants by 16% for Fe metal, 29% for Mn metal and 35% sulfur for 20 days after being given treatment.

1. Introduction

Aquatic plants have an important role in geochemical processes in artificial wetlands based on the active and passive transport of their elements [1]. The process of absorption of heavy metals from soil and sediment through plant roots is known as phytoremediation. Phytoremediation techniques can be a good choice in wastewater management systems because they are relatively inexpensive and friendly as a suitable treatment option for developing countries [2].

Plants have a very important role in removing pollutants. The absorption of heavy metals by plants in the wetland system is leaves and roots. Submerged rooted plants have the potential to extract metals from sediment and water, whereas rootless plants can only extract metals from water [3]. The rate of metal removal by plants varies widely, depending on the rate of plant growth and the concentration of heavy metals in plant system. Metal uptake rates per unit area of wetland are often much higher for herbaceous plants. In the case of uptake heavy metals in leaves, it is passive movement in the water phase through cracks in the cuticle or through the stomata to the cell wall and then the plasmalemma [3].

Sediment pH, the context of organic matter and plant genotype, can have a marked effect on metal uptake. Metal bioaccumulation is based on root uptake but also plants can accumulate relative amounts of metals by uptake of leaves from atmospheric deposits in plant leaves [3].

Phytoremediation has a role in a constructed wetland system by using water plants to accumulate metals in water and soil. The use of plants that are able to associate with bacteria or mycorrhizae will greatly assist in the process of managing acid mine drainage. With the right design, a passive treatment system can have a lifespan of more than 20 years which is useful for reducing the acid mine drainage is also very important to support rehabilitation and reclamation activities of ex-mining areas [4].

Phytoremediation is an alternative in conventional wastewater treatment in developing countries, as suggested by previous researchers (Ghosh and Singh, 2005). Successful application of the technique depends on identifying plant species with the appropriate set of characteristics [5]. In assessing the phytoremediation potential of a species, researchers often emphasize the very high biomass production, bioaccumulation, and tolerance to heavy metals by plants [6]. Hyperaccumulator plants with the capacity to hold very high amounts of heavy metals (usually 0.1% to 3% of dry weight) are even more desirable. In this regard, the *Vetiveria zizanoides* plant shows considerable potential as a candidate for phytoremediation, as it exhibits relatively high accumulation and translocation capabilities of heavy metals in artificial wetlands. In addition, this plant is also able to tolerate low water pH as well as the influence of dissolved metal concentrations such as iron, manganese, and sulfur. This can be seen from during the 20 days of the research process, there were still some plants that had no signs of wilting.

2. Materials and methods

This study used an experimental method by conducting an experiment planting *Vetiveria zizanoides* in an experimental reactor. The materials used in this research are:

a) Plastic tub used as experimental adhesive;
b) Acid mine drainage;
c) Soil as a media planting;
d) The organic material used is bokashi fertilizer;
e) Aerators are used to provide treatment so that the water experiences movement (only used in ponds with treatment 4).

The experimental reactor has dimensions of 60 cm x 47 cm x 40 cm with the experimental design used consisting of 1 control pond and 3 ponds with different treatments which can be seen in (Figure 1). The treatment arrangement in this experiment was a combination of organic substrates and *Vetiveria zizanoides* consisting of:

a) Without treatment (control), the amount of soil is 10 kg and the volume of Acid Mine Drainage (AMD) is 28 liters.
b) Treatment 1 (AMD + soil + 20 stems of *Vetiveria zizanoides*), with 10 kg of soil and 28 liters of AMD volume.
c) Treatment 2 (AMD + soil + organic substrate + 20 stems of *Vetiveria zizanoides*), with 10 kg of soil, 10 kg of organic substrate and 28 liters of AMD volume.
d) Treatment 3 (AMD + soil + organic substrate + 20 stems of *Vetiveria zizanoides*), with 10 kg of soil, 10 kg of organic substrate and 28 liters of AMD volume. In treatment 4, the experimental pond was given an aerator which functions to provide treatment of water movement in the pond.

Observations were made on the development of *Vetiveria zizanoides* plants for 20 days after being given the treatment. Then an analysis of the effect of *Vetiveria zizanoides* on the improvement of the quality of acidic water in the experimental reactor was also carried out for 20 days. Then on the 20th day after giving the treatment, samples of *Vetiveria zizanoides* were taken for laboratory analysis. With the parameters tested include the metal content of iron, manganese and sulfur.

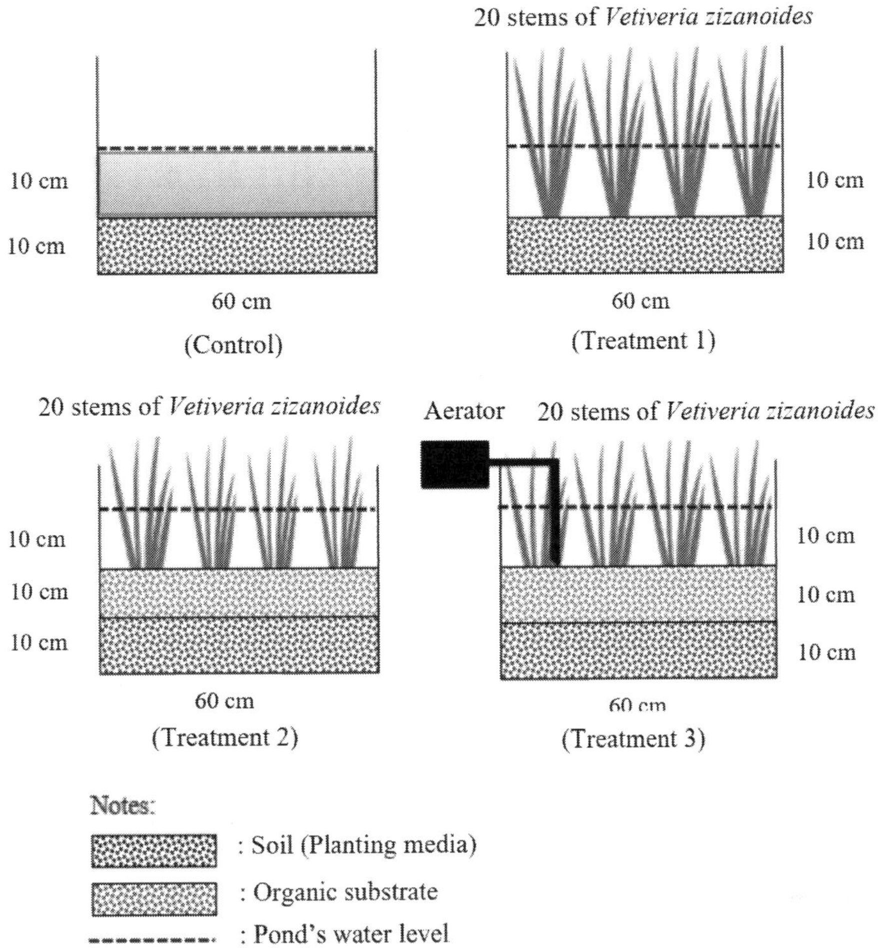

Figure 1. Experimental reactor design

3. Results and discussion

3.1. Vetiveria zizanoides plant survival

At the time of the study, it was seen that in treatment 1 starting on the 12th day after the plants were put in water there were 14 *Vetiveria zizanoides* stems experiencing signs of wilting and 1 stem experiencing death, while on the 20th day there were 17 stems experiencing signs of wilting and 1 stem experienced death. So, from a total of 20 *Vetiveria zizanoides* stems given in the first treatment, only 2 stems survived without any signs of wilting. The death of *Vetiveria zizanoides* plants in treatment 1 was due to the tolerance limits of plants to heavy metals such as iron and manganese which were absorbed by *Vetiveria zizanoides* so that they affected the condition of plants.

In treatment 2, it began to appear on the 12th day after the plants were put in water, there were 14 stems experiencing signs of wilting, then on the 20th day there were 19 stems withering. So that out of a total of 20 plants given in this second treatment, only 1 stem still survives without any signs of wilting. Whereas in treatment 3, it was seen that on the 12th day after the plants were put in water, there were 16 stems that had wilted, then on the 20th day there were 17 stems that had wilted. So that from a total of 20 plants given in this 3rd treatment, only 3 stems still survive without any signs of wilting.

In treatment 2 and 3, *Vetiveria zizanoides* were able to survive because of the availability of nutrients derived from organic matter. Another thing can also be caused by the stimulation of the growth of sulfate-reducing bacteria from organic substrates which helps in accelerating the deposition of heavy metals so that the heavy metal concentrations in the 3rd and 4th treatments are not too high. *Vetiveria zizanoides* survival conditions can be seen in Figure 2 and Figure 3.

(a) Treatment 1 (b) Treatment 2 (c) Treatment 3

Figure 2. The condition of *Vetiveria zizanoides* on the 0th day after treatment

(a) Treatment 1 (b) Treatment 2 (c) Treatment 3

Figure 3. The condition of *Vetiveria zizanoides* on the 20th day after treatment

3.2. Effectiveness of Vetiveria zizanoides in heavy metal absorption

Plants are an important component in the remediation process in wetland environments. The role of plants is to provide sites for microbial attachment, remove oxygen from their roots and provide a source of organic material for heterotrophic microbes. So that the reduction and oxidation processes of sulfate, iron and manganese will be more prevalent in the root area because of the high abundance of microorganisms, including reducing bacteria [7].

In this study, laboratory testing was carried out on *Vetiveria zizanoides* plants to determine the effectiveness of the ability of *Vetiveria zizanoides* in absorption of heavy metals (iron, manganese and sulfur) in the management of acid mine drainage. The results of laboratory analysis of plants before and after treatment can be seen in Table 1.

Table 1. Concentrations of sulfur, iron, and manganese in *Vetiveria zizanoides*.

Type of Testing	Test Result			
	Before Treatment	Treatment 1	Treatment 2	Treatment 3
Iron (ppm)	552.10	8,230.51	710.51	657.35
Manganese (ppm)	139.00	232.85	168.43	195.65
Sulfur (%)	0.33	0.58	0.50	0.51

From Table 1 above, it can be calculated the effectiveness of increasing the concentration of iron, manganese and sulfur metals. A measure that states how far the target (quantity, quality and time) has been achieved is called effectiveness [8]. The greater the percentage of targets achieved, the higher the effectiveness. Therefore, in analyzing the effectiveness of increasing the concentration of iron, manganese and sulfur metals. can use the following equation [9].

$$\text{\% Effectiveness Increase} = 1 - \frac{Ca}{Ct} \text{ x } 100\ \% \quad (1)$$

$$\text{\% Decreasing effectiveness} = 1 - \frac{Ct}{Ca} \text{ x } 100\ \% \quad (2)$$

Description:
Ca = initial concentration
Ct = final concentration

Table 2. Average effectiveness of increasing iron, manganese and sulfur concentration in *Vetiveria zizanoides*.

Type of Testing	Test Result		
	Treatment 1	Treatment 2	Treatment 3
Iron (%)	93%	22%	16%
Manganese (%)	40%	17%	29%
Sulfur (%)	43%	34%	35%

The results showed that the concentration of iron, manganese and sulfur was higher in *Vetiveria zizanoides* plants found in treatment 1. That is, with an increase in efficiency of 93% for ferrous metals, 40% for manganese and sulfur metals by 43% for 20 days after being given. treatment. This is because in treatment 1 there is no addition of an organic substrate so that the active role in the absorption of iron, manganese and sulfur metals is only in *Vetiveria zizanoides* plants. The high concentration of metals contained in these plants causes the plants to wither and experience death.

The efficiency of increasing metal concentration in treatment 2 was 22% for iron, 17% for manganese and sulfur by 34% for 20 days after treatment. And in treatment 3 has an increase efficiency of 16% for iron, 29% for manganese and sulfur by 35% for 20 days after being given treatment. The concentrations of iron, manganese and sulfur metals contained in the *Vetiveria zizanoides* plant in treatment 2 and treatment 3 were not too high due to the influence of the organic substrate so that they can stimulate sulfate-reducing bacteria which will react with dissolved metal ions. The high concentration of iron and manganese in plant tissue is influenced by the high concentration of these elements in the organic substrate. The existence of composting with the addition of alkaline materials can be an alternative in reducing the concentration of iron and manganese [10].

3.3. Effectiveness of increasing water pH and decreasing heavy metals

The effectiveness of increasing pH and decreasing heavy metals was influenced by the treatment using *Vetiveria zizanoides* and organic substrates. Based on the results of the calculation of the effectiveness of each treatment, the average effectiveness can be seen in Table 3.

From Table 3, it can be seen that in treatment 1 by planting *Vetiveria zizanoides* in the experimental reactor without a combination of organic substrates has not been able to improve the quality of acid mine drainage. The table shows that the effectiveness of increasing the pH of water in treatment 1 is only 17%, meaning that the increase in water pH is very small and the resulting pH still does not meet the quality standards of coal mining waste. Likewise, for the effectiveness of decreasing the manganese concentration, in treatment 1 there was an increase in the manganese concentration by

49% from the initial concentration. This is thought to be due to conditions of low water pH, because manganese deposition depends on pH and Fe^{2+} concentration, where manganese will form a stable precipitate at high pH and very low Fe2+ concentration. This phenomenon can also be caused by AMD at low pH resulting in an increase in metal desorption and mineral solubility which results in an increase in the concentration of toxic heavy metals [11]. And in treatment 1 this does not use a combination of organic substrates so that the only active role in decreasing the manganese concentration is absorption by the plants themselves.

Table 3. Average effectiveness of each treatment.

Parameter	Treatment			
	Before Treatment	Treatment 1	Treatment 2	Treatment 3
pH	2%	17%	58%	58%
Iron	9%	73%	94%	97%
Manganese	9%	-49%	76%	97%

Treatment 3 = Highest effectiveness

For the results, the average effectiveness in treatment 2 and treatment 3 was not too different, but there was a significant difference in the increase in the concentration of manganese. The increase in manganese metal concentration was seen more quickly in treatment 3. This is because in this treatment the aeration process was experienced in the experimental reactor. So that it causes the supply of oxygen to the roots of the plant to take place properly. With a good oxygen supply, the development of microorganisms in this system will be better so that the decrease in manganese concentration can be seen more quickly.

In bacterial metabolism on organic substrates, microbial-driven reduction of sulfates and iron is a naturally occurring process in wetland sediments that facilitates the removal of metals from acid mine drainage through an increase in pH, which in turn results in the deposition of metals as either hydroxides or sulfides [12].

4. Conclusion

a.) In the study, it was shown that *Vetiveria zizanoides* plants can thrive in acidic water pH conditions. *Vetiveria zizanoides* is able to survive in the experimental reactor with treatment conditions 1, 2 and 3. In treatment 1, it can be seen that of the 20 *Vetiveria zizanoides* stems only 1 plant has died, meaning that the plant is still able to survive in acidic water conditions even without the influence of the substrate. organic. This proves that *Vetiveria zizanoides* is very tolerant of acid mine drainage.

b.) Treatment 2 and treatment 3 are treatments that have a high effectiveness value in managing the quality of acid coal mine water. The high concentration of sulfur, iron and manganese metals in *Vetiveria zizanoides* is influenced by the high concentration of these elements in the organic substrate. This can be seen in treatment 1, the concentration of Fe and Mn metals in *Vetiveria zizanoides* is very high compared to plants found in treatment 2 and treatment 3, because there is no addition of organic substrates in treatment 1. So that the ability of plants to absorb heavy metals is proportional to the amount of material. organics added to the growing medium.

Acknowledments

This drafting paper cannot be separated from the support of various parties, especially magister study program of Mining Engineering UPN "Veteran" Yogyakarta. The writer also expresses his gratitude to LPPM UPN "Veteran" Yogyakarta who has funded this research.

References

[1] Bose S, Vedamati J, Rai V and Ramanathan A L 2008 Metal uptake and transport by Tyaha angustata L. grown on metal contaminated waste amended soil An implication of

phytoremediation. Geoderma **145** (1-2) 136-142
[2] Ghosh M and Singh SP 2005 A review of phytoremediation of heavy metals and utilization of itsbyproducts Appl Ecol Environ Res **3**(1) 1–18
[3] Sheoran A S and Sheoran V 2006 Heavy metal removal mechanism of acid mine drainage in wetlands: a critical review Minerals engineering **19**(2) 105-116
[4] Ziemkiewicz P F, Skousen J G and Simmons J 2003 Long-term performance of passive acid mine drainage treatment systems. Mine water and the Environment, **22**(3) 118-129
[5] Massa N, Andreucci F, Poli M, Aceto M, Barbato R and Berta G 2010 Screening for heavy metalaccumulators amongst autochtonous plants in a polluted site in Italy Ecotoxicol Environ Saf **73** 1988–1997
[6] Anning A K, Korsah P E and Addo-Fordjour P 2013 Phytoremediation of wastewater with Limnocharis flava, Thalia geniculata and Typha latifolia in constructed wetlands International journal of phytoremediation **15**(5) 452-464
[7] Skousen J, Rose A, Geidel G, Foreman J, Evans R and Hellier W 1998. Handbook of technologies for avoidance and remediation of acid mine drainage National Mine Land Reclamation Center, Morgantown **131**
[8] Gobel A P 2018 Efektifitas pemanfaatan fly ash batubara sebagai adsorben dalam menetralisir air asam tambang pada settling pond penambangan banko PT. Bukit Asam (Persero) Tbk Jurnal Mineral, Energi, dan Lingkungan, **2**(1) 1-11
[9] Schermerhorn Jr and John R 1986 Management for Productivity New York John Willey & Sons
[10] Sandrawati A 2012 Management of Acid Mine Water through Artificial Swamps Based on In Situ Materials in Coal Mining (Case Study at the Sambarata Mining Site PT Berau Coal Berau Regency East Kalimantan Province) Thesis Bogor Agricultural University Postgraduate
[11] Yunus R and Prihatini N S 2018 Phytoremediation of Fe and Mn Acid Water from Coal Mine with Water Hyacinth (Eichornia crassipes) and Purun Tikus (Eleocharis dulcis) in artificial wetland systems at PT JBG South Kalimantan Sainsmat Scientific Journal of Natural Sciences **7**(1) 73-85
[12] Kalin M 2001 Biogeochemical and ecological consideration in designing wetland treatment systems in post-mining landscapes Waste Management **21** 191–196
[13] Haghollahi A, Fazaelipoor M H and Schaffie M 2016 The effect of soil type on the bioremediation of petroleum contaminated soils Journal of environmental management **180** 197-201
[14] Hallberg K B and Johnson D B 2005 Microbiology of a wetland ecosystem constructed to remediate mine drainage from a heavy metal mine, Sci Total Environ **338** 53–66
[15] Rout G R and Sahoo S 2015 Role of Iron in Plant Growth and Metabolism Review in Agricultural Science **3** 1-24
[16] Russell R A, Holden P J, Wilde K L and Neilan B A 2003 Demonstration of the use of Scenedesmus and Carteria biomass to drive bacterial sulfate reduction by Desulfovibrio alcoholovorans isolated from an artificial wetland. Hydrometallurgy **71**(1-2) 227-234

FTIR analysis of polyethylene glycol treated bacterial cellulose pellicle

J Maulana[1], H Suryanto[1,2], B D Susilo[1], U Yanuhar[3], A Aminnudin[1], Y R A Pradana[1] and R D Bintara[1]

[1]Department of Mechanical Engineering, Faculty of Engineering, Universitas Negeri Malang, Jl. Surabaya 6, Malang, East Java, Indonesia
[2]Center of Excellence for Cellulose Composite (CECCom), Department of Mechanical Engineering, Universitas Negeri Malang, Jl. Semarang 5, Malang 65145, Indonesia
[3]Department of Waters resources Management, Faculty of Fisheries and Marine Sciences, Universitas Brawijaya, Jl. Veteran Malang, East Java, Indonesia

heru.suryanto.ft@um.ac.id

Abstrak. This research identified the effect of additional polyethylene glycol (PEG) in bacterial cellulose made from pineapple peel waste. The produced pellicles were soaked in PEG solution with 0%, 1%, 2.5%, and 5%. Then, they were homogenised using an ultrasonic homogeniser for 15 minutes before dried using the freeze-drying method. The functional groups were observed using FTIR. The functional groups showed the presence of cellulose I and cellulose II at 439.08 cm^{-1}, 2895.15 cm^{-1}, 1963.53 cm^{-1} 1633.71 cm^{-1}, 1315.45 cm^{-1}, 839.03 cm^{-1} wavenumbers. Additional PEG caused an increase of the C-H bond at 1963.53 cm^{-1} wave number.

1. Introduction

Nowadays, people become aware of waste management, be it biodegradable or non-biodegradable waste. Indonesia is a country that produces the largest pineapple globally besides Thailand, the Philippines, Malaysia, Hawaii, Ivory Coast, Kenya, Brazil, Taiwan, Australia, India and South Africa [1]. High pineapple production means high pineapple peel waste. An effort to manage pineapple peel waste is by using it as a culture to develop bacterial cellulose (BC). Pineapple peel has fructose, glucose, and sucrose that can be used as bacterial cellulose cultivate.

Bacterial cellulose is a polymer network measuring 20–50 nm and has a 3-dimensional structure derived from 3-dimensional hydrogel bonds [2]. Bacterial cellulose has the same structure as plants cellulose. Besides, it also has high mechanical strength, crystallinity, and hydrophilicity [3]. It also has purity, high biocompatibility, and degraded properties [4–6]. Hence, cellulose is often applied to packaging, gas barriers, novel drug delivery systems, antibacterial membranes, composite membranes, implants, wound dressings, and magnetic films [7–14]. Bacterial cellulose can also be used as an air filter without generating non-recyclable waste.

Producing bacterial cellulose membrane for filter application requires attention in the drying method. The freeze-drying method [15–19] with a perforated surface of the specimen is suitable for air filter applications. Besides the drying process, the material as a surfactant also requires attention because it can influence the bond characteristic of bacterial cellulose and other metal. Infrared spectroscopy is a method to analyse a bond produced by a material, bacterial cellulose included. Thus, a further review of the additional surfactant effect on bacterial cellulose bond characteristic was needed using infrared spectroscopy. This research studied the effect of additional surfactant on bacterial cellulose properties using additional Poly Ethylene Glycol (PEG) 4000 with 0%, 1%, 2.5%, and 5% variations. This study aimed to observe those additional PEG as surfactant on bacterial cellulose pellicle bonds using the Fourier Transform InfraRed (FTIR).

2. Materials and methods

2.1. Materials
Pineapple peels were obtained from a plantation in Blitar, East Java, Indonesia. The bacteria to make the cellulose were *Acetobacter xylinum*, Glucose, NaOH, and Poly Ethylene Glycol (PEG) 4000, and they were obtained from CV Makmur Sejati, Malang, East Java, Indonesia.

2.2. Bacterial cellulose synthesis
300 gram of pineapple peel was washed, crushed, and extracted into juice. The extraction used 2 L of water, 10% (w/v) sugar, and acetic acid to achieve a 4.5 pH. Then, 20% of the *A. xylinum* bacteria was added into the culture before fermentation for ten days at 30°C temperature. The produced pellicles were transparent and floated on the solution. After harvesting, the pellicles were rinsed with distilled water to remove bacteria and other impurities. The pellicles were soaked in 1% NaOH for 2 hours, stirred using a magnetic stirrer at 350 rpm speed, and rinsed with distilled water until they reached 7.0 pH.

2.3. Mixing process
The celluloses were cut in 5 cm x 5 cm dimensions, soaked in the PEG solutions (0%, 1%, 2.5%, and 5%), then put into ultrasonic homogeniser for 15 minutes in 300 W power 20 kHz frequency.

2.4. Drying process
The bacterial cellulose pellicles were put into vacuum freeze-drying (Berkley Scientific, China) at 1 Pa and were condensed at -30°C for two days. The dried samples were stored at 58% humidity and 25°C temperature.

2.5. FTIR analysis
The bacterial cellulose functional group was analysed using the Fourier Transform Infrared spectrometer/FTIR (Shimadzu IR Prestige-21). The spectrums were recorded in 4 cm^{-1} resolution within the 400–4000 cm^{-1} range.

3. Results and discussion

3.1. Synthesis results
Researchers succeeded in synthesising bacterial cellulose using pineapple peel waste. Fibre cellulose, as the production of *Acetobacter xylinum* bacteria, had a different structure than plant fibre. Bacterial cellulose has a higher purity of up to 100% than cellulose in plants containing hemicellulose and lignin [20]. However, bacterial cellulose also contained impurities such as *Acetobacter xylinum* bacteria, nucleic acid, and protein from pineapple skin extract that needed to be removed using NaOH. Figure 1a shows the wet cellulose after cleaned with NaOH and cut into 5 cm x 5 cm. Figure 1b displays the cellulose after soaked in PEG 4000.

Figure 2 exhibits the plasticisation mechanism of bacterial cellulose with PEG 4000. During the drying process, the bacterial cellulose chain bond tissue increased the hydrogen bonds, limiting the relative movement between the fibres. This condition caused bacterial cellulose to have robust and brittle properties [21]. PEG 4000 not only acted as surface coverage but also penetrated the bacterial cellulose tissue. As a result, the hydrogen interactions between bacterial cellulose were reduced, and the relative movement between the fibres became freer [21].

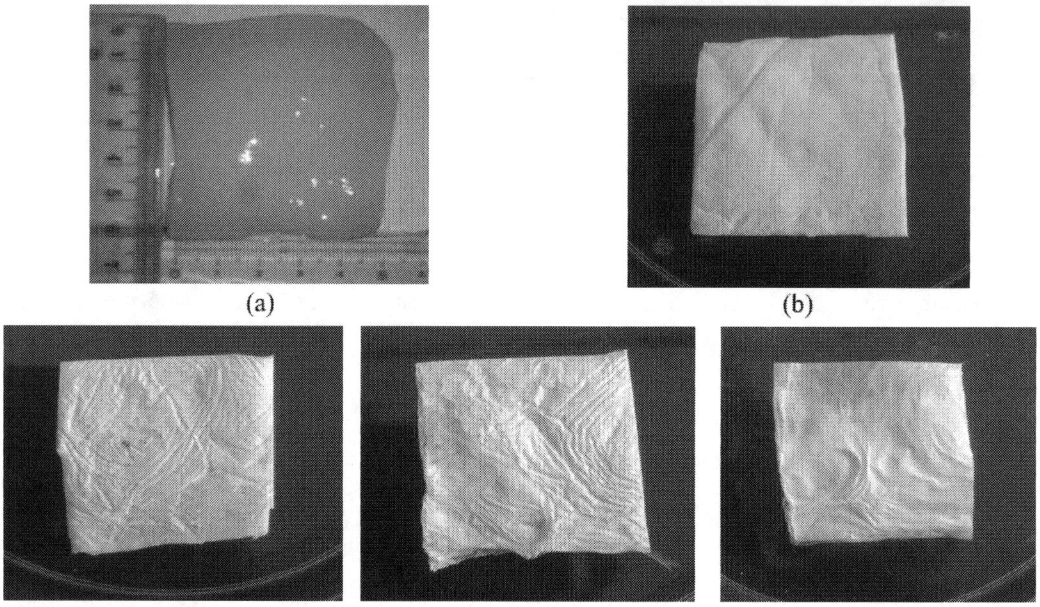

(a) (b)
(c) (d) (e)

Figure 1. Bacterial cellulose synthesis results (a) freeze dry without PEG (b) with additional 1% PEG (c) with additional 2.5% PEG (d) with additional 5% PEG

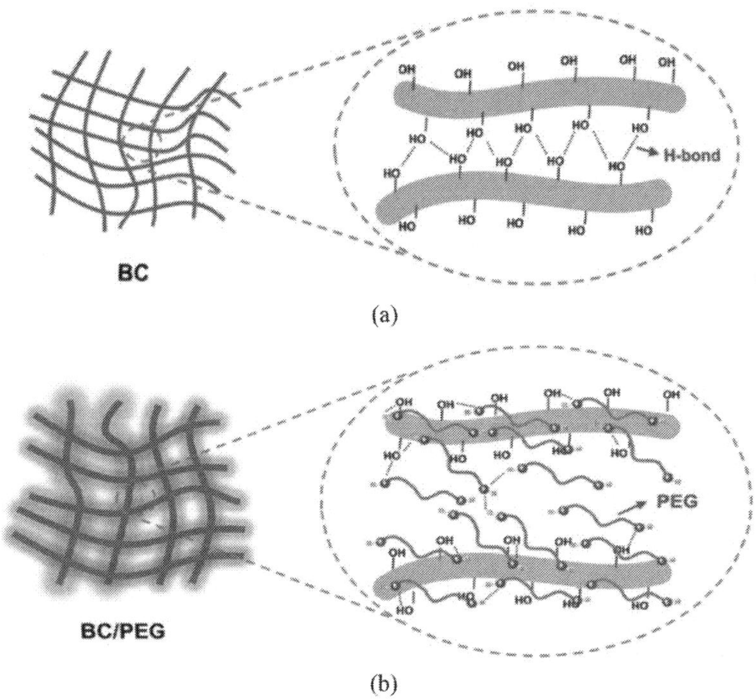

Figure 2. Bacterial cellulose tissue illustrations (a) without PEG (b) with PEG [21]

3.2. FTIR analysis

Table 1 and Figure 3 show the FTIR results of pure bacterial cellulose specimen and additional PEG 4000. Based on the images, before and after additional PEG 4000 bacterial celluloses were dominated with O-H bond at around 3439.08 cm^{-1}. It indicated a cellulose type I. There was the C-H bond at 2895.15 cm^{-1} wavenumbers, which indicated cellulose type II. There was an intensity increase of the C-H bending bond. The O-H characteristics of water absorption are also shown in the appearance of the C-O bond at 1633.71 cm^{-1}. The wavenumber of 839.03 cm^{-1} saw a transmittance peak following the percentage increase of PEG 4000.

In the O-H bond at 3459 cm^{-1} wavenumbers, the more curved graph indicated the change in intensity value. The graph identified the crystallinity of cellulose I and cellulose II on the O-H bond [22]. Following the additional PEG 4000, the hydroxyl group bond decreases. The O-H graph change between 3750–3200 cm^{-1} and intensity increase in C-H bending bond at 1936.53 cm^{-1} showed the additional PEG 4000 identity. Adding PEG 4000 did not produce a new bond but intensity improvement in some bonds. This occurrence was because the functional groups of cellulose (($C_6H_{10}O_5$)n) and PEG 4000 (H(OCH_2CH_2)nOH) consist of mostly the same bonds and do not have much difference in wavelengths [23].

Table 1. Bacterial cellulose functional group analysis

Wavenumber (cm^{-1})	Bond	Component
3439.08	O-H bond	Cellulose I and Cellulose I
2895.15	Alkanes C-H bond	Cellulose I
1963	C-H	Polyethylene Glycol
1633.71	C-O bond	Cellulose I and Cellulose II
839.03	C-H	Cellulose I

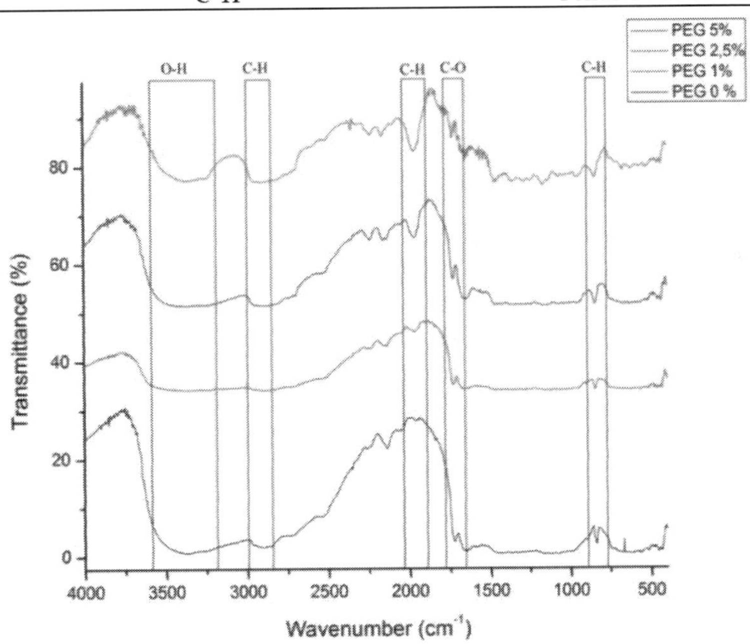

Figure 3. FTIR analysis result of bacterial Cellulose with additional PEG

4. Conclusion

Researchers succeeded in adding PEG 4000 into bacterial cellulose bond using immersion with ultrasonic homogeniser method. Additional PEG 4000 did not change the initial bond of bacterial cellulose, only influenced the intensity values of the bonds. Particularly in C-H bending bond and signed by deeper intensity valley graph at 1963.52 cm^{-1} wave number. In the future, this material from pineapple peel waste will be developed as filtering material for air and water treatment.

Acknowledgement
A great appreciation was delivered to the DRPM Dikti for the grant of research scheme of Penelitian Kompetitif Nasional, Fundamental Research 2021.

References
[1] Siti Roha A M, Zainal S, Noriham A and Nadzirah K Z 2013 Determination of sugar content in pineapple waste variety N36 *Int. Food Res. J.* **20** 19413.
[2] Song H J, Li H, Seo J H, Kim M J and Kim S J 2009 Pilot-scale production of bacterial cellulose by a spherical type bubble column bioreactor using saccharified food wastes *Korean J. Chem. Eng.* **26** 1416.
[3] Anicuta S, Dobre L, Stroescu M and Jipa I 2010 Fourier transform infrared (ftir) spectroscopy for characterization of antimicrobial films containing chitosan *Analele Univ. din Oradea Fasc. Ecotoxicologie, Zooteh. si Tehnol. Ind. Aliment.* 123440.
[4] Mahsuli T, Suryanto H, Larasati A and Muhajir M 2019 Mechanical properties of bacterial nanocellulose membrane from pineapple peel waste after homogenization process *AIP Conf. Proc.* **2120.**
[5] Maneerung T, Tokura S and Rujiravanit R 2008 Impregnation of silver nanoparticles into bacterial cellulose for antimicrobial wound dressing *Carbohydr. Polym.* **72** 4351.
[6] Hosseini H, Zirakjou A, Goodarzi V, Mousavi S M, Khonakdar H A and Zamanlui S 2020 Lightweight aerogels based on bacterial cellulose/silver nanoparticles/polyaniline with tuning morphology of polyaniline and application in soft tissue engineering *Int. J. Biol. Macromol.* **152** 5767.
[7] Abdul Khalil H P S, Davoudpour Y, Islam M N, Mustapha A, Sudesh K, Dungani R and Jawaid M 2014 Production and modification of nanofibrillated cellulose using various mechanical processes: A review *Carbohydr. Polym.* **99** 64965.
[8] Luo H, Xiong G, Hu D, Ren K, Yao F, Zhu Y, Gao C and Wan Y 2013 Characterization of TEMPO-oxidized bacterial cellulose scaffolds for tissue engineering applications *Mater. Chem. Phys.* **143** 3739.
[9] Phutanon N, Motina K, Chang Y H and Ummartyotin S 2019 Development of CuO particles onto bacterial cellulose sheets by forced hydrolysis: A synergistic approach for generating sheets with photocatalytic and antibiofouling properties *Int. J. Biol. Macromol.* **136** 114252.
[10] Mohammedi Z 2017 Structure, Properties and Medical Advances for Biocellulose Applications: A Review *Am. J. Polym. Sci. Technol.* **3** 89.
[11] Zheng Y, Yang J, Zheng W, Wang X, Xiang C, Tang L, Zhang W, Chen S and Wang H 2013 Synthesis of flexible magnetic nanohybrid based on bacterial cellulose under ultrasonic irradiation *Mater. Sci. Eng. C* **33** 240712.
[12] Sureshkumar M, Siswanto D Y and Lee C K 2010 Magnetic antimicrobial nanocomposite based on bacterial cellulose and silver nanoparticles *J. Mater. Chem.* **20** 694855.
[13] Suryanto H, Muhajir M, Sutrisno T A, Yanuhar U and Bintara R D 2020 Effect of disintegration process on the properties of bacterial cellulose foam *Key Eng. Mater.* **851 KEM** 8691.
[14] Sandberg L I C 2011 Porous filters of nanofibrillated cellulose
[15] Zhou Y, Guo X, Li X, Fu J, Liu J, Hong F and Qiao J 2020 In-situ growth of CuO/Cu nanocomposite electrode for efficient CO2 electroreduction to CO with bacterial cellulose as support *J. CO2 Util.* **37** 18894.
[16] Pawcenis D, Chlebda D K, Jędrzejczyk R J, Leśniak M, Sitarz M and Łojewska J 2019 Preparation of silver nanoparticles using different fractions of TEMPO-oxidized nanocellulose *Eur. Polym. J.* **116** 24255.
[17] Almasi H, Mehryar L and Ghadertaj A 2019 Characterization of CuO-bacterial cellulose nanohybrids fabricated by in-situ and ex-situ impregnation methods *Carbohydr. Polym.* **222**
[18] Suryanto H, Muhajir M, Sutrisno T A, Mudjiono, Zakia N and Yanuhar U 2019 The Mechanical Strength and Morphology of Bacterial Cellulose Films: The Effect of NaOH Concentration

IOP Conf. Ser. Mater. Sci. Eng. **515** 18.

[19] Sardjono S A, Suryanto H, Aminnudin and Muhajir M 2019 Crystallinity and morphology of the bacterial nanocellulose membrane extracted from pineapple peel waste using high-pressure homogenizer *AIP Conf. Proc.* **2120.**

[20] Sun Y, Meng C, Zheng Y, Xie Y, He W, Wang Y, Qiao K and Yue L 2018 The effects of two biocompatible plasticizers on the performance of dry bacterial cellulose membrane: a comparative study *Cellulose* **25** 5893908.

[21] Fengel D 1992 Characterization of Cellulose by Deconvoluting the OH Valency Range In FTIR Spectra *Holzforschung* **46** 2838.

[22] Agustin T A and Putra A 2019 The Effect of Addition of Polyethylene Glycol (PEG) on Biodegradable Plastic Based on Bacterial Cellulosa from Coconut Water (Coconus Nucifera) *Int. J. Progress. Sci. Technol.* **17** 507.

The change of land use patterns and cover on the surface run-off in Krueng Meuraksa sub-watershed

A Savitri[1], A Achmad[2] and N Fadhly[3]

[1]Student, Department of Civil Engineering, Syiah Kuala University, Banda Aceh, Indonesia
[2]Department of Architectural and Planning, Syiah Kuala University, Banda Aceh, Indonesia.
[3]Department of Civil Engineering, Syiah Kuala University, Banda Aceh, Indonesia

noer6637@unsyiah.ac.id

Abstract. The increasing population and the development of Lhokseumawe city from 2008-2018 resulted in changes in land use, therefore the type of land cover will be a benchmark for how much the surface run-off will occur. In this study there are six categories of land use types, thus are shrubs, plantations, settlements, open land, dryland farming and rice field. Krueng Meuraksa sub-watershed is in Lhokseumawe, North Aceh Regency, Aceh Province. The area of Krueng Meuraksa sub-watershed is 9,958.56 ha. This study aims to observe how the pattern of land use change and run-off in 2008–2018. This study uses secondary data collection methods obtained from related resources. The result shows that there is a change in land use where the area of land cover continues to increase in the residential area, on the contrary the area of land cover continues to decrease in dryland farming area. These changes lead to an increase in the coefficient of surface run-off in Krueng Meuraksa sub-watershed in 2008–2018, where in 2008 it was 0.726 and in 2018 it was 0.787. Based on the run-off coefficient, rainfall intensity and sub-watershed in Krueng Meuraksa, the amount of run-off was 333.403 m^3/d in 2008 and 361.765 m^3/d in 2018. The increase in the amount of run-off occurs as the run-off coefficient increases every year

1. Introduction

Lhokseumawe is an area which is potential for disasters such as earthquakes, tsunamis, floods, abrasion, and tidal waves. The growth population in Lhokseumawe from 2008 to 2018 has increased. The average growth rate of the population of Lhokseumawe is 8%. Lhokseumawe has an area of 181.06 km^2 with various land used. The highest portion used as a residential area with amount of 2,861.79 ha [1].

The increasing number of residents and the development of Lhokseumawe have resulted in the conversion of land functions, therefore, the type of land cover will be a benchmark for how much surface run-off will occur. Land cover types such as open land areas will have small surface run-off because they have a large infiltration capacity. Changes in land use damage the natural vegetation cover [2], if the open land is converted into a built-up area, it will cause the reduction of infiltration capacity, as a consequence, the water will flow more easily into rivers and the volume of river water will rise faster and will cause the flood. Considering that, it is necessary to do a research on the affect of the changes in land use on the surface run-off in Lhokseumawe. There are 3 (three) sub-watersheds in Lhokseumawe thus are Krueng Geukuh sub-watershed, Krueng Meuraksa sub-watershed and Krueng Pase sub-watershed. In this study took a place in the Krueng Meuraksa sub-watershed which is the largest sub-watershed covering the city of Lhokseumawe.

Concerning the surface run-off, the type of land cover will be the measurement of how much surface run-off will occur. For instance, forest areas have small surface run-off due to their large infiltration capacity. If the forest area is converted into open land or settlements, it will reduce the infiltration capacity due to soil surface compression. As consequence water will flow more easily into the river, the water volume in the river will rise faster and will cause a flood [3].

This research will be focused in Lhokseumawe, Krueng Meuraksa sub-watershed, where no research has been conducted regarding the change of land use on the surface water run-off before. So the authors are interested to study in detailed about the effect of surface water run-off due to land conversion in 2008 to 2018 by using map data in the form of a shapefile and based on the hydrological aspect. The results show the increase amount of run-off occurs as the run-off coefficient increases every year. This information can be useful to mitigate the flood by realizing the right RTRW for Lhokseumawe.

2. Materials and methods

2.1. Study location

This research conducted in Krueng Meuraksa sub-watershed. It located near urban areas, thus the distribution of land use types in the Krueng Meuraksa sub-watershed is varied. Meuraksa sub-watershed located in Lhokseumawe and North Aceh District, Aceh Province. The area of Krueng Meuraksa sub-watershed is 9958.56 ha. Astronomically, the Krueng Meuraksa sub-watershed is located in 5 ° 12'57.14 "U - 97° 6'59.23" E and 5 ° 3'36.27 "U - 97 ° 10'50.08" E. The following can be seen the Krueng Meuraksa sub-watershed map in Figure 1.

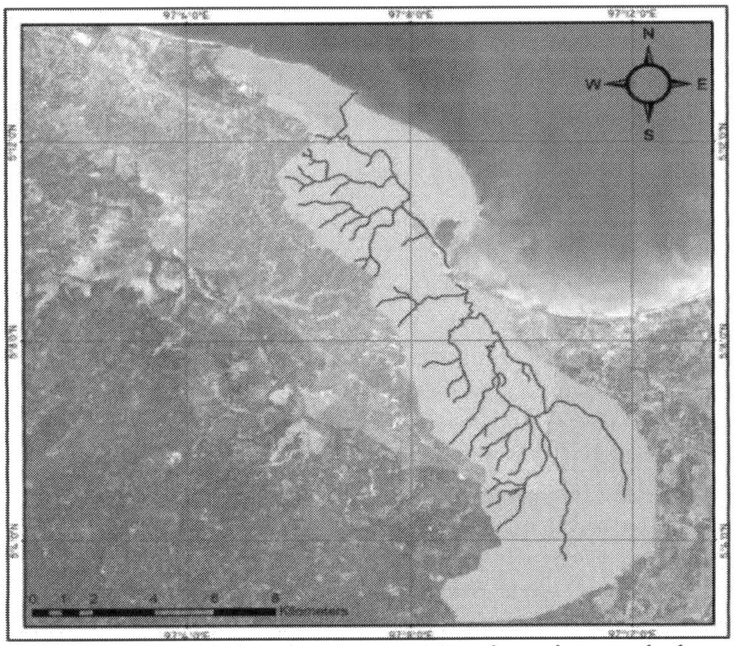

Figure 1. Study location: Krueng Meuraksa sub-watershed

2.2. Research methodology

This research is a case study of the change of land use and surface run-off in the Krueng Meuraksa sub-watershed. The data collection method used in this research is secondary data, where:

- Secondary data is obtained through literature study and collecting secondary data or information from various resources and related agencies, such as: topographic maps, both printed and digital maps, aerial photographs, administrative boundary maps or regional maps, land use maps, data climate, hydrological data and previous research journals.
- Map data in the form of shapefiles obtained from Krueng Aceh Watershed and Protected Forest Management Center (BPDAS-HL).
- Data of Hydrology is taken from rainfall data from Malikussaleh Meteorological Station in North Aceh and the Lhokseumawe BPBD Rain Post.
- Population data is obtained from the Central Statistics Agency (BPS) of Aceh Province.

The data used to determine the use of land cover consists of a map in the form of a shapefile obtained from the Krueng Aceh Watershed and Protected Forest Management Center (BPDAS-HL). The type of land use is the most dominant land use [4] it consists of shrubs, plantations, settlements, open land, dryland farming and rice fields. The Calculation of rainfall used the amount of daily rainfall and the amount of monthly rainfall which collected from Malikussaleh Meteorological Station in North Aceh from 2008–2018.

3. Results and discussions

3.1. Results

3.1.1. The change in land use

The change in land use is a transistion process from previous land use to another. It caused by humans adaptation to the conditions that change overtime. Land use change analysis carried out with the help of ArcGIS 10.5. By comparing the results of the land use analysis in 2008–2018, the changes in land use area are as shown in Table 1:

Table 1. The change of land use in sub-watershed in Krueng Meuraksa

2008			2018			Change of Land 2008–2018	
Types of Land Cover	Ha	%	Types of Land Cover	Ha	%	Ha	%
Shrubs	314.771	3.161	Shrubs	18.076	0.182	-296.695	-2.979
Plantation	26.752	0.269	Plantation	2,127.880	21.367	2,101.129	21.099
Settlement	2,080.292	20.889	Settlement	3,257.241	32.708	1,176.950	11.818
Open Land	14.742	0.148	Open Land	6.450	0.065	-8.293	-0.083
Dryland Farming	5,107.438	51.287	Dryland Farming	2,074.901	20.835	-3,032.536	-30.452
Rice Field	1,583.109	15.897	Rice Field	1,659.277	16.662	76.169	0.765
Water Body and Pond	831.455	8.349	Water Body and Pond	814.732	8.181	-16.723	-0.168
Total	9,958.558	100.000	Total	9,958.558	100.000	0.000	0.000

Source : BPDAS-HL Krueng Aceh, 2020 [5]
Description : Negative (-) shows land reduction
: Positive (+) Shows land addition

Based on Table 1, it can be seen that in the period of 2008–2018 there was a change in land cover which is an increase in plantation area by 2,101.129 ha, settlements by 1,176.950 ha and Rice fields by 76.169 ha. On the other hand, there was a reduction in the area of dryland farming by 3,032.536 ha, shrub by 296.695 ha and open land by 8.293 ha. Significant changes that occur are the increase amount of plantation area and the reduction of dryland farming area. The picture of the change of land use in 2008 and 2018 can be seen in Figure 2. This is the same as in the Keduang watershed where the most striking changes in this period is the addition of dry fields and the reduction of mixed gardens. In percentage terms, the largest area increase was in the cover/use of open land (739.13%) [6].

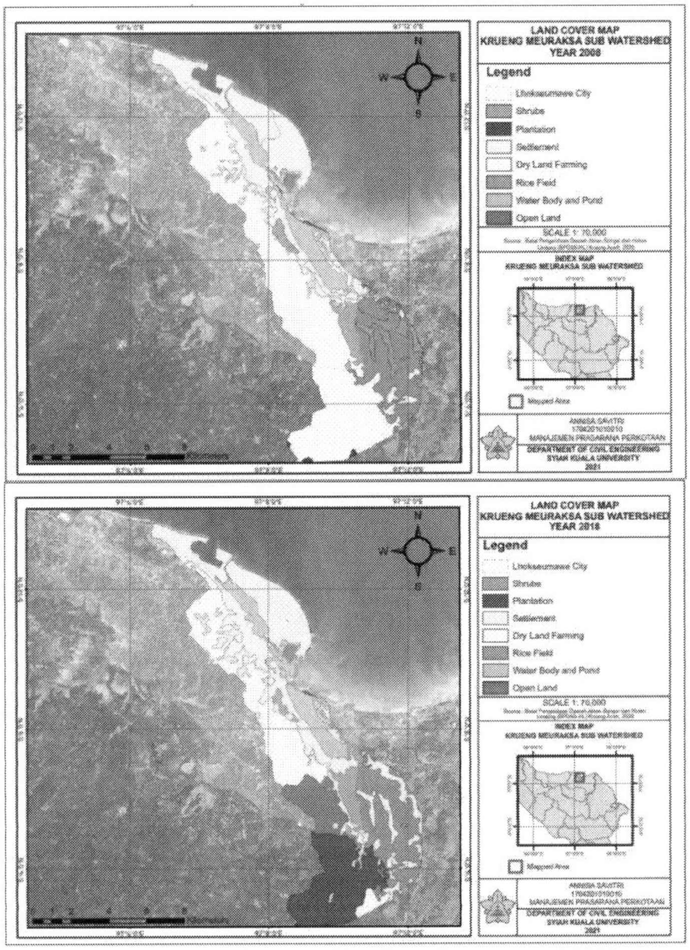

Figure 2. The change of land use in 2008 and 2018

3.1.2. Rain intensity

The intensity of rain used to calculate the run-off in this study is the plan rain intensity for 10 years. The average amount of rainfall in the Krueng Meuraksa sub-watershed for 10 years from 2008 to 2018 can be seen in Table 2 as follows:

Table 2. Average monthly rainfall in sub-watershed Krueng Meuraksa from 2008–2018

Month	Average Rainfall (mm)									
	2008	2009	2010	2011	2013	2014	2015	2016	2017	2018
Jan	6.7	13.6	25.6	6.8	8.0	3.9	12.2	15.8	11.8	4.7
Feb	1.0	3.5	8.3	9.0	8.0	19.0	14.8	9.7	8.5	12.7
Mar	12.2	12.4	13.8	25.2	15.2	39.8	1.5	6.4	7.7	8.8
Apr	8.7	13.9	24.7	5.3	22.8	16.7	5.1	8.0	6.0	12.6
May	6.3	10.0	11.6	9.3	12.8	13.2	9.1	7.1	8.8	12.1

Table 2. Average monthly rainfall in sub-watershed Krueng Meuraksa from 2008–2018 (contd.)

Month	Average Rainfall (mm)									
	2008	2009	2010	2011	2013	2014	2015	2016	2017	2018
June	2.0	29.3	25.6	7.3	16.7	5.8	9.4	9.0	11.3	9.1
July	13.3	8.4	5.9	9.3	13.0	10.4	6.0	7.4	7.1	22.6
August	12.8	12.4	12.9	13.5	9.8	17.8	9.8	2.8	11.5	16.2
Sept	8.0	5.7	11.3	3.8	6.3	9.5	12.9	6.5	8.1	4.8
Oct	6.3	7.3	7.1	17.5	8.1	13.9	6.0	2.8	16.1	13.5
Nov	23.8	26.8	15.5	12.0	4.3	13.1	21.7	13.1	10.0	14.7
Dec	12.6	10.2	13.5	21.8	12.6	30.1	9.6	10.3	18.6	4.5
Total	113.6	153.5	175.6	140.7	137.4	193.1	117.9	98.9	125.7	136.2

Source: Malikussaleh Station Meteorology, North Aceh, 2011 [7]

The rainfall plan is obtained from rainfall data from the Malikussaleh Station Meteorology, North Aceh. The selection of distribution used is the log person 3 distribution, after the the test of Ck and Cs. The concentration time (tc) is 5.023 hours. The return period used is 10 years. The peak flow rate estimation model generally used in the rain interval of 10 years [8]. The results of the rainfall plan calculation using the Log Person III method can be seen in Table 3 as follows:

Table 3. The value of rainfall plan log-pearson III method

No	Reset Period	Xi	Kt	S	R_{24} (mm)
1	2 yrs.	110.75	-0.003	0.122	106.817
2	5 yrs.		0.841		135.313
3	10 yrs.		1.284		153.177

3.2. Discussion

3.2.1. The increasing and decreasing rate of land cover use

The change of land use is a form of transition from previous land uses to another. Data of land use in this study obtained from the Center for Watershed Management and Protection Forest (BPDAS-HL) Krueng Aceh, the type of land use and change in land use area from 2008 to 2018 in sub-watershed Krueng Meuraksa can be seen at Figure 3 as follows:

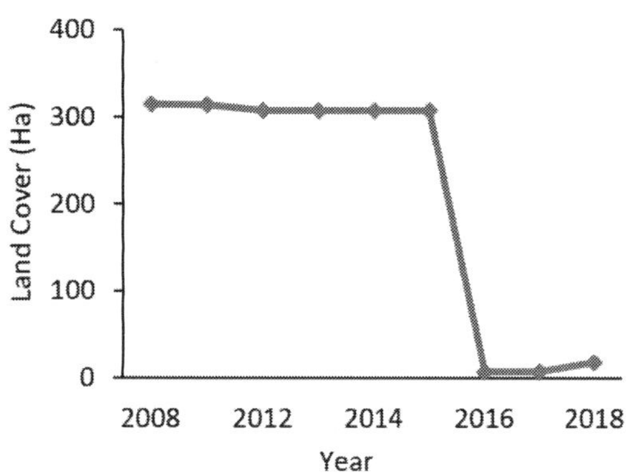

(a) Shrubs

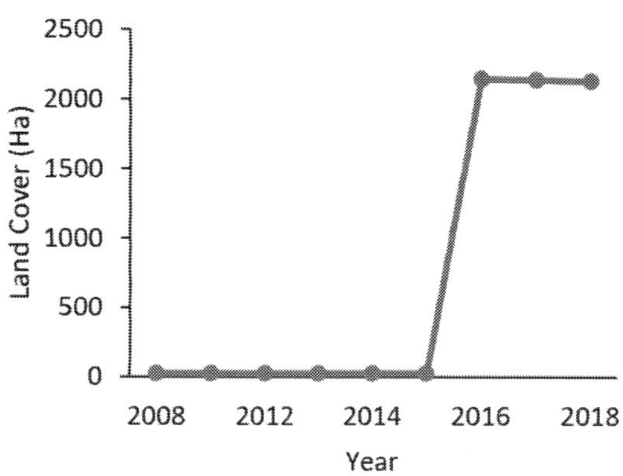

(b) Plantation

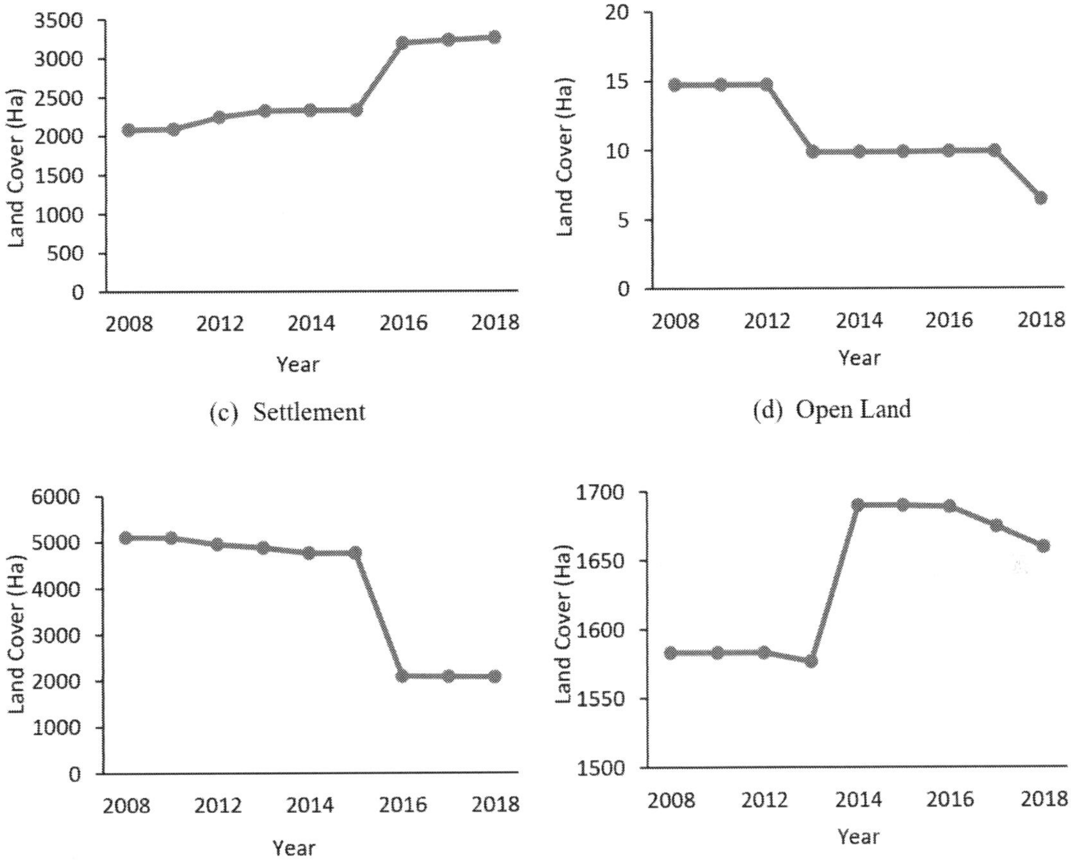

(c) Settlement (d) Open Land

(e) Dryland Farming (f) Rice Field

Figure 3. The change of land use area from 2008–2018

There was an additional amount of shrub land area as much as 296.695 ha from 2008 to 2018. The shrub land change into the development of residential land area. The reduction in plantation land from 2008 to 2018 as much as 2,101.129 comes from the changes in dryland farming. Residential land used from 2008 to 2018 was also increased by 1,176.950 ha. The addition of this residential land increase every year along with the increase in population, it comes from changes in dryland farming, ponds, rice fields and shrubs. The use of open land is decreasing from 2008 to 2018 as much as 8.293 ha. The reduction of the open land area is because it changes to pond. The reduction use of dryland farming from 2008 to 2018 is 3,032.536 ha. The reduction take place in 2015–2016 because it changes to residential land, plantations, and rice fields. The use of rice fields from 2008 to 2018 was increase in the rice field area as much as 76.169 ha. The addition of rice fields comes from changes in dryland farming, ponds and shrubs.

From Figure 3, there was a reduction in the shrub and dryland farming area in 2015 to 2016. It changed into residential land development, and in that year, there was also a reduction in dryland farming into plantation land. In 2012 to 2013 and 2017 to 2018 there was a reduction of open land into ponds. In 2013 to 2014 there was an increase in rice fields due to the reduction of land from dryland farming, ponds and shrubs.

The change of land use and land use categories indicate that land use is dynamic, closely related to space and time. The change depends on direct human activities and through regulations [9]. From Figure 23, we can see that the area of land cover continues to increase is in residential land use, on the contrary, the area of land cover that continues to decrease is found in the use of dryland farming. In general, land use changes from low vegetation cover to high vegetation cover resulted in an increase in surface run-off that will occur [10].

3.2.2. Krueng Meuraksa sub-watershed run-off

The decrease in land use area where the land surface becomes a watertight surface resulted in an increase of run-off value [11]. The calculation used in surface run-off is the Rational Method. It calculated based on the run-off coefficient (C) in 2008 which was 0.726. The intensity of the 10-year plan (I) was 18.106 mm/hour. The area of the Krueng Meuraksa sub-watershed in 2008 was 9,127.103 Ha, the correction factor for the hectare area was 0.00278. With these data, the surface run-off obtained in 2008 was 333.403 m^3/d. While in 2018 the run-off coefficient (C) in 2018 was 0.787. The intensity of the 10-year plan (I) was 18.106 mm/hour. The area of the Krueng Meuraksa sub-watershed in 2018 was 9,143.826 Ha, with 0.00278 factor correction for the hectare area, the value of surface run-off was 361.765 m^3/d.

[12] also observes the effect of changes in land use in the Batang Arau watershed in 2006 and 2012. In this study, it can be seen that there was a change in the coefficient of surface run-off in 2006 and 2012. In 2006 the coefficient of surface run-off was 0.39 and in 2012 it was 0.41. Based on the run-off coefficient, the rainfall plan intensity and the area of the Batang Arau watershed, the run-off in 2006 was 327.20 m^3/d and in 2012 was 339.51 m^3/d. This study also sees the relationship between land use changes that affect surface run-off.

4. Conclusion

This study observes the change in Krueng Meuraksa sub-watershed land use by monitoring the land cover area using ArcGIS 10.5 in 2008 to 2018.Next step is to identify the relationship between land use changes and run-off. There are six categories of land use types thus are shrubs, plantations, settlements, open land, dryland farming and rice fields. Here are the conclusions from this research:

- In the period 2008–2018 there were changes of land use in Krueng Meuraksa sub-watershed, thus are:
  a. Reduction of shrub land area by 296.695 ha, open land by 8.293 ha and dryland farming by 3,032.536 ha;
  b. The addition of plantation land area is 2,101.129 ha, settlements is 1,176.950 ha and rice fields is 76.169 ha. In this case, the area of land cover that continues to increase every year is the residential land use, on the contrary the area of land cover continues to decrease every year is the use dryland farming.
- There was a change in the surface run-off coefficient of Krueng Meuraksa sub-watershed in 2008–2018, where in 2008 it was 0.726 and in 2018 it was 0.787. Based on the run-off coefficient, rainfall intensity and area of the Krueng Meuraksa sub-watershed, the surface run-off value in 2008 was 333.403 m^3/d and in 2018 it was 361.765 m^3/d. The increase in the amount of run-off occurs as the run-off coefficient increases.

References

[1] Badan Perencanaan Pembangunan Daerah Kota Lhokseumawe 2018 *Kajian Rasio Kemandirian Daerah Kota Lhokseumawe* vol 53 (Lhokseumawe: Bappeda Kota Lhokseumawe)

[2] Yaşar Korkanç S 2018 Effects of the land use/cover on the surface run-off and soil loss in the Niğde-Akkaya Dam Watershed, Turkey *Catena* **163** 233–243

[3] Cindy Harifa A C, Sholichin M and Prayogo T B 2017 Analisa Pengaruh Perubahan Penutupan Lahan Terhadap Debit Sungai Sub Das Metro Dengan Menggunakan Program Arcswat *J. Tenik Pengair.* **008** 1–14

[4] Fadhly N, Program M, Perencanaan D, Pascasrjana S, Matondang A R, Doktor P, Wilayah P, Doktor P, Wilayah P, Saleh S M, Teknik F, Sipil J and Syiah U 2016 Hubungan struktur ruang terhadap bangkitan perjalanan di kota banda aceh *Pros. Forum Stud. Transp. Antar*

Perguru. Tinggi **4** 336–344
[5] Aceh BPDAS-HL Krueng 2020 Perubahan Penggunaan Lahan di Sub DAS Krueng Meuraksa
[6] Widyaningsih I W 2008 Pengaruh perubahan tata guna lahan di sub DAS keduang ditinjau dari aspek hidrologi *Msc Thesis* Universitas Sebelas Maret, Surakarta, Indonesia
[7] Malikussaleh BMKG Stasiun Meteorologi 2011 Rata-Rata Curah Hujan di sub DAS Krueng Aceh
[8] Arsyad S 2010 *Konservasi Tanah dan Air* (Bogor: IPB Press)
[9] Achmad A, Irwansyah M, Nizamuddin N and Ramli I 2019 Land Use and Cover Changes and Their Implications on Local Climate in Sabang City, Weh Island, Indonesia *J. Urban Plan. Dev.* **145** 04019017
[10] Fu Q, Li B, Hou Y, Bi X and Zhang X 2017 Effects of land use and climate change on ecosystem services in Central Asia's arid regions: A case study in Altay Prefecture, China *Sci. Total Environ.* **607–608** 633–646
[11] Syauqi M B 2017 Studi Pengaruh Perubahan Penggunaan Lahan Terhadap Perubahan Infiltrasi Dan Limpasan Permukaan Di Sebagian Wilayah Desa Maguwoharjo, Depok, Sleman, D.I. Yogyakarta *J. Bumi Indones.* **6(4)**
[12] Nurhamidah N, Junaidi A and Kurniawan M 2018 Tinjauan Perubahan Tata Guna Lahan Terhadap Limpasan Permukaan. Kasus : DAS Batang Arau Padang *J. Rekayasa Sipil* **14(2)** 131

Planning of fishing residential in Kuta Village (supporting the Mandalika special economic zone) based on green concept

R S Saptaningtyas[1], P K Utomo[2], S Hilyana[3] and S N R Anwar[4]

[1]Department of Architecture, University of Mataram, Mataram, Indonesia
[2]Department of Architecture, Mulawarman University, Samarinda, Indonesia
[3]Department of Fishery, University of Mataram, Mataram, Indonesia
[4]Department of Civil Engineering, Universitas Negeri Malang, Malang, Indonesia

rinisaptaningtyas@unram.ac.id

Abstract. A Fishing housing in Kuta Village in Central Lombok Regency supports the Mandalika Special Economic Zone (SEZ). Housing with environmental issues and inadequate infrastructure in the fishing village of Kuta village is a fact. On the other hand, this area is a potential tourist destination in West Nusa Tenggara Indonesia. The aim is to produce a reference document for developing fisherman housing in Kuta Village based on a sustainable environment. This methodology uses a descriptive qualitative approach. The conclusion is that the planning of fishing settlement areas must be developed with a sustainable development model. Where in this development must adopt materials that are friendly and adaptive to the environment. The development of sustainable tourism may not be fully appreciated by stakeholders, but there is a responsibility to maintain regional development, especially in coastal areas. Sustainability should be considered by developers and policymakers to create a more desirable tourism future for coastal areas.

1. Introduction

Fishing residential is the one of the basic human needs concerning the feasibility and welfare of a fishing community in an area. The house not only functions as a place to live or shelter, more than that; the house also has a strategic function in its role as a center for family education, cultural conformity and quality improvement for future generations. As a part special economic zones of Mandalika, where the active development is the development of coastal tourism, the existence and plans for the settlement or fishing village of Kuta beach, especially in the west of the area, must be integrated with the plan to develop a tourist area in a broad manner and form an integral part that will be become one of the tourist destinations itself. Thus, the development of housing and fishermen residence for Kuta village is not based solely on physical development but must be linked to the social, economic and cultural dimensions that support community life in a sustainable manner.

2. Literature review

Sustainability is a part of tourism development because three are three aspects of sustainable tourism: environment, social, and economic [1,2]. The development of coastal areas must be designed carefully because it considers the future, especially environmental conditions using the past to inform the present in providing recreation [3]. Discussion about the relationships between land and sea, natural and non-natural, human and ecosystem, need to be understood through economic perspective and landscape management through local cultural interactions. [4,5].

Ecotourism can be defined as any recreational activity that explicitly involves the viewing or extraction of other living beings in the wild [6]. The planning and development that can be used as a concept in the long term is ecotourism. As a specialization, nature-based tourism and community sustainability transformed into ecotourism, being an alternative that represents the discourse of sustainability. At this point, development does not only depend on the developers, but is also influenced by the community and policymakers.

The dynamics of coastal tourism cannot be separated from the economic factors and policy dimensions that affect the evolution of tourism in general in a socio-cultural framework [7]. Special Economic Zones (SEZs) are evidence of this dynamic, where they emerge from national policies aimed at increasing the economy on a local to international scale. The Mandalika area in Central Lombok is one of the areas designated as an SEZ based on Government Regulation 52/2014 on the Mandalika Special Economic Zone.

The awareness of sustainable tourism is necessary for stakeholders [8]. At this point, stakeholders must take notice to the basic needs of the community. Settlement is one of the most priority aspects. Quality of the settlement environment will increase the benefits for community [9], although the quantity of attraction is another one potential approach to develop tourism. Coastal destinations should be able to provide cultural attractions and events [10]. The community development factors of human resource development such as skill enhancement and the community readiness in development change are aspects of the development process [11].

In arranging the Kuta Fishing Village, it is very important to pay attention to the aspects that affect the sustainability of people's lives in the area. This is because fishing settlements have their own uniqueness, namely socio-economic problems (communities with high levels of density and number of family members as well as low levels of income, savings and education), and physical environmental problems (limited or unavailability of sanitation and solid waste infrastructure [12]. According to [13], the physical arrangement of fishermen settlements can be done by exploring (a) accessibility /achievement; (b) zoning; (c) the existence of public, supporting and social facilities; (d) design, house; and (e) residential utilities. Meanwhile, according to [14], the arrangement of fishermen's settlements must pay attention to infrastructure (in the form of roads, drainage, clean water and electricity), and facilities (in the form of open spaces, schools, places of worship, and fish auction centers).

A settlement can be said to be a quality settlement when it is able to accommodate the needs of its inhabitants. Strategies that can be carried out in achieving this goal include adjusting building dimensions, building structures, roof coverings, building walls, building floors, distances between buildings, and building orientation/direction [15]. According to [16], a strategy that can be applied is to pay attention to environmental adaptation, socio-economic dynamics, sanitation and the environment of sustainable development.

The arrangement of Kuta Fishermen Village can be viewed from two approaches, namely the regional approach (macro) and the architectural approach (micro). The main thing in area planning is road arrangement as the main key to area accessibility. Roads are the basic infrastructure needed to support access and networks in settlements [17]. Provision of area support facilities is also a further priority, in the form of drainage networks, sanitation and open spaces. The existence of Green Open Space (RTH) is one of the dominant preferences in residential areas [18], so that in the arrangement of the Kuta Fishermen Village, RTH must be provided in an appropriate proportion.

In addition, coastal residential areas can also adopt the concept of ecological and environmentally friendly architecture with proper zoning [19]. Buildings that are categorized as public facilities must be placed in a location that is affordable and easily accessible. As an area where almost all of the people work as fishermen, the existence of the fish market is the main economic support. It should be emphasized here that the existence of a fish auction center (TPI) is very important because market imbalances which are the cause of the low economy of the fishing community are weakening the fish marketing network due to the malfunctioning of the TPI [20].

3. Research method

This research approach uses a descriptive-qualitative approach. So, what is produced from the survey in the field as primary data about how the existing condition is, will be narrated in writing so that it is easy to understand. Then for secondary data obtained from related agencies supported by direct observations in the field to obtain internal and external data on the tourist area of the fishing village of Kuta.

Further analysis, related to how to find out the tourism potential, can be done by planning and rearranging the fishing settlement area of Kuta. This is done to create a fishing village that is environmentally friendly and sustainable.

4. Result and discussion

Based on Law Number 1 of 2011 concerning Housing and Settlement Areas, a residential area is part of the environment outside a protected area, either urban or rural, which functions as a residential or residential environment and a place for activities that support life and livelihoods. Referring to the same law, Housing and Settlement Areas are spatial entities that cannot be separated from the community as residents. Figure 1 shows the area of fishing village in Kuta (Mandalika).

Settlements are composed of housing, environment, and infrastructure. The smallest unit in the context of a residential area is a house. A house is a building that functions as a place to live that is suitable for habitation, a means of fostering a family, a reflection of the dignity of its inhabitants, as well as an asset for its owner. There are several types of houses, including a) Independent houses; b) Public houses; c) Special homes; d) Country Houses. Each type of house has its own characteristics and specifics.

Settlements in the Kuta Village area consist of self-help houses. Housing development in this area has been going on for decades. Since the 1970s, the community has been occupying the land in this area for generations. They moved their house several times, both in Kuta 2 sub village and to the area around Kuta 2 sub village. Some of the people who started living in this area have also sold their land and moved out of Kuta Village. In a period of decades, the residents of Kuta 2 sub village tend to be homogeneous, there has been no significant change in the composition of the community in the last 50 years. Figure 2 shows the building and environment condition of fishing residential in Kuta (Mandalika).

4.1. *Building condition*

From the regularity of the buildings, the houses in the kuta fishing village area are very crowded and disorganized. The distance between buildings is very tight, which is an average of 1.5 meters. There is no clear building mass pattern. This is because the community builds without paying attention to the land lots and the orientation of the buildings on it. Due to the irregularity of the buildings in this area, the face of the area tends to be messy.

The condition of housing units in this area is not very feasible. From the physical aspect of the building, some houses still use non-permanent materials. Some who use permanent materials still have not applied the correct structural and architectural rules. Due to the condition of the housing units, this area has become slum and unfit for habitation.

4.2. *Waste management*

One of the most crucial problems in the Kuta Fishing Village area is the problem of solid waste. In Kuta village, the area around the fishing village is the area with the least waste facilities. Waste management in this area is also the least feasible compared to other areas in Kuta village.

Due to the absence of trash can facilities, residents tend to dispose of garbage in places where it is not supposed to be. One of the places where garbage is dumped is on the shoreline around the fishing boat parking area. As a result of the garbage in this area, the view becomes unattractive, and the smell of garbage disturbs the comfort of people who are active around. This condition also damaged the image of the tourism area around Kuta Fishing Village

The Kuta fishing village area requires solid waste management that can accommodate the volume of residents' waste while being designed with a neat design so that it looks clean. Garbage collection by managers must also consider special access so as not to interfere with activities or damage the comfort of residents.

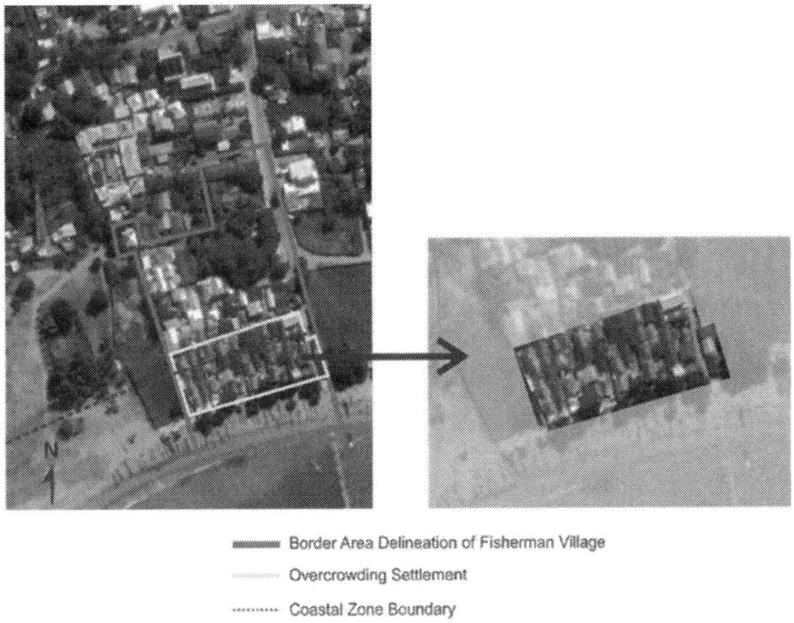

Figure 1. Order of building

Figure 2. House building and waste conditions

4.3. *Discussion*

With a profession as a fisherman, most residents in the Kuta fishing village area own a boat. This is an advantage because these boats can support tourism activities. By boat, residents can offer services between tourists and for exploration of the surrounding sea. This potential if managed properly will attract both local and foreign tourists.

The current condition of the area is not suitable for a tourism area. The settlement is slum, and the infrastructure is inadequate. This is a challenge for the development of this area to increase tourism. In terms of their livelihoods as fishermen, the income of residents in fishing village is far from sufficient.

Regarding the rules and regulations stating that permanent buildings cannot be located in the Coastal Line Area, residents can understand this and are willing to be relocated to a place that does not violate the rules. The availability of them to be relocated is an important factor in the arrangement of the area because it requires a design that does not conflict with the prevailing regulations. In addition, moving residents will facilitate the design of housing that is more feasible for them, so that it will provide benefits for them.

The area of the Kuta fishing village has received enough attention from the government, from the central level to the regional level. Various programs in physical form have been carried out in this area. One of the newest programs in 2017 is the arrangement of the Directorate General of Human Settlements of the Ministry of Public Work and Spatial Planning and the Transportation Service of Central Lombok Regency. These programs are an indication that this area can attract programs from stakeholders in its development.

Likewise, ITDC (International Tourism Development Cooperative) as a state-owned company that has a strong interest in the development of the SEZs Mandalika, also pays attention to the management and arrangement of the Kuta Fishermen Village area. With the role of ITDC, structuring opportunities in order to increase tourism can be boosted more optimally. This will benefit ITDC as well as benefit citizens in terms of their welfare.

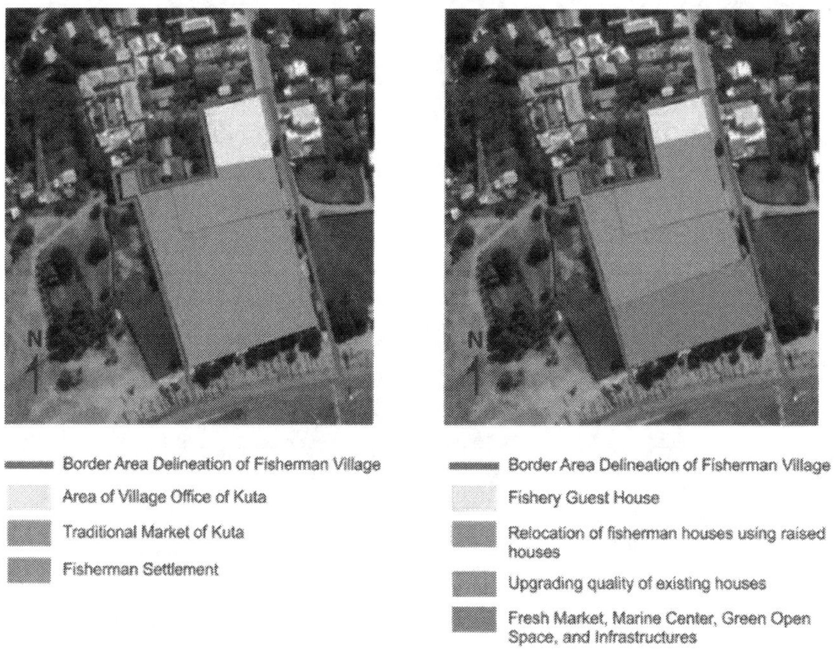

Figure 3. Sustainable development concept of Kuta Village (Mandalika)

The causes of slum, among others, are the improper aspects of the settlement as mentioned in the settlement analysis section. In overcoming this problem, associated with tourism, architecture is the solution. The fact that the people living in the fishing village area are willing to be reorganized and relocated has become an entry point for the role of architecture. This is an opportunity to design more feasible residential units and supporting facilities to support tourism activities

With the relocation and construction of new houses for residents of the Kuta Fishing Village area, local wisdom can be embedded in the arrangement. Especially if it is related to the condition of the location which is on the coast, local wisdom will emerge as a solution to the challenge of how the fishermen's house takes shape. The next challenge is the use of building materials, structural considerations, which in the end are all packaged in aesthetics to give birth to a typical fisherman housing concept that is worthy of its residents as well as being a magnet for tourists.

The condition of Kuta fishing village currently does not have a proper green open space. Residents still gather in sporadic places and children play in places that do not guarantee their comfort. In fact, a settlement really needs open space to improve the quality of the area. The absence of adequate green open space in Kuta fishing Village has made this area degraded from an environmental point of view.

The density of buildings in the Kuta fishing village area is the cause of the lack of quality open spaces. This density needs to be broken down so that the solid-void area becomes better. In this open space, facilities for supporting community activities can be built. Thus, the livelihoods of the people living in this area will be of higher quality. In addition, the existence of green open space will be able to support the potential that exists in the area, for example tourism potential.

Figure 4. Iconic fonts area of Kuta (Mandalika)

Figure 5. Planning of fishermen's settlements in Kuta (Mandalika)

Based on facts above, the planning of fishing village in Kuta beach (Mandalika) is proposed as Figure 3 until Figure 5. Figure 5 illustrates that a healthy environment can be planned by designing environmentally friendly infrastructure. For example: for fishermen's settlements, houses are designed on stilts so that if sea water rises, water will not enter the house. The building materials are also earthquake-friendly, both from the walls and the roof. while the drainage system uses an environmentally friendly drainage system in the sense that every 20 m distance is infiltrated so that not all the water that flows is wasted into the sea but some seeps into the ground. This is done so that the quality of soil is maintained.

5. Conclusion

Fishermen's residential areas must be preserved and developed with a sustainable development model – which in this development must be adopted materials that are friendly and adaptive to the environment. It is done so the impact of this development to improving the quality of the environment will also increase the community's economy and continue towards a healthy and prosperous housing environment. The development of sustainability tourism may not be fully appreciated by stakeholders [21], but there is a responsibility to maintain regional development, especially in coastal areas. Sustainability must be considered by developers and policymakers to create a more desirable tourism future for coastal areas [22].

Acknowledgment

Thanks to the Indonesia Tourism Development Cooperation (ITDC) which has funded this research in supporting the development of the area around SEZ Mandalika in West Nusa Tenggara of Indonesia.

References

[1] Fennell D A 2003 *Ecotourism: an Introduction* (New York: Routledge)
[2] Baker S 2006 *Sustainable Development* (New York: Routledge)
[3] Cetin M 2016 Sustainability of urban coastal area management: A case study on Cide *J Sustain For.* **35** 527–41
[4] Chakraborty S, Saha S K and Ahmed S S 2020 Recreational services in tourism dominated coastal ecosystems: Bringing the non-economic values into focus *J Outdoor Recreat Tour* **30** 100279
[5] Urquhart J and Acott T A 2014 Sense of place in cultural ecosystem services: The case of Cornish fishing communities. *Soc Nat Resour.* **27** 3–19.
[6] Johnson A F, Gonzales C, Townsel A and Cisneros-Montemayor A M 2019 Marine ecotourism in the gulf of California and the baja California peninsula: Research trends and information gaps *Sci Mar.* **83** 177–85.
[7] Clavé S A and Wilson J 2017 The evolution of coastal tourism destinations: a path plasticity perspective on tourism urbanization *J Sustain Tour.* **25** 96–112.
[8] Bernini C, Emili S and Vici L 2020 Are mass tourists sensitive to sustainability? *Tour Econ.*
[9] Sesotyaningtyas M and Manaf A 2015 Analysis of sustainable tourism village development at Kutoharjo Village, Kendal Regency of Central Java *Procedia - Soc Behav Sci.* **184** 273–80
[10] Valle P O Do, Guerreiro M, Mendes J and Silva J A 2011 The cultural offer as a tourist product in coastal destinations: The Case of Algarve, Portugal *Tour Hosp Res.* **11** 233–47
[11] Ahmad P, Yusof F and Abdullah F 2013 Local economic growth and community sustainability *Procedia - Soc Behav Sci.* **101** 437–44
[12] Christiawan P I and Budiarta I G 2017 Entitas permukiman kumuh di wilayah pesisir *J Ilmu Sos dan Hum.* **6** 179.
[13] Anisa A, Septiawan T, Rahmah G L N, Kadeli K, Putro S A and Kurnia T 2018 Eksplorasi kondisi fisik dan non fisik perikanan tradisional di Kampung *J Langkau Betang* **5** 55–67.
[14] Aldi R H M, Trisnandari L and Ikaputra I 2019 Karakteristik dan pola kampung nelayan. *Tesa Arsit.* **17** 115–26.
[15] Purwanto E and Darmawan E 2018 The adaptation strategy of dwelling in the riverside settlement of the Arut River in Pangkalan Bun City, West Kotawaringin Regency, Central Kalimantan. *IOP Conf Ser Earth Environ Sci.* **213**
[16] Wulandari E, Nasution B, Djamaludin M and Sabila F 2020 Proses tumbuh kembang

permukiman perdesaan Pantai Muara Sungai di pesisir barat Aceh dan adaptasi keberlanjutannya *J Arsit Zo.* **3** 195–208.
[17] Sharvina A N, Faqih M and Santosa H R 2017 Sustainable fisherman settlement development *IPTEK J Proc Ser.* **0**
[18] Syafrina A, Tampubolon A C, Suhendri, Hasriyanti N and Kusuma H E 2018 Preferensi masyarakat tentang lingkungan perumahan yang ingin ditinggali *Rev Urban Archit Stud.* **16** 32–45.
[19] Yuliani S, Setyaningsih W and Winarto Y 2018 Strategi penataan kawasan Pantai Klayar Pacitan sebagai destinasi pariwisata berkelanjutan dengan prinsip arsitektur ekologis *Rev Urban Archit Stud.* **16** 1–12.
[20] Cahaya A 2015 Fishermen community in the coastal area: A note from Indonesian poor family *Procedia Econ Financ.* **26** 29–33.
[21] Hengky S H, Kikvidze Z 2021 Sustainable coastal tourism in Tanjung Kelayang, Indonesia. *Tour Plan Dev.* **18** 365–70.
[22] Andersen I M V, Blichfeldt B S and Liburd J J 2018 Sustainability in coastal tourism development: an example from Denmark *Curr Issues Tour.* **21** 1329–36.

IOP Conf. Series: Earth and Environmental Science **847** (2021) 012010 doi:10.1088/1755-1315/847/1/012010

Indoor temperature in perimeter spaces on warm humid tropical courtyard housing model

I Defiana[1]

[1]Architecture Department. Sepuluh Nopember Institute of Technology, Indonesia

may.d@arch.its.ac.id

Abstract. In general, Indonesia housing provides open space at front and fence. Open space inside that expose to sky is predicated as inner courtyard. Housing provide open or green space on side rarely. In humid tropics, high temperatures most of the year have an effect on thermal comfort. To fulfill thermal comfort, a housing model with inner courtyards on side and fence is proposed. In this case, representation of warm humid tropic is Surabaya. Simulation method is used to predict indoor temperature of spaces in perimeter courtyard. Spaces located on inner courtyard perimeters are living room, bedrooms and kitchen. These rooms have duration of occupancy more than one hour. With the existence of these inner courtyards, inhabitants can connect with green open space. The result show, on the hottest month, side and fence of inner courtyard model could support indoor temperature. For 14 hours, living room and bedrooms in thermal neutral zone. Except kitchen, only for 4 hours in thermal neutral zone

1. Introduction

Global warming is an issue that has been widely discussed. It has an impact on rising temperatures and sea levels. Enhancement cooling energy use in urban areas as trigger global warming. Application of bioclimatic design is an effort to reduce the impact of global warming in urban area, in response to climate change. Bioclimatic design is a strategy for designing buildings in order to provide comfort for their occupants based on climatic conditions [1]. Degree of accuracy of climate analysis holds a key factor in the bioclimatic concept that is useful for designing buildings. Accuracy of climate analysis supports efforts for thermal comfort and energy efficiency. Climate analysis in bioclimatic should use microclimate data. Bioclimatic chart principally uses some climate elements [1]. It is still valid today with some adjustments [2]. On the bioclimatic chart, there is a comfort zone, where the user feels comfortable [3]. The comfort zone can be expanded by providing wind flow [4].

Several studies of bioclimatic have been conducted. Bioclimatic concept has been applied to the proposed design in the fishing settlement. The design aims to anticipate sea level rising [5]. Design considers climatic elements, namely temperature, wind speed and rainfall. The application of bioclimatic principles to building envelope materials can reduce cooling energy consumption, including the implementation of stack effect strategy can provide better thermal conditions [6]. Mapping of distinct climate zone in Madagascar region show thermal comfort range from 20^0C to 26^0C [7]. Bioclimatic designs are generally aimed to achieve thermal comfort or physiological cooling and energy savings.

Thermal comfort is a condition where residents feel comfortable with the surrounding temperature [4]. Humans are in thermal equilibrium if they are in normal body temperature, which is around 37^0C,

IOP Conf. Series: Earth and Environmental Science **847** (2021) 012010 doi:10.1088/1755-1315/847/1/012010

while the body's skin surface temperature is around 31-34^{0}C. In terms of comfort, temperature is the main factor in determining thermal comfort. Outside temperatures influence indoor temperature in building spaces. Neutral temperature is the basis that can be used to range of thermal comfort. Thermal comfort range is +/- 2.5K from neutral temperatures [4]. Previous research show that thermal comfort for Surabaya is at 24.5 ^{0}C – 28.9^{0}C [8]. Thermal comfort range for Thailand is in range 27^{0}C - 31^{0}C [9]. In general, buildings in dry tropics have courtyard, as a center of mass orientation. Courtyard building as a building that has an indoor or semi-outdoor open space [10] and [11]. The characteristic of courtyard is that one side of courtyard faces the sky. In Iran, courtyard support non-physical comfort for inhabitants [12]. In this study, courtyard is referred as inner courtyard.

Microclimate has a role in modifying outdoor thermal comfort. This is important so that users feel comfortable doing activities in the courtyard [11]. Research on indoor temperature has been carried out on 5 types of boarding house which have courtyards [13]. The results show that the indoors temperature are 4.9-7.3^{0}C lower than the outdoor temperature for 12 hours. But at nighttime, it is higher around 0.8-1.3^{0}C for 5 hours.

Research objective was to observe indoor temperature in perimeter spaces. These spaces are spread across the two-inner courtyard in a humid tropical area. Thermal conditions in these spaces are categorized as comfortable or not, especially when they are used. If the space is passively comfortable, the cooling energy usage can be reduced. Thereby, proposed model could participate in mitigation of climate change

2. Design method

Surabaya is a city located near the equator and has a high residential growth. Therefore, Surabaya is used as a representation of warm humid tropic urban housing. Field study was used as a tool to obtain characteristics of a model proposal, and typology of spaces in courtyard perimeter. It shows that housings have two inner courtyards, rarely. The application of courtyard is one of the responses to passive cooling strategies. Propose model has two inner courtyards, at side and fence, which are assumed to be in a grid patterned residential neighborhood. It has building covered ratio of 50%. Inner courtyard in this model has sky view factor in 0.7-0.8 or 70-80% open to sky. 20-30% of courtyard area shadowed by roof. Model canyon or height to width ratio is 0.4-0.6. This ratio has a higher opportunity of wind flow compared to another ratio [14]. Wind flow accelerates heat loss by convection process.

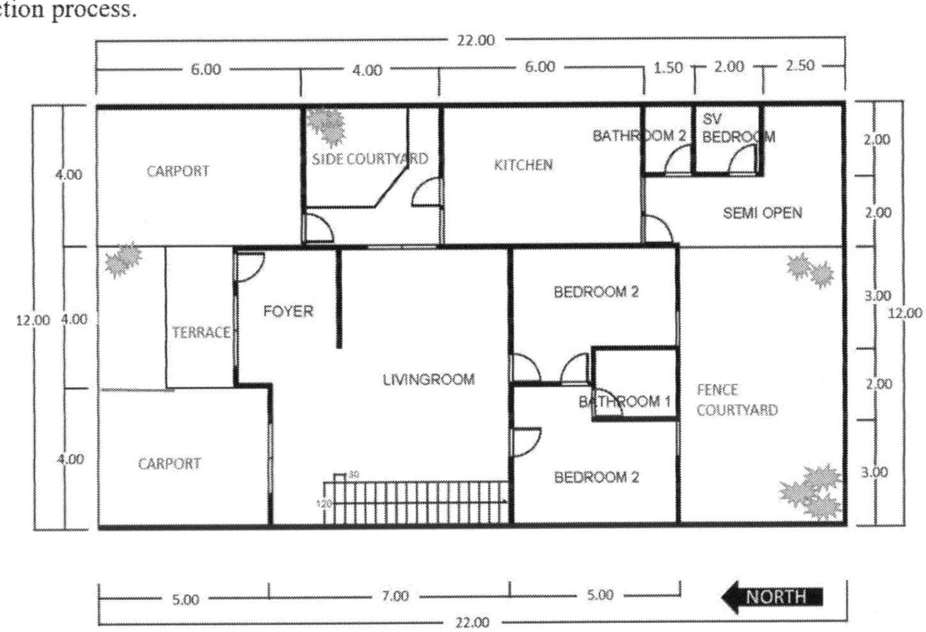

(a) First floor

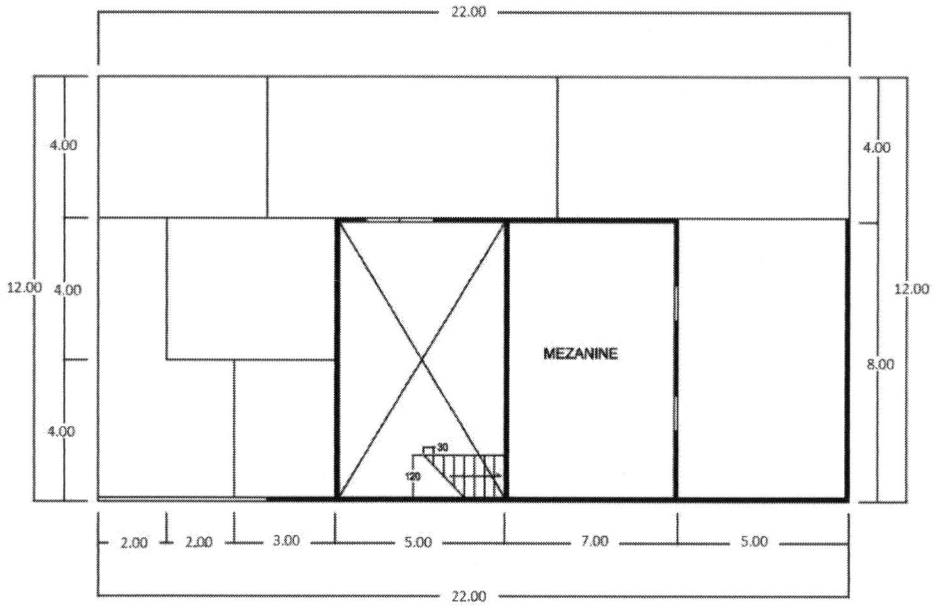

(b) Mezzanine

Figure 1. Plan of model

Simulation used design builder V4.1, for prediction indoor temperature. Simulations are carried out for 24 hours in the hottest month. The observed spaces are located on perimeter of inner courtyard. These rooms have usage duration more than 1 hour per day. The spaces on perimeter of side inner court are living room and kitchen. Living room and kitchen windows are oriented to side inner courtyard. Both of bedroom windows are oriented to fence courtyard. All spaces utilize natural ventilation. Model applies jalousie as type opening. Application of natural ventilation is expected to reduce energy use. designs that apply natural ventilation are expected to be sustainable design. The simulation assumed that wall on the model using plastered bricks, painted white. Roof material is dark red concrete tile. Model proposal can be seen in figure 1.

3. Result and discussion

3.1. Climate analysis

In this study, hottest month was determined based on climate data for the last 10 years. The climate element used to determine the hottest month is temperature. Based on data, November has highest temperature. It is over upper neutral temperature. Climate data is obtained from Meteorology, Climatology and Geophysics Bureau (BMKG), Juanda-Surabaya. Figure 2 below shows the average temperature over the last 10 years.

In Figure 2, the highest air temperature is November. The highest temperature shows 30.25^{0}C, while the lowest temperature is in July, around 26.9^{0}C. For over 10-year, the difference in highest and lowest temperature is less than 5K. Based on equation, neutral temperature is 27^{0}C [4]. Range of comfort zone is +/- 2.5. During January till September, outdoor air temperature has been in the upper limit of neutral temperature, except October-December.

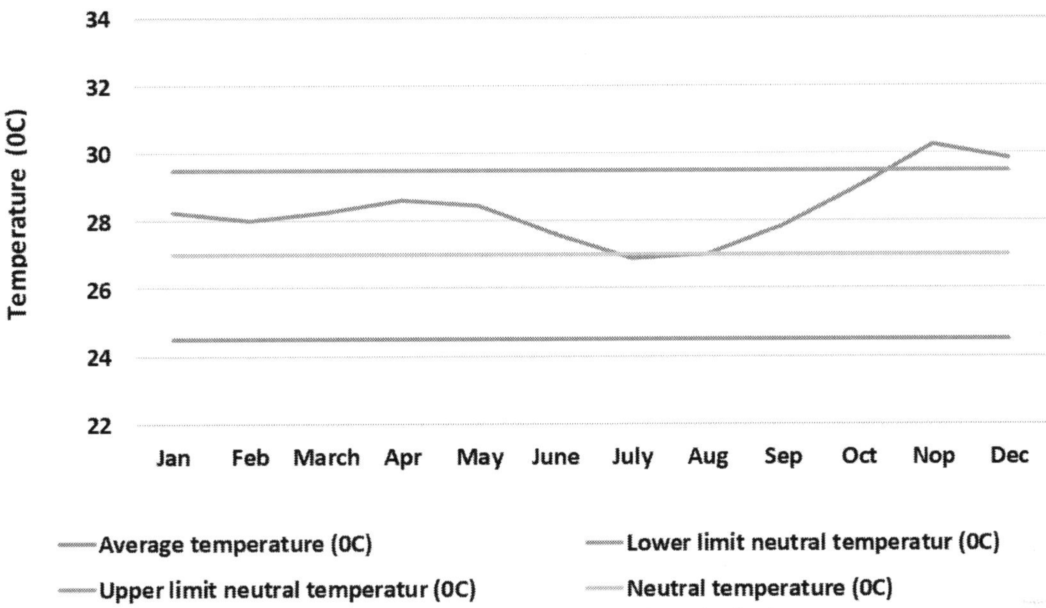

Figure 2. Average air temperature of Surabaya in the last 10 years

3.2. Indoor temperature on side inner courtyard spaces

The facade of the model is oriented to North. Short side wall of inner courtyard wall is oriented to North and South, while the long side wall is oriented to East and West. East side is bordered 9 meters wall. Wall on East side block solar radiation at 06.00-09.00 a.m. Window in living room is oriented to the east, has 30% window to wall ratio. The window at kitchen parallel to short side of courtyard wall. It has 20% window to wall ratio. When both of rooms are used at 05.00-07.00 a.m. and 4.00-10.00 p.m. all windows are opened. Apart from these hour windows are closed. Figure 3 show the result of indoor and outdoor temperatures of the model in the hottest month

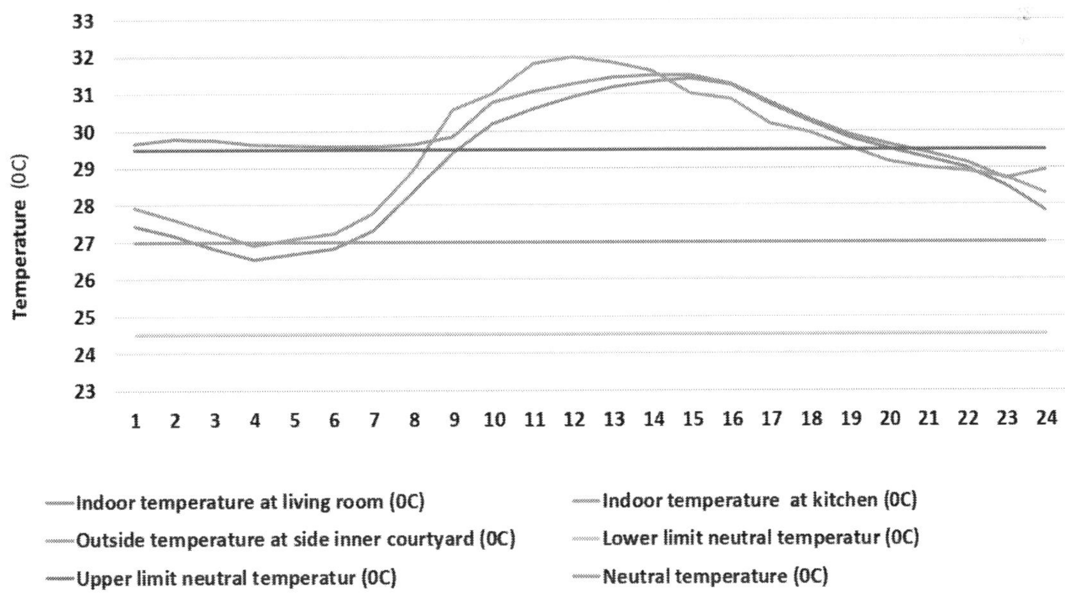

Figure 3. Indoors temperature in living room and kitchen, and outdoor temperature at side inner courtyard

The neutral temperature in this study is 27^{0}C. Range of comfort zone are 24.5-29.5^{0}C. Side of inner courtyard at 080.00 p.m. - 08.00 a.m. in comfort zone, but it is in the upper limit comfort zone. Except at 02.00-07.00 a.m., it is in the lower limit comfort zone. Outside temperature of side courtyard at 8 a.m. to 7 p.m. always above the comfort zone (see figure 3). Because the side courtyard is 70% open to sky. It means more exposure to solar radiation.

In the morning, living room is in a comfort zone. When inhabitants have activities in morning and at night, temperature in kitchen almost above comfort zone. It is caused by internal heat loads. It produces latent heats and sensible heats, such as stoves, rice cookers, dispensers. At 04.00-07.00 p.m., indoor temperature in living room is still above the comfort zone. This is influenced by outdoor temperature of side inner courtyard. Temperature is still above the comfort zone. From 08.00 p.m. till 09.00 a.m., living room is in the comfort zone. The volume of living room 3 times larger than kitchen volume, encourage slow down the heating process. At living room inhabitant can feel thermal comfort for 14 hours, or 60% of 24 hours, while at the kitchen feel comfort for 4 hours or 20% of 24 hours.

At nighttime outdoor temperatures of the side inner courtyard is in comfort zone. It provides an opportunity for heat loss convection from inside to outside building. Indoor temperature in the living room has always lower than the kitchen and side inner courtyard. The temperature difference between living room and inner courtyard approximately 0.5-1 K.

Around the side courtyard is a high wall, approximately 9 meter. On the west side is a brick wall painted dark gray. Time lag from materials previously exposed to the sun also contributes to the increase in the side courtyard temperature in the afternoon. Materials with high albedo and exposure to the sun can increase the temperature of the surrounding air during the day [15].

3.3. *Indoor temperature on fence inner courtyard spaces*

In contrast to the side inner courtyard, the long side of fence inner courtyard is facing to North-South, while the short side is East-West. East and South sides are shadowed by 9 meters high wall. During the day it receives heat from the West and South. This courtyard is almost 80% open to the sky

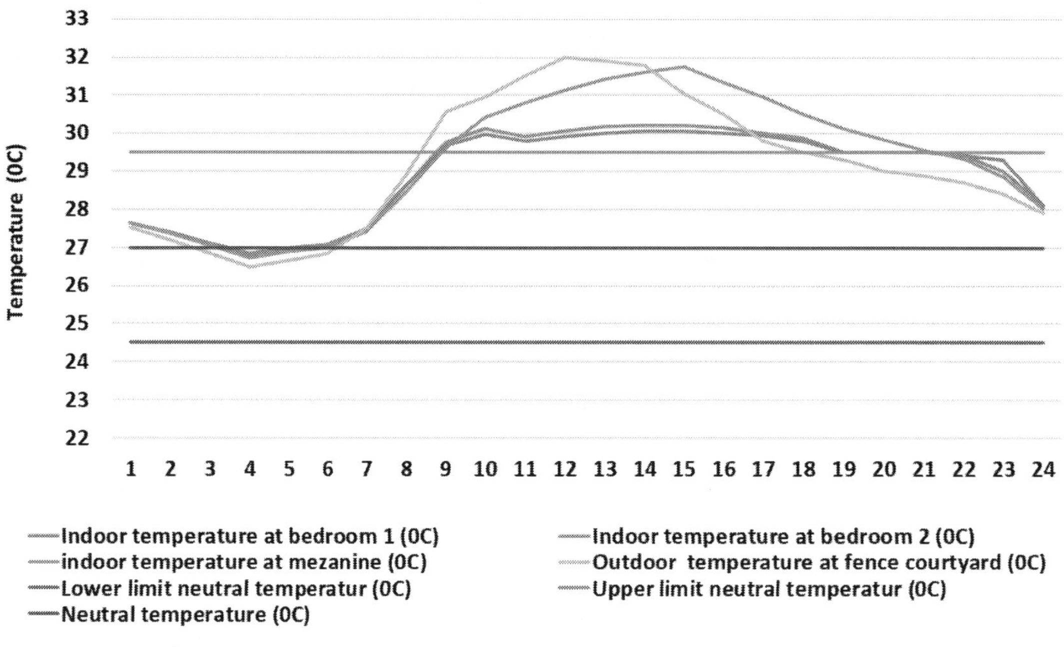

Figure 4. Indoor temperature at perimeter spaces of fence courtyard housing model

At 08.00 a.m.- 05.00 p.m., outdoor temperatures in the fence courtyard is above comfort zone. The highest temperature at 12.00 as shown in figure 4. Fence inner courtyard is assumed to be covered by plants, with a cover percentage of around 40%. Low albedo material and medium green spaces was able to reduce the temperature by around 2.1-5.30K compared to the use of low albedo material and zero greenery [16]. This means that the use of plants is able to reduce environmental temperature.

This model has two bedrooms. Two bedrooms have 30% window to wall ratio. Windows are facing to South. Inhabitants have daily activities in bedroom from 10 p.m. to 7 a.m. Its is assumed that both rooms are used all windows are opened. Apart from these hours the windows are closed. Figure 4 shows the bedrooms are in the neutral temperature zone from 7 p.m. to 8 a.m. or for 14 hours. If calculated as is 60%.

4. Conclusion remark

Open space has an important role in the fulfilment of thermal comfort. This is shown by the proposed model in this study. The model is designed to have two inner courtyards. With two inner courtyards located in different positions, it provides opportunities for spaces on the perimeter to be connected to the outer space. The spaces on the perimeter are bedrooms, family rooms that have a high space occupancy duration. The kitchen is also placed on the perimeter of the inner courtyard to get fresh air. The simulation results show that the three rooms located on the perimeter of the inner courtyard can meet the neutral temperature of 14 hours, except for the kitchen. When bedroom and living room are used, temperature in a neutral temperature zone. Another thing to note is the characteristics of the inner courtyard. It is open approximately 70-80% and gives a greater chance of sun exposure. It will raise the temperature, unless the floor material in the inner courtyard is covered with green plants.

Acknowledgments
The authors gratefully acknowledge financial support from the Institut Teknologi Sepuluh Nopember for this work, under project scheme of the Publication Writing and IPR Incentive Program (PPHKI).

References

[1] Olgyay and Victor 1963 Design with Climate, Bioclimatic approach to architectural regionalism, Princenton University Press, Princenton, New Jersey
[2] Daemei, Baghaei A et al. 2019 Bioclimatic design strategies: A guideline to enhance human thermal comfort in Cfa climate zones, *Journal of Building Engineering* **25** 100758
[3] Givoni, Baruch 992 Comfort, climate analysis and building design guidelines *Energy and Buildings* **18** 11-23
[4] Szokolay SV 2004 Introduction to Architectural Science: The Basis of Sustainable Design *Architectural Press* Oxford
[5] Defiana and Ima 2017 Open building concept for fisherman housing on the north coastal, Dimensi, *Journal of Architecture and Built of Environment* **44** 1 15-20
[6] Franco, Carvalho L, Mendes J C, Costa L C B, Pira R R and Peixoto R A F 2019 Design and thermal evaluation of a social housing model conceived with bioclimatic principles and recycled aggregates *Sustainable Cities and Society* **51** 101275
[7] Attia, Shady, Lacombea, Rakotondramiaranac H T, FrançoisGarded and Roshan G R 2019 *Analysis tool for bioclimatic design strategies in hot humid climates. Sustainable Cities and Society* 45 8-24
[8] Santosa and Mas 1993 Climatic Design for Warm Humid, Proceeding Seminar and Workshop Tropical Architecture ed: R.Aynsley, Australian institute of Tropical Architecture (AITA) Townsville, Australia
[9] Khedari and Joseph 2000 Thailand ventilation comfort chart *Energy and Building* **34** 245-249
[10] Koch-Nilsen H 2002 Stay Cool: A Design for the Built Environment in Hot Climates. London: Earthscan Dunstan House EC1N 8XA.
[11] Ghaffarianhoseini, Amirhosein, Berardi U and Ghaffarianhoseini A 2015 Thermal performance characteristics of unshaded courtyards in hot and humid climates *Building and environment* **87** 154-168

[12] Soflaeia, Farzaneh, Shokouhianb M and Zhu W (2017) Socio-environmental sustainability in traditional courtyard houses of Iran and China *Renewable and Sustainable Energy Reviews* **69** 1147–1169

[13] Nugroho A M 2020 courtyard as tropical hot humid design passive strategy: case study of Indonesian contemporary boarding houses in Surabaya Indonesia *Journal of Design and built environment* **20** 2

[14] Defiana and Ima 2012 Canyon design as potential for night ventilation on warm humid tropic housing estate *International Journal of Academic Research* **5** 5 101-106

[15] Erell, Evyatar, Pearlmutter D, Boneh D and Kutiel P B 2014 Effect of high-albedo materials on pedestrian heat stress in urban street canyons *Urban Climate* **10** 367-386

[16] Yuan, Jihui; Emuraa K and Farnhama C 2017 Is urban albedo or urban green covering more effective for urban microclimate improvement? A simulation for Osaka *Sustainable Cities and Society* **32** 78-86

Effect of addition polyethylene glycol on morphology and functional groups of composite films bacterial nanocellulose - graphite nanoplatelets

B D Susilo[1], H Suryanto[2, 3] and A Aminnudin[2]

[1]Master Program of Mechanical Engineering, Graduate School, Universitas Negeri Malang, Jl. Semarang 5, Malang 65145, Indonesia
[2]Center of Excellence for Cellulose Composite (CECCom), Department of Mechanical Engineering, Universitas Negeri Malang, Jl. Semarang 5, Malang 65145, Indonesia
[3]Centre of Advanced Material for Renewable Energy (CAMRY), Universitas Negeri Malang, Jl. Semarang 5, Malang 65145, Indonesia

heru.suryanto.ft@um.ac.id

Abstract. Polyethylene Glycol (PEG) existence on the composite film bacterial nanocellulose (BNC) reinforced with graphite nanoplatelets (GNP) has been analyzed in this research. Pineapple peel waste extract was used to synthesis the bacterial cellulose (BC) pellicle. A 1% NaOH solution was used to remove impurities from the BC pellicle. Mechanical disintegration and a high-pressure homogenization method were used to synthesize BNC. The high-pressure homogenizer (HPH) runs for five cycles at 150 bar before being filtered with a vacuum filter. The composite films are synthesized using a mixture of BNC, PEG 1%, and GNP 2.5%. Then the mixed slurry was dried using an oven for 14 hours at a temperature of 80 degrees Celsius. The morphology of composite BNC film was observed by using SEM. The addition of PEG causes the BC structure to become more porous. GNP is not well distributed in the BNC matrix but still acts as a dispersed filler. FTIR analysis is conducted to see its functional groups. The addition of GNP caused the hydroxyl group to be significantly reduced.

1. Introduction

Eco-friendly polymer composites should be biodegradable and renewable. These polymer composites are simple to compost and do not harm the environment [1,2]. Cellulose is an eco-friendly polymer, and it is an abundant natural polymer, producing many tons per year for paper and textile production [3,4]. Bacterial cellulose (BC) is pure cellulose produced by gram-negative bacteria such as *Acetobacter xylinum* and *Pseudomonas aeruginosa* [5,6], naturally formed through the bottom-up approach [7] and can produce cellulose with beneficial physical properties [8]. More than 90% of BC being water. It is made as a white hydrogel and consists of a tangled network of three-dimensional cellulose nanofiber [9] in the form of a band connected through a hydrogen bond tissue [10]. BC's chemical structure is similar to plant cellulose, but it differs in three-dimensional structure, degree of polymerization, and physicochemical characteristics [11]. Because of the cellulose chain characteristics, BC has a high crystallinity (60–80 percent). It also possesses outstanding mechanical characteristics as well as the capacity to be tuned by an in situ process [12]. Pure BC, on the other hand, is devoid of antimicrobial characteristics, optical transparency, conductivity, magnetism, and hydrophobicity [13]. BC has been utilized as a precursor for the synthesis of carbon nanofibers (CNF), which are used as an electrode material in supercapacitors and alkaline metal batteries, in certain prior research [9]. BC is also used as a binder on hybrid nanocomposites [14], and as a reinforcement agent. As a result, there are some possibilities for it to use as a binder such as electrodes of lithium battery [15,16].

As a biodegradable synthetic material, polyethylene glycol (PEG) is a biocompatible and non-toxic polymer. It's soluble in a wide range of organic solvents as well as water. [17]. The presence of polyethylene glycol (PEG) on the BC network has been shown by Liu et al. [18]. Biocompatibility and biodegradability are great in composites made of cellulose, graphene, and PEG. This research will provide a win-win method for developing composite-based environmentally friendly components [19].

The Graphite Nanoplatelet (GNP) is a 2D nano-sized carbon material composed of graphene sheets [20]. GNP can be used to alter thermal properties [21], electrical [22], mechanical [4] and electrochemical [23] in BC films [10]. In this experiment, GNP was used as a filler on the BNC matrix with PEG as a surfactant to produce a composite film of bacterial nanocellulose. Pineapple peel extract was used as the medium for synthesizing the BC pellicle. BNC was synthesized using the mechanical disintegration of BC. Then continue the homogenization process, conducted using a high-pressure homogenizer (HPH) and vacuum filtration. BNC, PEG, and GNP are dissolved and mixed by using a distillate water media. The purpose of this study was to see how adding PEG to BNC, and BNC/GNP composites affected their morphology and functional groups.

2. Materials and methods

2.1. Materials

Pineapple peel extract is used as a medium of BC culture. Other materials in the manufacture of culture media, supplied from CV. Makmur Sejati Malang, are glucose (C6H12O6), PEG 4000, urea, and solid NaOH. *A. xylinum* bacteria used as a fermentation agent obtained from the Applied Technology Lab of Muhammadiyah University of Malang, Indonesia. GNP was provided from CV. Gamma Scientific Biolab, Malang, East Java, Indonesia.

2.2. Synthesis of BC

Suryanto et al. method is used in this BC synthesis [24]. A high-speed blender was used to mix 300 grams of rotting pineapple peel with up to 1 liter of water for 1.5 minutes. Additional water was poured till the level reached 2 liters. Then mix for 1 minute in the blender. After that, strain the juice to remove the extract and heat until it boils. After boiling, the mixture was mixed with 7.5 wv% glucose and 0.5 wv% urea, then allowed to cool to room temperature. After cooling, 1 vv% *A. xylinum* bacterial starter was introduced and cultured for 10-14 days. The BC pellicle was collected and then cleaned with water before being cooked for up to 2 hours at 90°C with a NaOH 1 percent solution. Finally, it was rewashed until it was completely neutral.

2.3. Synthesis of BNC

The steps to synthesize BNC refer to Sardjono et al. [25]. Approximately about 50 gr of pellicle with 300 ml of distilled water was mechanically disintegrated using a high-speed blender for 1.5 minutes. Then 700 ml of distilled water was added and continue to blend for 2 minutes. BC solution homogenized using HPH at 150 bars as much as five cycles. It is then filtered using a vacuum filter.

2.4. Synthesis of composite BNC-PEG films

Ten grams of BNC dissolved using distilled water up to 200 ml and then stirred for 20 minutes. Next, PEG with a percentage of 1% was added. The solution then stirred using a magnetic stirrer for 25 minutes, continuing with the homogenization process in an ultrasonic homogenizer for 20 minutes. The sonication results are placed on a glass mould coated in aluminium foil and dried using an oven for 14 hours at 80°C [17]

2.5. Synthesis of composite BNC-PEG-GNP films
Ten grams of BNC dissolved using distilled water up to 200 ml volume. It was then stirred using a magnetic stirrer for 20 minutes. Next, PEG 1 wt% and GNP 2.5 wt% were added to the solution. Then continue to stir for 25 minutes. The solution is then sonicated using an ultrasonic homogenizer for 20 minutes. Finally, results are placed on a glass mould coated in aluminium foil and dried using an oven for 14 hours at 80°C [17].

2.6. Morphological analysis
The morphological observation was conducted using SEM, type Inspect-S50. Specimens were coated with 10 nm gold before it was observed.

2.7. FTIR analysis
The function group was analyzed at Advanced Mineral and Material Laboratory, Faculty of Science, State University of Malang, Indonesia, using the FTIR Shimadzu IR Prestige-21 Spectrum.

3. Results and discussion

3.1. Morphology
SEM observations are shown in Figure 1. The disintegration and HPH process already has resulted in nano-sized BNC fibers. Figure 1 (a) shows the morphology of BNC at magnification 50,000 times. This treatment can get 30-50 nm in diameter of BNC [25]. Also, the cellulose contact surface becomes wider and strengthens the possibility of interaction between cellulose in BNC and carbon in GNP.

Figure 1 (b) shows the results of an SEM micrograph from a BNC with a PEG 1 wt% film sample. The surface morphology, shown in Figure 1 (b) and (c), is a rough structure and severe wrinkles. PEG PEG makes the BC structure more porous because it increases the distance between the fibers. Figure 1 (c) shows the non-uniform dispersion of GNP in the BNC matrices. This is because GNP has a hydrophobic feature, while PEG is a hydrophilic polymer. Therefore, GNPs are challenging to be dispersed in PEG because of the weak interaction between GNPs and PEG [26].

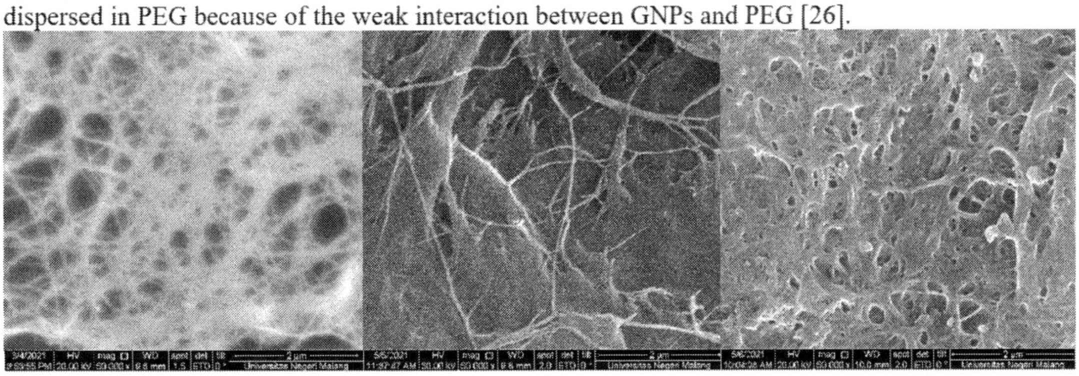

(a) (b) (c)

Figure 1. SEM observations of BNC-PEG-GNP composite film micrographs; (a) pure BNC films; (b) BNC with PEG 1 wt%; (c) composite films of BNC with PEG 1 wt% and GNP 2,5 wt%.

As a dispersed filler, GNP is distributed in the BNC matrix. GNP is hydrophobic, so GNP tends to agglomerate when dispersed in a water-based solution and forms a thicker layer on the interface [25]. The development of cellulose-PEG interaction interactions, which are favored owing to the high concentration of hydrogen bonds, might promote agglomeration of BNC particles [4]. Furthermore, hydrogen bonding between molecules and intramolecular hydrogen bonds between three hydroxyl groups of recurring units cause BNC fibers to cluster [27].

3.2. FTIR analysis

FTIR analysis begins by analyzing the FTIR spectrum graph of the BC/PEG/GNP composite film. The hydroxyl group (O-H) consists of intramolecular and intermolecular hydrogen bonds at wavenumber of 3200-3400 cm^{-1} [28]. Figure 2 shows that with the addition of PEG and GNP, the number of O-H bonds detected at the peak of 3412 cm^{-1} decreases. It indicates a reduction in the number of hydrogen bonds when the GNP content in the composite increases. Decreasing the number of hydrogen bonds can reduce the mechanical properties of composite films [4]. Then at 2899 cm^{-1}, a peak belongs to the C-H stretch due to the presence of CH and CH_2 groups in cellulose. A peak appears at 1647 cm^{-1}, corresponding to the –OH bond [29]. O-H and C=O groups strongly decreased. It may be caused by the cross-linking reaction between the O-H group of GNP, BNC, and PEG [1,31]. The C-O cluster that appears at the peak of 974 cm^{-1} tends to decrease and shift slightly because of the existence of GNP, and the thermal stability of the composite film will increase [13]. With the increasing temperature of the composite film during the drying process, rapid oxygen discharge occurs in CO and CO_2 gases [31]. The –OH functional group did not change and only shifted slightly after PEG was added to BC, while the addition of PEG caused the amount of –OH to decrease significantly.

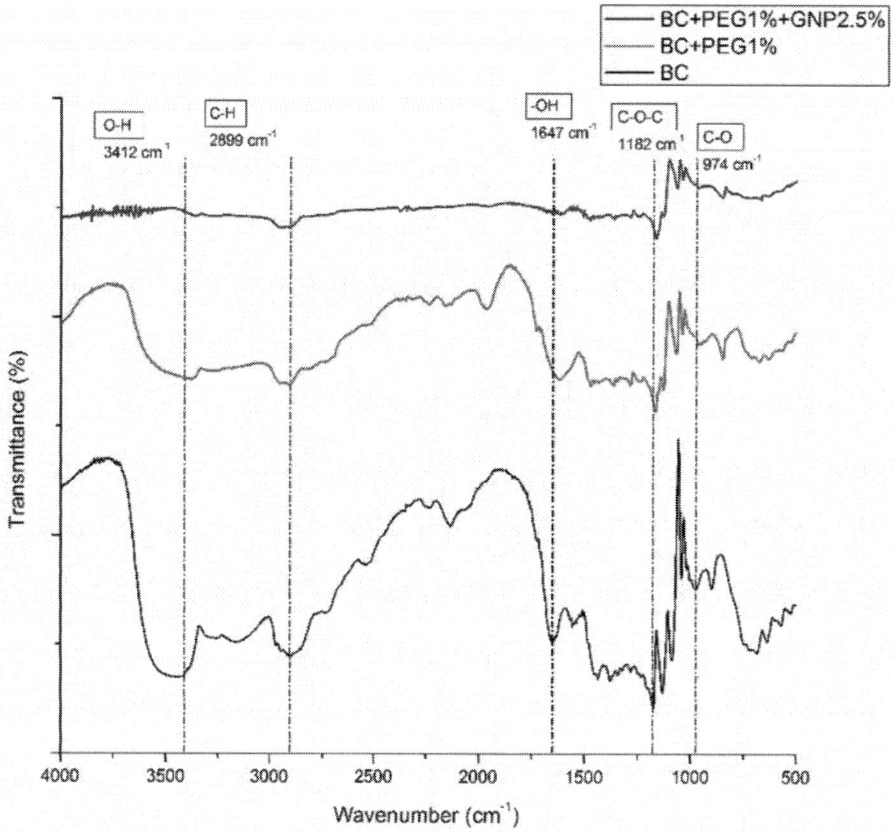

Figure 2. FTIR spectroscope of BNC/PEG/GNP composite films.

GNP is a hydrophobic material. It tends to agglomerate when dispersed in a water-based solution. It also could form a thicker layer on the interface [32]. The O-H group created high surface energy on the surface of nanocellulose, which that surface energy leads to insufficient wetting and weak interface adhesion with hydrophobic polymers [33]. So, it can be predicted that there was no chemical bonding between BNC and GNP. GNP just merged into the BNC matrix and made the structure become conductive film [2].

4. Conclusion

The composite film from combination BNC/PEG/GNP had been produced successfully. GNP agglomeration may cause by the hydrogen bonds that occur between BNC fibres. Since GNP was hydrophobic, most of them agglomerated while dissolved in water solvents. GNP was difficult to distribute throughout the composite film because of the hydrophilic properties of BNC. Low molecular weight PEG addition can increase the porosity of BNC. The addition of GNP caused the hydroxyl group to be significantly reduced.

Acknowledgement
A great appreciation was delivered to LP2M UM through the PNBP-PUI research grant 2021.

References

[1] El-Sakhawy M, Kamel S, Salama A and Tohamy H-A S 2018 Preparation and infrared study of cellulose based amphiphilic materials *J. Cellul. Chem. Technol.* **52** 193–200
[2] Tohamy H-A S, Anis B, Youssef M A, Abdallah A E M, El-Sakhawy M and Kamel S 2020 Preparation of eco-friendly graphene oxide from agricultural wastes for water treatment *Desalin. WATER Treat.* **191** 250–262
[3] Mahsuli T, Suryanto H, Larasati A and Muhajir M 2019 Mechanical properties of bacterial nanocellulose membrane from pineapple peel waste after homogenization process *AIP Conf. Proc.* **2120**
[4] Yi Z, Ren Y, Jaffer S, Tjong J, Sarshar Z and Sain M 2021 Facile fabrication of electrically conductive graphitic cellufoil for lighting and energy devices *Compos. Part B Eng.* **207** 108593
[5] Illa M P, Sharma C S and Khandelwal M 2021 Catalytic graphitization of bacterial cellulose–derived carbon nanofibers for stable and enhanced anodic performance of lithium-ion batteries *Mater. Today Chem.* **20** 100439
[6] Iguchi M, Yamanaka S and Budhiono A 2000 Bacterial cellulose - a masterpiece of nature's arts *J. Mater. Sci.* **35** 261–270
[7] Klemm D, Kramer F, Moritz S, Lindström T, Ankerfors M, Gray D and Dorris A 2011 Nanocelluloses: A new family of nature-based materials *Angew. Chemie - Int. Ed.* **50** 5438–5466
[8] Agustin T A and Putra A 2019 The Effect of Addition of Polyethylene Glycol (PEG) on Biodegradable Plastic Based on Bacterial Cellulosa from Coconut Water (Coconus Nucifera) *Int. J. Progress. Sci. Technol.* **17** 50–57
[9] Illa M P, Khandelwal M and Sharma C S 2018 Bacterial cellulose-derived carbon nanofibers as anode for lithium-ion batteries *Emergent Mater.* **1** 105–120
[10] Tayeb A, Amini E, Ghasemi S and Tajvidi M 2018 Cellulose Nanomaterials—Binding Properties and Applications: A Review *Molecules* **23** 2684
[11] Melliawati R and Djohan A C 2013 Analisis Karboksimetil Selulosa Dari Bakteri Acetobacter xylinum dan Acetobacter sp. RMG-2 *Ber. Biol.* **12** 335–344
[12] Römling U and Galperin M Y 2015 Bacterial cellulose biosynthesis: diversity of operons, subunits, products, and functions *Trends Microbiol.* **23** 545–557
[13] Erbas Kiziltas E, Kiziltas A, Rhodes K, Emanetoglu N W, Blumentritt M and Gardner D J 2016 Electrically conductive nano graphite-filled bacterial cellulose composites *Carbohydr. Polym.* **136** 1144–1151
[14] Juntaro J, Pommet M, Mantalaris A, Shaffer M and Bismarck A 2007 Nanocellulose enhanced interfaces in truly green unidirectional fibre reinforced composites *Compos. Interfaces* **14** 753–762
[15] Dutta S, Kim J, Ide Y, Kim J H, Hossain M S A, Bando Y, Yamauchi Y and Wu K C-W 2017 3D network of cellulose-based energy storage devices and related emerging applications *Mater. Horizons* **4** 522–545

IOP Conf. Series: Earth and Environmental Science **847** (2021) 012011 doi:10.1088/1755-1315/847/1/012011

[16] Sabo R, Yermakov A, Law C T and Elhajjar R 2016 Nanocellulose-enabled electronics, energy harvesting devices, smart materials and sensors: A review *J. Renew. Mater.* **4** 297–312
[17] Serag E, El Nemr A and El-Maghraby A 2017 Synthesis of highly effective novel graphene oxide-polyethylene glycol-polyvinyl alcohol nanocomposite hydrogel for copper removal *J. Water Environ. Nanotechnol.* **2** 223–234
[18] Liu H F, Feng Y H, Yan J K, Zhang D La and Li J C 2012 Influence of adding different poly (ethylene glycol) in the biosynthesis of bacterial cellulose *Advanced Materials Research* vol 418 (Trans Tech Publ) pp 589–592
[19] Tohamy H A S, El-Sakhawy M and Kamel S 2021 Carboxymethyl Cellulose-Grafted Graphene Oxide/Polyethylene Glycol for Efficient Ni(II) Adsorption *J. Polym. Environ.* **29** 859–870
[20] Aritonang H F, Wulandari R and Wuntu A D 2020 Synthesis and Characterization of Bacterial Cellulose/Nano-Graphite Nanocomposite Membranes *Macromol. Symp.* **391** 1900145
[21] Perets Y, Aleksandrovych L, Melnychenko M, Lazarenko O, Vovchenko L and Matzui L 2017 The Electrical Properties of Hybrid Composites Based on Multiwall Carbon Nanotubes with Graphite Nanoplatelets *Nanoscale Res. Lett.* **12**
[22] Zhou T, Chen D, Jiu J, Nge T T, Sugahara T, Nagao S, Koga H, Nogi M, Suganuma K, Wang X, Liu X, Cheng P, Wang T and Xiong D 2013 Electrically conductive bacterial cellulose composite membranes produced by the incorporation of graphite nanoplatelets in pristine bacterial cellulose membranes *Express Polym. Lett.* **7** 756–766
[23] Li Y, Sun Z, Liu D, Lu S, Li F, Gao G, Zhu M, Li M, Zhang Y, Bu H, Jia Z and Ding S 2020 Bacterial Cellulose Composite Solid Polymer Electrolyte With High Tensile Strength and Lithium Dendrite Inhibition for Long Life Battery *ENERGY Environ. Mater.* eem2.12122
[24] Suryanto H, Sutrisno T A, Muhajir M, Zakia N and Yanuhar U 2018 *MATEC Web of Conferences* **204** 05015
[25] Sardjono S A, Suryanto H, Aminnudin and Muhajir M 2019 *AIP Conference Proceedings* **2120** 080015
[26] Wei X, Jin X zheng, Zhang N, Qi X dong, Yang J hui, Zhou Z wan and Wang Y 2021 Constructing cellulose nanocrystal/graphene nanoplatelet networks in phase change materials toward intelligent thermal management *Carbohydr. Polym.* **253** 117290
[27] Long L-Y, Weng Y-X and Wang Y-Z 2018 Cellulose Aerogels: Synthesis, Applications, and Prospects *Polymers (Basel).* **10** 623
[28] Hishikawa Y, Togawa E and Kondo T 2017 Characterization of Individual Hydrogen Bonds in Crystalline Regenerated Cellulose Using Resolved Polarized FTIR Spectra *ACS Omega* **2** 1469–1476
[29] Sun Y, Meng C, Zheng Y, Xie Y, He W, Wang Y, Qiao K and Yue L 2018 The effects of two biocompatible plasticizers on the performance of dry bacterial cellulose membrane: a comparative study *Cellulose* **25** 5893–5908
[30] El-Sakhawy M, Kamel S, Salama A, Youssef M A, Elsaid W and Tohamy H 2017 Amphiphilic cellulose as stabilizer for oil/water emulsion *Egypt. J. Chem* **60** 181
[31] Qin H, Zhou Y, Huang J, Zhang C, Wang B, Jin S and Zhou Q 2017 Catalytic Graphitization Strategy for the Synthesis of Graphitic Carbon Nanocages and Electrochemical Performance *Int. J. Electrochem. Sci* **12** 10599–10604
[32] Alberts M, Kalaitzidou K and Melkote S 2009 An investigation of graphite nanoplatelets as lubricant in grinding *Int. J. Mach. Tools Manuf.* **49** 966–970
[33] Dufresne A and Belgacem M N 2013 Cellulose-reinforced composites: from micro-to nanoscale *Polímeros* **23** 277–286

IOP Conf. Series: Earth and Environmental Science **847** (2021) 012012 doi:10.1088/1755-1315/847/1/012012

Inoculation of *Bacillus cereus* enhance phytoremediation efficiency of *Pistia stratiotes* and *Eichhornia crassipes* in removing heavy metal Pb

N Z Zahari[1], P M Tuah[1] and S A Rahim[1]

[1]Faculty of Science & Natural Resources, Universiti Malaysia Sabah, UMS Road, 88400, Kota Kinabalu Sabah, Malaysia.

zaidazahari@ums.edu.my

Abstract. This study investigates the effects of inoculation rhizhosperic microbes namely *Bacillus cereus*-NMeHI-Cr2 towards the roots of two locally aquatic macrophytes in removing heavy metals Pb from simulated wastewater. Water lettuce (*Pistia stratiotes*) and water hyacinth (*Eichhornia crassipes*) were selected in this study due to their absorption capability and acts as biological filters by accumulating heavy metals in their tissue's bodies. The inoculation of *B. cereus*-NMeHI-Cr2 onto the roots of *P. stratiotes* and *E. crassipes* was conducted in non-continuous horizontal flow (NCHF) reactor with addition of 1.5mg/L of Pb comprising of T2: treatment with inoculation of *B. cereus*-NMeHI-Cr2, T4: Non inoculation of *B. cereus*-NMeHI-Cr2 and T6: no plant growth in NCHF reactor (control). After 15 days of cultivation, the treatment integrated with inoculation of *B. cereus*-NMeHI-Cr2 resulted 83.7% removal efficiency of Pb followed by T4 (70.4%) and T6 (2%) respectively. Furthermore, 63.8 ug/g of Pb was observed in T2 treatment with similar distribution pattern as roots > leaves > stalks in *E. crassipes* and roots> leaves in *P. stratiotes*. The physical changes of both plants were also monitored to show any toxicity symptoms experienced as a result of heavy metal Pb. This study validates that the removal of Pb was done via rhizofiltration mechanism as the value of enrichment factor was found below than 1000. The identified plants species with inoculation of *Bacillus cereus*-NMeHI-Cr2 provides a simple and cost-effective method for application of metal polluted water.

1. Introduction

Water pollution comes in various forms with its associated effects, depending on the type of pollutant entering the environment [1]. Some heavy metal for example lead is widely used due to its physical and chemical properties such as malleability and ductility. Thus, it can be found in many industries and places. Sources of lead includes leaded gasoline, smelting of lead, lead based paint, as well as ingestion of water obtained from leaded pipes. The emission of lead from these activities can be bioaccumulated in the bones of living things that can subsequently be released into the blood during pregnancies, menopause, or lactation, causing deleterious effects (anaemia, kidney damage, fertility reduction and decrease cognitive function) [2]. Besides that, the excessive uptake of lead in plants will induce stunted growth, the blackening of roots, as well as chlorosis [3]. Phytoremediation can be defined as a low cost, energy-friendly, as well as aesthetic method of using green plants to restore, detoxify, and reclaim places that are polluted with varying levels of environmental pollutants [4]. This method required simple operation and maintenance, and at the same time is efficient in the removal of various types of pollutants such as organic matter, nutrients, and heavy metals [5]. The combination with rhizoremediation technique is one of the bioremediation processes through which root are inoculated with beneficial microorganisms to increase the efficiency of contaminants in the extraction process [6]. This process works as win-win situation whereby the attachment of microorganisms would increase the availability of the compounds; thus, the roots of plants help in the removal of compound in contaminated water. Previous studies on phytoremediation had shown that the inoculation of beneficial microorganisms towards roots of plants have advantages to host plants against heavy metal stress [7]. Therefore, this study aims to highlight the rhizoremediation effect of both plants with inoculation of *B. cereus*-NMeHI-Cr2 in removing heavy metal Pb from simulated wastewater. The distribution patterns of lead in both plant tissues will also discussed in this paper.

2. Materials and method

2.1. Aquatic macrophytes sampling
Two locally indigenous aquatic macrophytes *Pistia stratiotes* (water lettuce) and *Eichhornia crassipes* (water hyacinth) were selected to access their ability in removing metal Pb from simulated wastewater under laboratory scale. *P. stratiotes* and *E. crassipes* (Figure 1) were sampled from unpolluted sites (Kg. Bolong Lama, Tuaran and Kg. Kibabaig, Penampang), Sabah and transported to the laboratory using 20L tank. Both plants were rinsed with tap water and distilled water to remove waste and any undesirable sediment. Cleaned plants were placed into a 40L tank and were acclimatized for 2 weeks in UMS lake water at the laboratory for prior experiments. Light was provided by fluorescent light in the lab and cultivation at room temperature.

(a) (b)

Figure 1. Selected plants: (a) *P. stratiotes* and (b) *E. crassipes* growing in natural habitat used as phytoremediation of heavy metal Pb.

2.2. Experimental set-up
Non-continuous horizontal flow (NCHF) reactor was designed with 1 glass tanks being set up horizontally. The measurements glass of the tank is (30 x 60 x 30 cm) as illustrated in Figure 2. A bacterial suspension with 10^8 cell mL^{-1} was prepared and submerged in the roots of plants for 10 minutes as described by [8]. The details of experiments conducted were presented in Table 1.

Figure 2. Experimental set-up on the removal of Pb using Non-Continuous Horizontal Flow (NCHF) reactor

Table 1. Different types of treatment on inoculation of *B. cereus*-NMeHI-Cr2 onto roots of *P. stratiotes* and *E. crassipesin* using NCHF reactor

Treatment (T)	Experiment
T2	50g of *P. stratiotes* and *E. crassipes* were inoculated with rhizospheric bacteria *B. cereus*-NMeHI-Cr2. Approximately 1.5 mg/L of Pb was added at 1.5 mg/L concentration in 40L Lake UMS water.
T4	50g of *P. stratiotes* and *E. crassipes* were cultivated in 40L Lake UMS water with 1.5 mg/L concentration of Pb was added in the NCHF reactor.
T6 (Control)	There are no plants cultivated in 1.5 mg/L concentration of Pb in the NCHF reactor.

2.3. Sample analysis

The simulated wastewater was sampled at 15 days intervals throughout the experiment. The concentration of Pb in the samples were collected in duplicate each day with the use of clean 50 mL polyethene bottles. Prior to sampling, the simulated wastewater was gently stirred to ensure the collection of a homogenous sample. For heavy metal analysis in the plant samples, the roots, leaves, and stalks were collected at Day 0 and day 15 in duplicate. The samples obtained were placed into a clean aluminium container and dried in an oven at 75°C for 24 hours. The samples were digested using Nitric acid as according to USEPA method 3050b. The final solution was poured into an acid - washed polyethene bottle and placed in refrigeration at below 4°C prior to heavy metal analysis using Inductively Coupled Plasma Spectroscopy (ICP – OES). Standard calibration curve for metal Pb was established to determine their concentrations.

2.3.1. Removal efficiency and distribution of metals Pb by both aquatic plants

Heavy metal removal efficiency was used to determine how well both plants to remove Pb from simulated wastewater [9]. The formula used to calculate heavy metal removal efficiency is shown below:

$$Removal\ (\%) = \frac{Ti-Tf}{Ti} x100 \quad (1)$$

where:

T_i = Total Pb initial (mg/L)

T_f= Total Pb final (mg/L)

$$Uptake\ of\ Pb\ in\ plant\ parts\ (ug/g) = \frac{Cf - Ci}{Dw} x\ T \quad (2)$$

where:
C_f = Concentration final in plant parts (ug/mL)
C_i = Concentration initial in plant parts (ug/mL)
T_V = Total volume (mL)
D_W = Weight of biomass (g)

2.3.2. Phytoremediation mechanism uptakes of P.stratiotes and E. crassipes by determining the Enrichment Factors (EF) and Translocation Factors (TF).

The enrichment factors and translocation factors of heavy metal Pb in simulated wastewater and plant parts were calculated. The enrichment factors (EF) measuring the degree of enrichment and transfer of metals into both plant samples from simulated wastewater was evaluated by the ratio of the metal in plants (ug/g) to the metal in the simulated wastewater (mg/L). The translocation factors (TF) which refers to the transfer of metals (ug/g) to the shoot from the roots (ug/g). The EF [10] and TF [11] formulae was calculated as below:

$$EF = \frac{C\ plants}{C\ water} \quad (3)$$

where:
EF = Enrichment factors
C_{plants} = concentration in plants
C_{water} = concentration in water

$$TF = \frac{C\ shoot}{C\ root} \quad (4)$$

where:
TF = Translocation factors
C_{shoots} = concentration in shoots
C_{roots} = concentration in roots

3. Results and discussion

3.1. Effect of inoculation on removal efficiency of Pb by both aquatic plants in NCHF reactor

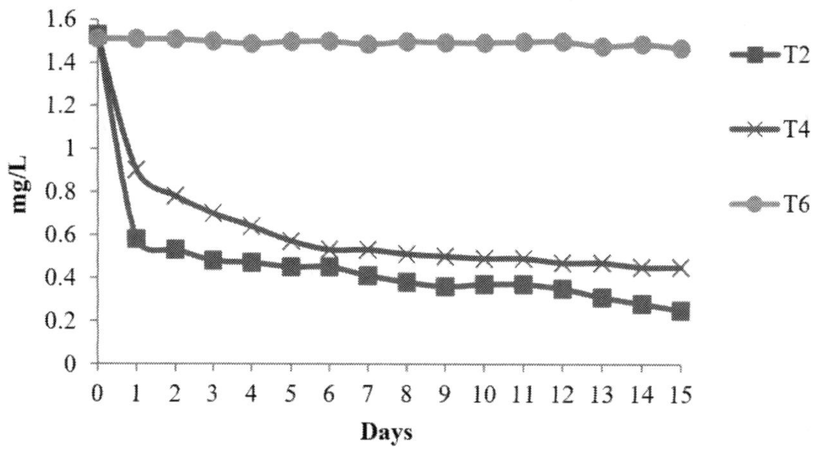

Figure 3. Pb removal efficiency by *P. stratiotes* and *E. crassipes* in NCHF reactors at different treatments: T2: inoculation of *B. cereus*-NMeHI-Cr2; T4: Non inoculation of *B. cereus*-NMeHI-Cr2 and T6: Non cultivation of plants in NCHF reactor (Control)

Figure 3 shows the residual concentration of Pb at different times upon different exposure experiments in NCHF reactors. On the contrary, treatment T6 (without plants) in control reactors do not evidence significant changes. The data shows that experiment containing *P. stratiotes* and *E. crassipes* with inoculation of *B. cereus*-NMeHI-Cr2 (T2) has a sudden drop on the first days of the experiment and the removal efficiency was continuing to decrease until day 15. On the other hand, T4 with non-inoculation of rhizospheric microbes shows that the removal of Pb was slightly decrease. Overall, after 15 days of the experiment the removal of Pb was highest in treatment T2 with 84% followed by T4 (70%) and control; T6 (2%) respectively. The results indicates that the effects of rhizospheric microbes towards both plants had increased the removal efficiency of Pb. This finding agrees with [12] stated that inoculation of beneficial microorganisms in the roots had aid in Pb removal process.

The characteristics of *B. cereus* -NMeHI-Cr2 which is Gram positive bacteria that contains of teichoic acids and acids associated to the cell wall, plays an important role in the uptake of Pb. These results are in line with [13], whereby the treatment of contaminated soil with heavy metal by *Trofoliumrefens* had shown that removal efficiency of metals is highly increased with inoculation of *B. cereus* as compared to control treatment. He found that the associated acids Indole acetic acid (IAA) was produce by *B. cereus* during the treatment process thus increase the bioaccumulation ability exhibited by the bacterial cells. The data proved that the inoculation of *B. cereus* -NMeHI-Cr2 towards roots of *P. stratiotes* and *E. crassipes* had increased the amount of Pb removal by both plants. Along the experiment, the physical changes of both plants were observed indicates that toxicity symptoms might occur due to toxic effect of Pb. Based on the observation, both plants in treatment T2 were recorded healthy within the exposure times of experiments. However, a physical change on *P. stratiotes* was observed in T4 whereby the leaves of this plant are wilting and yellowing after day 10 of exposure. This is due to the toxic effect of Pb whereby the capability to tolerate to this high concentration of Pb are decreased as exposure time is increased. Table 2 shows the summary observation that have been recorded during 15 days of the experiment.

Table 2. Pb removal efficiency by *P. stratiotes* and *E. crassipes* using NCHF reactors and toxicity symptoms observed during 15 days of treatment

Treatments	Removal (%)	Toxicity symptoms								
		Roots				Leaves				Stalks
		Fragile		Decay		Yellowing		Wilting		Wilting
		P	*E*	*P*	*E*	*P*	*E*	*P*	*E*	*E*
T2 - Inoculation of *B. cereus*-NMeHI-Cr2	**83.7**	-	-	-	-	√	-	√	-	-
T4 - Non inoculation of *B. cereus*-NMeHI-Cr2	**70.4**	-	-	-	-	√	-	√	-	-
T6 - Control (No cultivation of plants)	**2.0**	-	-	-	-	√	-	√	-	-

*P = P. stratiotes, *E = E. crassipes* * √= positive physical changes observed

3.2. Uptake and distribution of Pb in plant tissues of P. stratiotes and E. crassipes

Figure 4 showed the accumulation of Pb in plant tissues of *P. stratiotes* and *E. crassipes* after 15 days of cultivation. The results showed that the highest uptake of Pb was observed in roots of *E. crassipes* inoculated with *B. cereus*-NMeHI-Cr2 with 25.6 ug/g whereas the stalks in T4 reactor showed the lowest uptake with 6 ug/g. It can be note that a similar pattern of distribution in both plants' parts can be found higher in roots as compared to leaves and stalks. This finding is similar to previous studies done by [14] mentioned that the amounts of metals are mainly deposited in roots rather than leaves and stalks. [14] stated that the roots is the major storage and collection of metals in plants followed by vacuoles of epidermis and bundle sheath of leaves respectively.

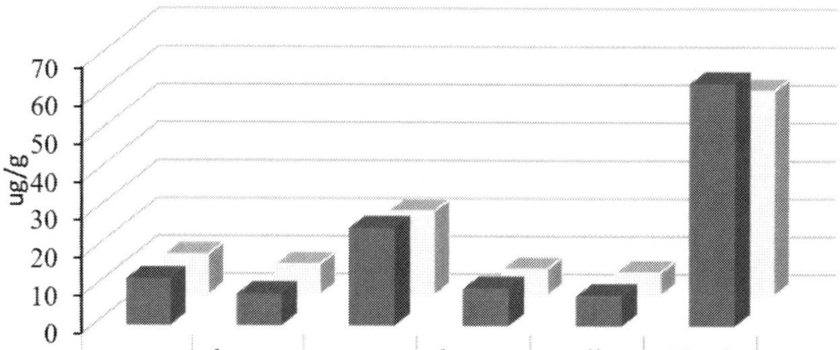

	roots	leaves	roots	leaves	stalks	Total PS+EC
T2	12.2	8.3	25.6	9.8	7.9	63.8
T4	10.66	8.1	22.1	6.9	6	53.9

Figure 4. Distribution of 1.5 mg/L of Pb in roots and leaves (*P. stratiotes*) and roots, leaves and stalks (*E. crassipes*)

The summary distribution patterns of these metals in plant tissues of *P. stratiotes* are in descending order from roots to leaves whereas for *E. crassipes* the pattern is as roots> leaves> stalks. It is important to note that the correlation of the amount of metal removed from simulated wastewater with the amount of metal uptake by both plants was significant. This is proved by the metal's uptake by both aquatic plants in NCHF reactor with the decreasing metals concentration after 15 days of the cultivation. It is important to note that the effects of rhizospheric microbes and roots of both plants had great influence in uptake Pb from simulated wastewater. In this study, both plants and *B. cereus*-NMeHI-Cr2 had shown synergistic effects whereby the plant provides bacteria with a specific carbon source, thus this induces bacteria to reduce the phytotoxicity of contaminated water. It is interesting to know that roots of *P. stratiotes* and *E. crassipes* form an association with *B. cereus*-NMeHI-Cr2 which increase the remediation activity thus stimulate the microbial community to boost metabolic activity in degrading Pb [15]. The present findings suggested that the effects of rhizospheric microbes (*B. cereus*-NMeHI-Cr2) increase both plant biomass and thereby stabilize the biochemical process [16]. Besides that, *B. cereus*-NMeHI-Cr2 have shown to produce auxin which in turn eliciting the plant growth, increase the metal uptake and enhance systematic resistance. This suggest that that both plants can be used as phytoremediation plants for removal of Pb in industrial wastewater.

3.3. Enrichment factor of Pb in P. stratiotes and E. crassipes

Figure 5 showed the enrichment factors of Pb in roots and shoots of *P. stratiotes* and *E. crassipes* cultivated in NCHF reactors. The EF of Pb in both plants are observed in range of 12.3–28.3. The highest EF can be found in *E. crassipes* in T2 with (28.3) whereas the lowest can be observed in *P. stratiotes* in T4 with (12.3). This finding shows that the roots of both plants play as a main role in close with simulated wastewater. As roots is the main source of nutrients for these plants, the inoculation of *B. cereus*-NMeHI-Cr2 in T2 had showed that both plant roots growth healthy with hairy branching roots thus makes it available in absorbed metals into roots as compared to shoots. This finding is in line with [17], whereby the effects of rhizospheric microbes towards both roots system had resulted in high EF of roots as compared to shoots in both plants studied.

According to [16] some bacteria had shown internal absorption of metal in certain part of bacteria cells or dispersed throughout the cell in nanoparticle and had a mechanism in protecting itself from acute toxicity effect by the shortening and forming wooly coat around the cells in order to decrease surface area from further contact with metal ions. However, both plants in this study are not considered as hyperaccumulator of Pb as the EF obtained in roots and shoots are below than 1000 [10]. Since the EF of Pb in both plants are in the range of 12.3 – 28.3, they are only best classified as rhizofiltration plants as suggested by [10].

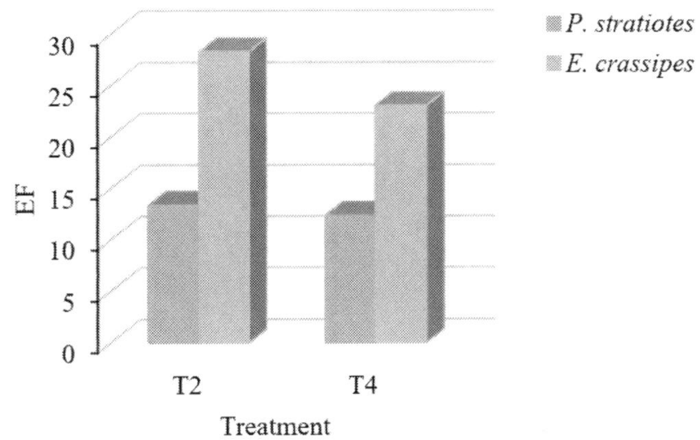

Figure 5. Enrichment factors of Pb in roots and shoots of *P. stratiotes* and *E. crassipes* at different treatments studied

3.4. Translocation factor of Pb in P. stratiotes and E. crassipes

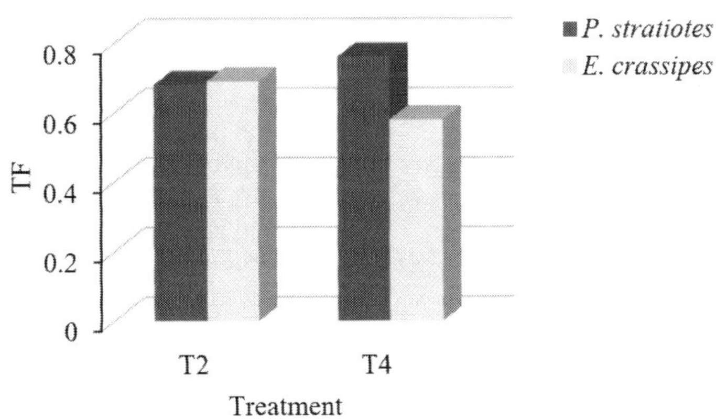

Figure 6. Translocation factors of roots and shoots in *P. stratiotes* and *E crassipes* cultivated in NCHF reactors

Translocation factor is an important indicator to mark the accumulation and storage of heavy metals from roots to aerial part. Figure 6 showed the translocation factors of Pb in roots and shoots of *P. stratiotes* and *E. crassipes*. Data on TF showed that the average value for by both plants studied are below than 1 which indicates that these plants have limited translocation of Pb to the aerial part. The highest TF value was found in *P. stratiotes* (T4) with 0.76 while the lowest was observed in *E. crassipes* (T4) with 0.58. [18] reported that the TF value which is higher than 1 indicates that the plants is accumulator species whereby this plant have an efficient metal transporter system to transport metal from roots to leaves [19]. Otherwise for TF value lower than 1, this plant generally known as metal excluder species. It is worth to know that the effects of rhizospheric microbes *B. cereus*-NMeHI-Cr2 towards roots of *P. stratiotes* and *E. crassipes* had displayed a low significance in translocating Pb into the aerial part. However, *P. stratiotes* and *E. crassipes* had perform as good rhizofiltration mechanism plants and can be used to remove heavy metal especially Pb from industrial wastewater.

4. Conclusion

The effects of rhizospheric microbes *B. cereus* -NMeHI-Cr2 onto the roots of *P. stratiotes* and *E. crassipes* had displayed higher removal efficiency of Pb with 83.7%. On the other hand, experiments with non-inoculation showed a lowest removal percentage with 13.3% differences. The data suggested that the inoculation of *B. cereus* -NMeHI-Cr2 had evidently form a synergistic effect for removal of Pb. The finding also reported that both plants have similar accumulation patterns whereby the distribution of Pb was higher in roots followed by leaves and stalks. Therefore, this study concludes that *P. stratiotes* and *E. crassipes* have a huge potential to remove metal polluted water. It is important to note that both plants must be harvested regularly to avoid the metals release back in to the environment.

References

[1] Masindi V and Muedi K L 2018 Environmental contamination by heavy metals *Heavy Metal* **7** 115–133

[2] Monisha J, Tenzin T, Naresh A, and Krishnamurthy N B 2014 Toxicity mechanism and health effects of some heavy metals *Interdisciplinary Toxicology* **7(2)** 60–72

[3] Sharma P and Dubey R S 2005 Lead toxicity in plants *Brazilian Journal of Plant Physiology* **17(1)** 35–52

[4] Bouzid N 2021 Phytoremediation: a suitable environmental technology for heavy metals decontamination *SN Applied Sciences* **3** 286

[5] Malik A 2007 Review article: Environmental challenge via an opportunity: The case of water hyacinth *Environmental International* **33** 122–138

[6] Yan A, Wang Y, Tan S N, Mohd Y M L, Gosh S, and Chen Z 2020 Phytoremediation: A promosing approach for revegetation of heavy metal-polluted land *Front. Plant Sci.* **11**

[7] Hoflich G and Metz R 1997 Interaction of plant microorganism accosiation in heavy metal containing soil from sewage farms *Bodenkultur* **48** 238–247

[8] Burd G I, Dixo D G, and Glick B R 2000 Plant growth promoting bacteria that decreases heavy metals toxicity in plants *Can. J. Microbiol.* **46** 237–245

[9] Mahardika G, Rinanti A, and Fachrul M F 2018 Phytoremediation of heavy metal copper (Cu^{2+}) by sunflower (Helianthus annuus l.) *Earth and Environmental Science* **106** 1–7

[10] Foluso O, Agunbiade Bamidele I, Olu-Owolabi Kayode O, and Adebowale 2009 Phytoremediation potential of *Eichhornia crassipes* in metal-contaminated coastal water *Biosource. Techno.* **4(11)** 4521–4526

[11] O'Connell K P, Goodman R M, Handelsman J 1996 Engineering the rhizosphere: Expressing a bias *Trends in Biotechnology* **14** 83–88.

[12] So L M, Chu L M and Wong P K 2003 Microbial enhancement of Cu^{2+} removal capacity of *Eichhornia crassipes* (Mart.) *Chemosphere* **52** 1499–1503.

[13] Azoom-Aguilar C and Barea J M 1981 Field inoculation of Medicago with VA mycorrhiza and Rhizobium in phosphate fixing agricultural soil *Soil Biol.Biochem.* **13** 19–22

[14] Hu P J, Qiu R L, Senthilkumar P, Jiang D, Chen Z W, Tang Y T, and Liu F J

2009 Tolerance, accumulation and distribution of zinc and cadmium in hyperaccumulator *Potentilla griffithii*. *Environ. Exp. Bot* **66** 317–325

[15] Jing Y, He Z, and Yang X 2007 Role of soil rhizobacteria in phytoremediation of heavy metal contaminated soils. *J Zhejiang Univ Sci B.* **8** 192–207

[16] Deepa C K, Dastager S G, and Pandey A 2010 Plant growth-promoting activity in newly isolated *Bacillus thioparus* (NII-0902) from Western ghat forest, India *World J Microbiol Biotechnol.* **4(12)** 2277–2283

[17] Chaudhry Q, Blom-Zandstra M, and Gupta S K 2005 Utilizing the synergy between plants and rhizosphere microorganisms to enhance breakdown of organic pollutants in the environment *Environ SciPollut Res.* **12** 34–48

[18] Yanqun Z, Yuan L, Jianjun C, Haiyan C, Li Q, and Schvartz C 2005 Hyperaccumulation of Pb, Zn and Cd in herbaceous grown on lead-zinc mining area in Yunnan, China *Environ. Int.* **31** 755–762

[19] Nur Zaida Z, Ng Sean F, Fera N C and Sahibin A R 2021. The potential of *Pistia stratiotes* in the phytoremediation of selected metals from simulated wastewater. *International Journal of Technology* **3** (613-624)

Countermeasures analysis of landslide in the cikeusal village of Tasikmalaya regency

A Salimah[1], Y Yelvi[1] and M Hazmi[2]

[1]Department of Civil Engineering, Politeknik Negeri Jakarta, Prof. Dr. G.A. Siwabessy, UI Depok, Indonesia
[2]Directorate General of Water Resources, Ministry of Public Works and Housing, 12110, Indonesia

aisyah.salimah@sipil.pnj.ac.id

Abstract. The construction of telecommunication towers aims to improve customer service. So that advances in the field of communication and information technology can be felt by residents who are far away in remote parts of the country. The construction of this tower must be planned as well as possible, especially for areas prone to landslides because towers are usually placed in higher areas. The landslide incident at the location of the telecommunication tower in Cikeusal Village, Tasikmalaya Regency can be used as an evaluation for other areas prone to landslides. The landslide can endanger the settlements under the tower. Field investigations are carried out as a basis for making plans to overcome landslides. The analysis was performed using Plaxis 2D software. The results of the strengthening analysis after the erosion prevention were carried out (with several stage), the factor of safety was 1.805 at section B and 3,228 at section C. The recommendations are given for landslide prevention are replacing the old retaining wall with the new one gradually, adding a column in the middle of the retaining wall, adding bored piles under the retaining wall, making drain pipe on the retaining wall to accommodate water flow from the ground surface, replacing the old stairs for access roads that were affected by landslides with new stairs and made with 2 applications so that the slope of the stairs is not too steep, as well as making a foundation for the stairs. It is hoped that the landslide occurrence and the results of the analysis of slope stability in Cikeusal Village, Tasikmalaya Regency can be used as evaluation material in the construction of telecommunication towers for landslide-prone areas.

1. Introduction

Landslides have been common in Indonesia, especially during the rainy season. Landslides are common on plateaus with steep slopes [1]. Based on InaRISK analysis, Tasikmalaya Regency, West Java Province is an area with a moderate to high hazard class category that is prone to potential flood and landslide hazards. There are 25 sub-districts in the medium to high hazard class for the potential flood hazard and 35 sub-districts in the landslide hazard [2] One of them is Cikeusal Village which is in Tanjungjaya District, Tasikmalaya Regency.

Cikeusal Village is one of the areas on the border between West Java and Central Java which has unstable soil and is prone to natural disasters. Regional conditions like this should be a concern for local governments in carrying out development and regional development. Just as there was a landslide in the telecommunication tower in Cikeusal Village. This tower was built to improve telecommunications services to consumers so that residents in this village area can reach the telecommunications network, especially in the digitalization era with the current pandemic condition. This incident also happened in Cisewu, Garut, West Java, and in Sukaresmi Village, Sukabumi Regency, West Java [3][4]. The type of landslide that occurred in the Cikesal area is a local landslide that can be handled more easily than other landslides. However, because of the landslide, it will still endanger the settlements of the population below it. Landslides that occurred in the Cikeusal area were caused by a layer of embankment soil at a depth of 0 to 2 meters that were not completely compacted and the occurrence of rain with high intensity for one month and the condition of the retaining wall which suffered a lot of damage. A retaining wall is a structure designed and built to withstand lateral (horizontal) earth pressure. An important factor in designing and constructing a retaining wall is to ensure that it does not move or collapse under the applied loads [5][6][7][8]. However, with the lack of friction on the retaining wall, rainwater entering the soil pores causes a decrease in soil adhesion and decreases soil shear strength. Water that enters the drainpipe is trapped in it causing the retaining wall to crack and unable to withstand its load and the moving soil load causing landslides.

Therefore, urgent treatment is needed to prevent ongoing landslides. First, an investigation is carried out in the field to collect data on conditions after the landslide which will later be used as a basis for analyzing preventing landslides. The landslide that occurs can be seen in Figure 1. Then an analysis is carried out using Plaxis 2 D software with various conditions needed to obtain the required factor of safety (FOS). In general, the value of FOS $\geq$ 1,25 is a normal design to provide an estimate of the FOS in slope stability analysis [9]. The parameters generated in the slope stability analysis are the failure plane shape and the FOS [10].

Figure 1. Landslide area at the edge of the Cikeusal – Tasik site fence

2. Methods

This research was conducted in Cikeusal, Tasikmalaya, West Java, where there is a telecommunications tower was threatened by landslides. The results of the investigation activities in the field obtained conditions as below.

IOP Conf. Series: Earth and Environmental Science **847** (2021) 012013 doi:10.1088/1755-1315/847/1/012013

2.1. Field Investigation Survey

2.1.1. Condition of foundation.

The condition of the foundation in the site area cannot be seen and predicted in detail, because it needs regular checks and special analysis such as horizontality checks. Horizontality is the horizontal state of a building, used to determine the continuous shift of a building or structure. From the results of field checks and based on Figure 2 and 3, the horizontality results for each pedestal can be explained in Table 1

Figure 2. Horizontally Tower

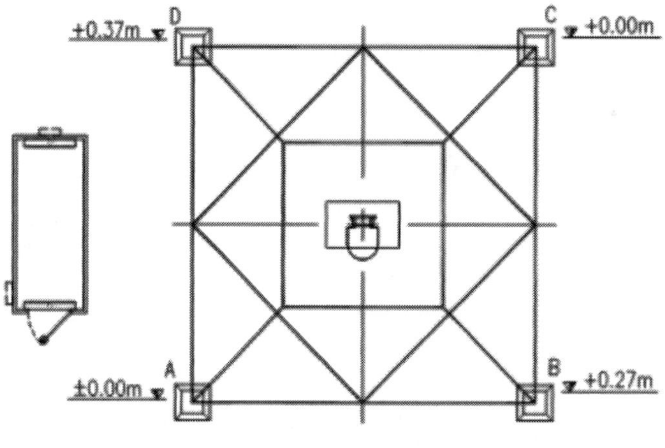

Figure 3. Horizontally Tower Position

Table 1. Horizontality Tower Check Results

Reference Point	Elevation (mm)	Difference (mm)	
A	0	0	Reference Point
B	270	+270	
C	0	+0	
D	370	+370	

2.1.2. Condition fence and access road.

The condition of the fence in the site area was partially damaged in the area affected by the landslide, the damage to the fence was caused by the damage to the retaining wall because the fence was supported directly on the retaining wall. The condition of the access road at the site is partially damaged, namely the steep descent of the stairs caused by a landslide at the site (Figures 4 and 5).

Figure 4. Condition of access road after a landslide

Figure 5. The condition of the fence after the landslide

2.1.3. Condition of retaining wall

Based on the results of the survey, the condition of the retaining wall with little difficulty causes rainwater that enters the ground to become groundwater pressure which imposes a burden on the retaining wall. Meanwhile, soil that is exposed to rainwater causes the adhesion between the soil to decrease and the soil moves to impose a retaining wall. The dimensions of the retaining wall are not able to withstand moving soil loads and withstand groundwater pressure, resulting in a landslide in the fenced area as shown in Figure 6.

Figure 6. Condition of Retaining wall after Landslide

Figure 6. Condition of Retaining wall after Landslide (contd.)

2.2. Slope Stability Reinforcement Analysis with Plaxis 2D
There are several stages for planning and countermeasures for landslides by applying several reinforcements. The first stage is replacing the existing retaining wall with a new retaining wall. The implementation of the work is to gradually remove the old retaining wall within 2.4 m to be replaced with a new retaining wall. The second stage with adding columns in the middle of the retaining wall to withstand the load so that those above it do not directly burden the retaining wall. The purpose adding piles under the retaining wall to withstand all forces on the retaining wall.

2.2.1. Material modelling.
The behavior of soil and rock under load is generally non-linear. This behavior can be modeled with various equations including the Mohr-Coulomb model, Hardening Soil Model, Soft Soil Model, and Soft Soil Creep Model. In this analysis, the Mohr-Coulomb model is used which requires 5 parameters, namely Cohesion(c), Shear Strength (φ), Modulus Young (E_{ref}), Poisson's Ratio (ν), Dry unit weight (γ_{dry}), and Saturated unit weight (γ_{sat})

Table 2. Material Design Parameters in landslide Simulation

Parameter		Layer 1	Layer 2	Layer 3	Retaining wall	Units
Model Material	Model	Mohr-Coulomb	Mohr-Coulomb	Mohr-Coulomb	Elastik Plastis	-
Type Material Behavior	Type	Undrained	Undrained	Undrained	Non-Porous	-
Saturated unit weight of Soil above the Phreatic Line	γ_{unsat}	16.33	17.5	21	23.5	kN/m³
Saturated unit weight of Soil below the Phreatic Line	γ_{sat}	18	19	22	23.5	kN/m³
Horizontal Permeability	k_x	0.00864	0.00864	0.00864	-	m/day
Vertical Permeability	k_y	0.00864	0.00864	0.00864	-	m/day
Young's modulus (constant)	E_{ref}	1400	3000	6750	20.000.000	kN/m²
Poisson's ratio	ν	0.3	0.3	0.3	0.15	-
Cohesion	c_{ref}	11.2	24	54		kN/m²
Friction	φ	32	54	40.937		°
Dilatancy angle	ψ	0	0	0	0	°

The cohesion value (c) and internal friction angle (φ) are obtained from the results of the Direct Shear soil test from laboratory analysis. Because the soil element has been deformed well past the

IOP Conf. Series: Earth and Environmental Science **847** (2021) 012013 doi:10.1088/1755-1315/847/1/012013

peak stress, so the remaining stress is the residual stress. In this case, the representative shear strength is the residual shear strength. Young's modulus (Eref) is obtained from testing the triaxial law method, for the preliminary design it can be used Young's Modulus (Eref) = 2 to 8qc [11]. Poisson's ratio value for clay is in the range 0.3 - 0.4 [12]. Using the Mohr-Coulomb model Poisson's ratio value is taken as 0.3. The value of the dilatancy angle (ψ) = 0°, for the shear angle depends on the type of soil layer in Table 2, an explanation of the soil parameters is given.

2.2.2. Model and simulation.

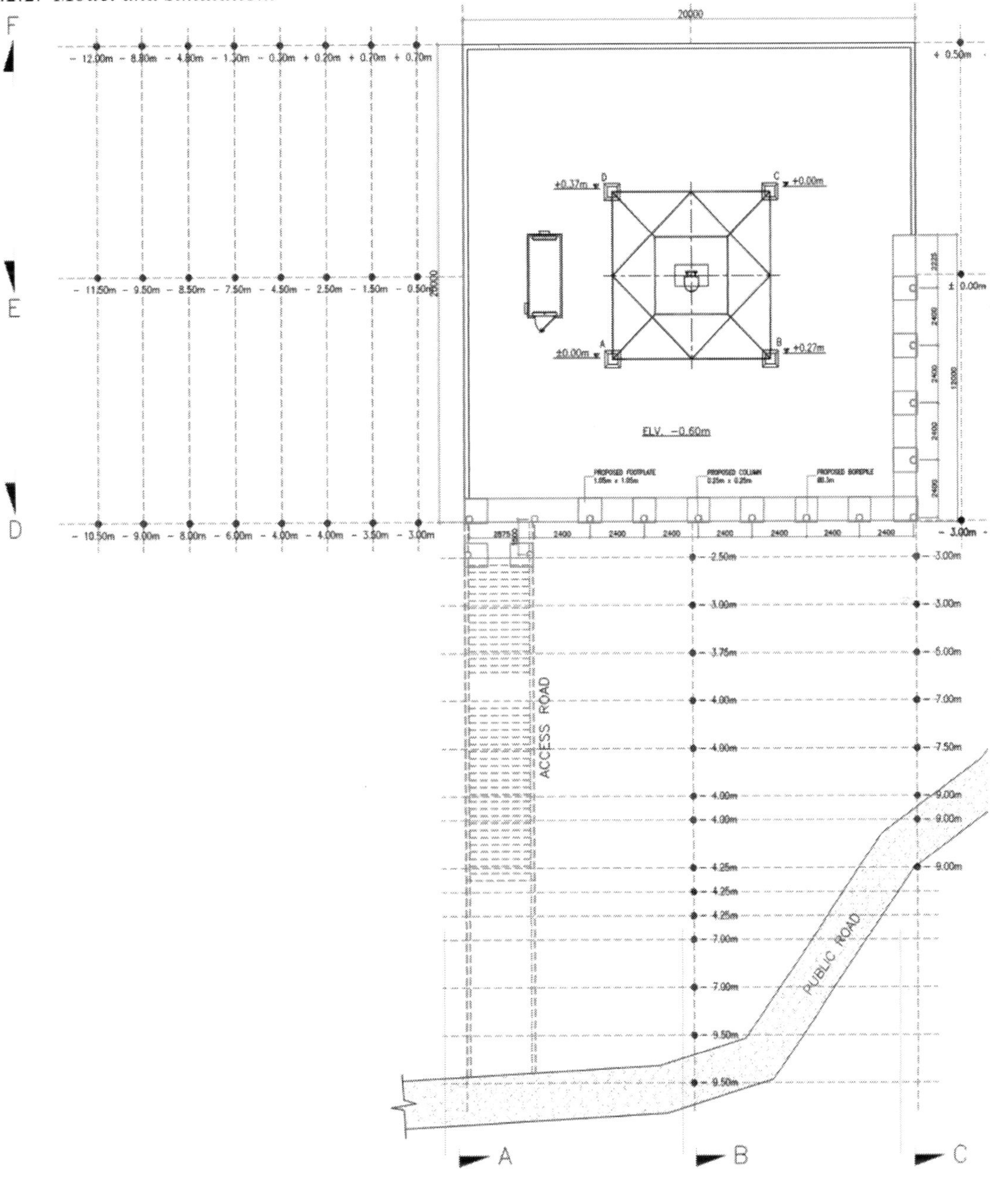

Figure 7. General Layout

The modeling is carried out using cross-sectional data (section profile) of slopes with reinforcement in the form of retaining walls. Figure 7 shows general layout of the location of the telecommunication tower. An overview of the design model for slope stability reinforcement with retaining walls and bored piles can be shown in Figure 8 and Figure 9, where the analysis is carried out in section B and section c as the impact of the landslide. The soil layer is modeled into 3 soil layers.

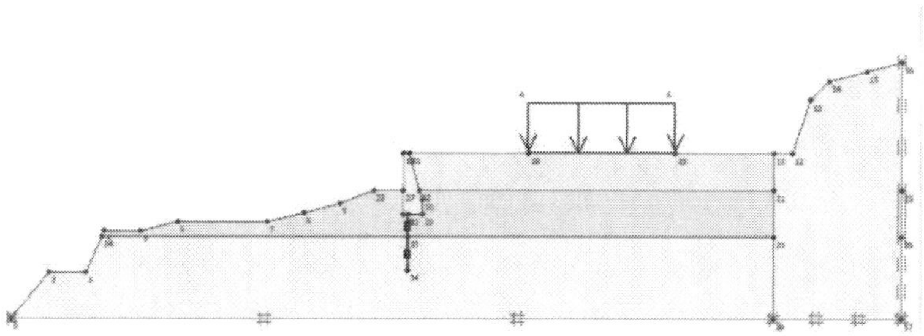

Figure 8. B-section Contour Design in Conditions after Landslide

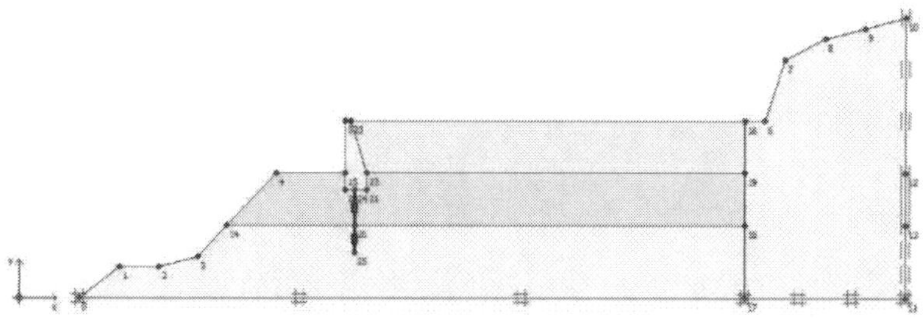

Figure 9. C -section Contour Design in Conditions After Landslide

2.2.3. . Slope Stability Analysis on Plaxis.

After the general description of the control for the existing condition and reinforcement has been completed, the next step is to enter the required soil parameters in the analysis as shown in Table 2 and Table 3 as well as in Figure 8 and Figure 9

Table 3. Material Design Parameters in Landslide Simulation

Parameter	Symbol	Pile	Units
Modulus of Elasticity	E	22.000.000	kN/m^2
Diameter	d	0.3	m
Pile length	L	3	m

2.2.4. Calculation input.

Slope formation material parameters, FOS calculation, and slope failure interpretation are the three main components of the slope stability study. [13]. The calculation stages are divided into four stages:

a. The initial phase is the default of the program (phase 0)
b. The gravity loading stage is the phase where the initial stress and strain due to the soil load itself and external loads are calculated (phase 1)
c. The multiplier stage will be selected after pressing the Define button. At this stage, there are many levels of variation.
d. The calculation phase of the factor of safety (FOS), which is the phase where the slope stability due to phases 1 and 2 is calculated.

Plaxis calculation for the 3 phases can be seen in Figure 10

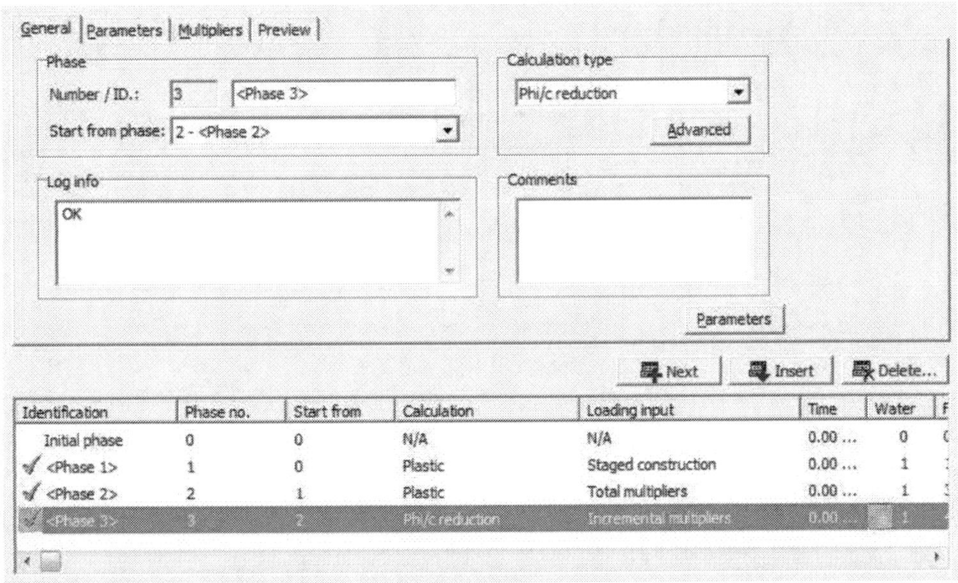

Figure 10. Calculation of Plaxis for 3 phases

3. Results and discussion

The first before in slope stability analysis, there are three important aspects to consider. The first concerns the slope-forming substance's material qualities. The second step is to calculate the factor of safety, and the third step is to define the slope failure definitions [13]. According to [5], FOS more than 1.25 is considered a rare landslide. Slopes with a lower safety factor have a higher risk of failure than slopes with a higher FOS [14].

3.1. Slope Stability Analysis Results with Plaxis

Retaining walls on slopes receive loads from the outside, namely tower loads, fence loads that support directly on the retaining wall, and soil loads that move to push the retaining wall. The input value of the load is 60 kN/m^2 as uniform load showed by blue arrow in downward direction in figure 11.a.

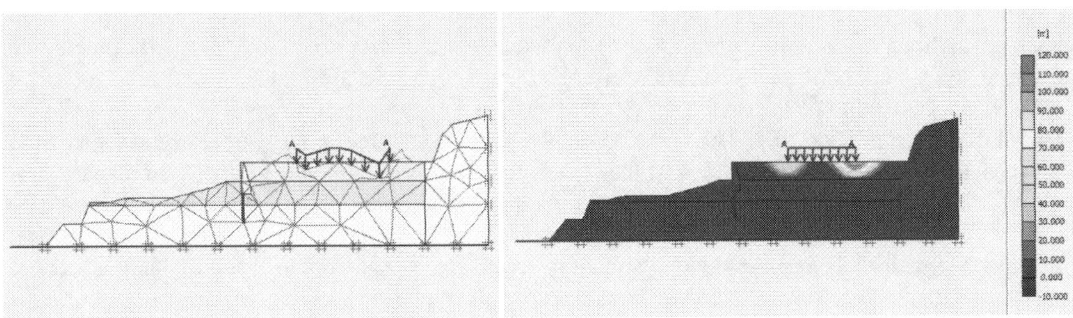

(a) (b)

Figure 11. (a) Displacement Strengthening Condition Section B, (b) Condition of failure

IOP Conf. Series: Earth and Environmental Science **847** (2021) 012013 doi:10.1088/1755-1315/847/1/012013

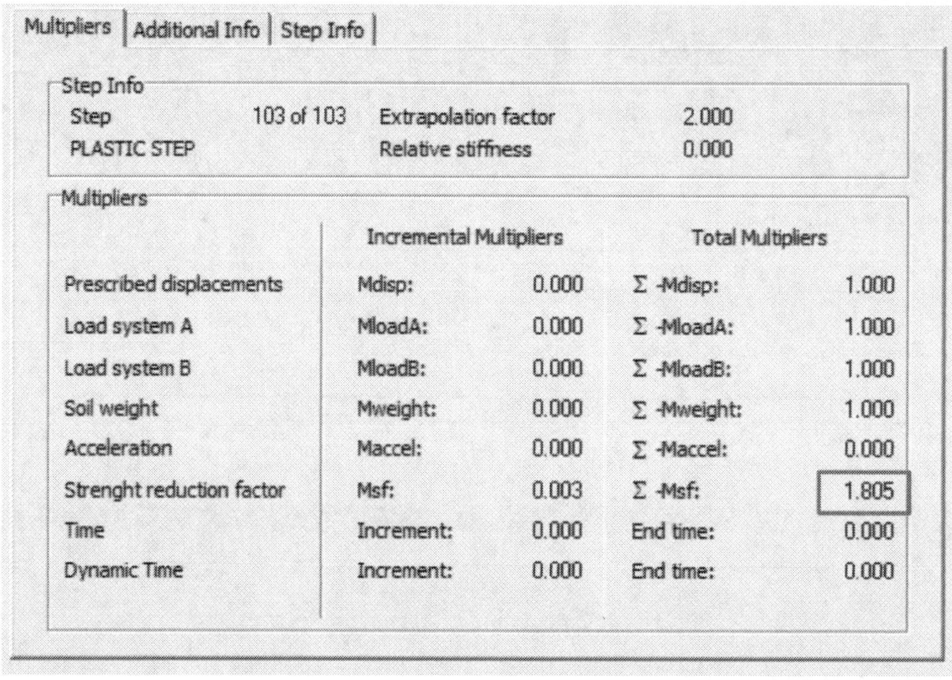

(c)

Figure 11. (c) FOS Strengthening

In Figures 11 and 12 the largest ground motion occurs in the built tower area. This can occur due to the high intensity of rain that affects the soil load and the condition of the dimensions of the retaining wall which is not able to withstand the load and the lack of difficulty in the retaining wall which causes an increase in load water pressure on the retaining wall. Figures 11.b and 12.b show the slipping zone's area potential, as well as the total regional displacement. At the reinforcement analysis stage, the slope FOS value for section B is 1.805, which is greater than the required 1.25 (Figure 11.c), while for section C, it is 3,228 (Figure 12.c). The function of the retaining wall is to withstand the earth pressure load in the tower area. While the function of the bored pile is to withstand the load transferred from the fence load and the retaining wall load. With conditions like the above, the condition of the slopes with reinforcement is in a safe condition

Based on the results of the analysis, obtained FOS from strengthening condition at section B is 1,805 and at section C is 3,288. This value is greater than the minimum required slope stability FOS (1,25). As a result, the safety condition at the slope's foot, could be achieved [5], [15]. In planning for reinforcement using retaining walls, a drainpipe is also needed on the retaining wall to accommodate the flow of water from the ground so that it does not enter the retaining wall and reduce the quality. The last stage with replacing the existing stairs for access roads that were affected by landslides with new stairs with dimensions of 30x20 cm steps with 2 stair traps made so that the slope of the stairs is not too steep, the stairs foundation is made in the form of columns and portals at the top of the stairs close to site area

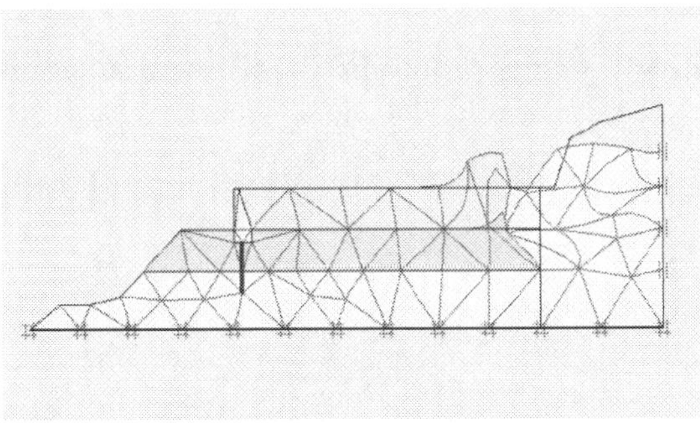

(a)

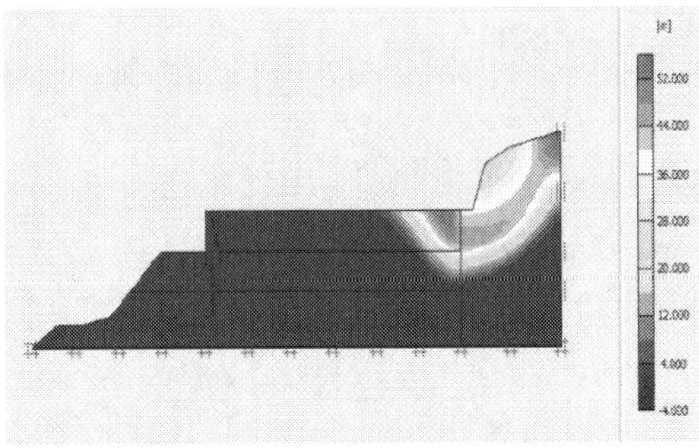

(b)

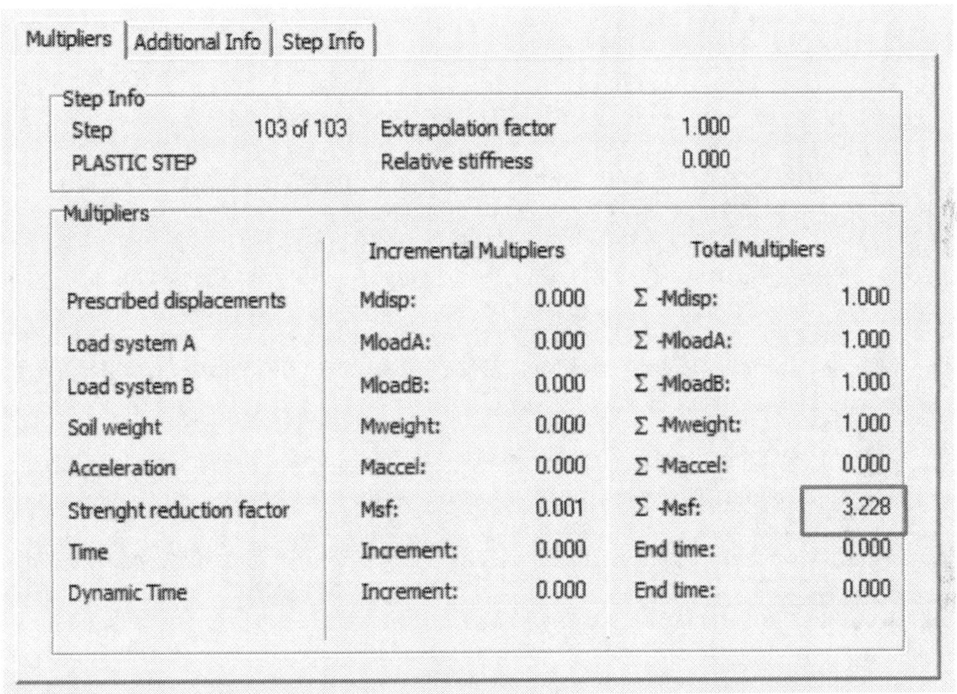

(c)

Figure 12. (a) Displacement Strengthening Condition Section C, (b) Condition of failure, (c) FOS Strengthening

4. Conclusion

Based on slope stability analysis using Plaxis 2D software, several recommendations are given to prevent landslides. The recommendations include making a new retaining wall, adding a column in the middle of the retaining wall, making drainpipe on the retaining wall, adding bored piles under the retaining wall, and replacing the existing stairs as an access road to the tower. These measures have the potential to strengthen the slope, with safety factor value more than 1,25. As a result, the safe condition could be achieved. To improve this research, it is necessary to carry out further analysis related to the effect of soil stability on extreme groundwater level conditions. So that it can be more anticipated during the heavy rainy season.

Acknowledgement
This study was carried out by extraordinary assistance from the surrounding local communities at Cikeusal, Tasikmalaya for the contribute during the research and Politeknik Negeri Jakarta for supporting the completion of the paper.

References
[1] Abidin H, Andreas H, Surono M, and Hendrasto M 2004 On the Use of GPS Survey Method for Studying Land Displacements on the Landslide Prone Areas 1–13
[2] BNPB Sejumlah Kerusakan Fisik Saat Banjir dan Longsor Terjang Tasikmalaya," https://bnpb.go.id/berita/sejumlah-kerusakan-fisik-saat-banjir-dan-longsor-terjang-tasikmalaya-.
[3] Salimah A, Ammar M, Rahmawati D, Yelvi, and S.Widya S T 2020 Landslide Analysis Study and Cisewu Countermeasures in District, Garut Regency, West Java *in IOP Conference Series: Materials Science and Engineering* **771**, 1
[4] Salimah A, Hasan M F R, Suripto S, Yelvi Y, and Sasongko I H 2019 Analisis Stabilitas Dan Perkuatan Lereng Menggunakan Plaxis2D Di Desa Sukaresmi Sukabumi, Jawa Barat *Jukung (Jurnal Tek. Lingkungan)* **5** 2 29–36
[5] Bowles J 1979 *Physical and Geotechnical Properties of Soils* (Tokyo: McGraw-Hill)
[6] D B M and S K 2014 Principle of Geotechnical Engineering
[7] Hardiyatmo H C 2002 *Teknik Fondasi 1* (Yogyakarta: Beta Offset)
[8] Hastuty I P and Prambudi R 2020 Analysis of slope stability by the planning of cantilever retaining wall reinforcement using the application of Plaxis (case study: Substation Panyabungan sta 0+060) *IOP Conf. Ser. Mater. Sci. Eng.* **851** 1 0–6, 2020.
[9] Ramadhan R, Munirwansyah M, and Sungkar M 2019 Faktor Keamanan Stabilitas Lereng pada Kondisi Eksisting dan Setelah Diperkuat Dinding Penahan Tanah Tipe Counterfort dengan Program Plaxis *Reka Buana J. Ilm. Tek. Sipil dan Tek. Kim* **5** 1 1
[10] Zheng H, Liu D F, and Li C G 2005 Slope stability analysis based on elasto-plastic finite element method *Int. J. Numer. Methods Eng* **64** 14 1871–1888
[11] Kézdi A 1974 *Handbook of Soil Mechanics* (Amsterdam: Elsevier Pub. Co)
[12] Marchenko, Mosicheva I, and Aniskin A 2018 *Estimation of Poisson's ratio of soil using stiffness of loose soils*
[13] Rocscience Inc 2001 *Application of the Finite Element Method to Slope Stability* no.1
[14] Mohammadinia A, Arulrajah A, Haghighi H, and Horpibulsuk S 2017 Effect of lime stabilization on the mechanical and micro-scale properties of recycled demolition materials *Sustain. Cities Soc* **30** 58–65
[15] Tsuchida T, Athapaththu A. M. R. G, Kawabata S, Kano S, Hanaoka T, and Yuri A 2014 "Individual landslide hazard assessment of natural valleys and slopes based on geotechnical investigation and analysis," *Soils Found* **54** 4 806–819

IOP Conf. Series: Earth and Environmental Science **847** (2021) 012014 doi:10.1088/1755-1315/847/1/012014

Calculation formula in determining recycling of disposable diapers waste as concrete composite materials

S Zuraida[1,2], B J Dewancker[3] and R B Margono[4]

[1]Doctoral Student, Graduate School of Environmental Engineering, The University of Kitakyushu, Japan
[2]Department of Civil Engineering, Institut Teknologi Sains Bandung, Indonesia
[3]Department of Architecture, The University of Kitakyushu, Japan
[4]Department of Architecture, Institut Teknologi Bandung, Indonesia

b0dbb409@eng.kitakyu-u.ac.jp

Abstract. Disposable diaper waste is a crucial environmental problem today. Population growth is one of the triggers for increasing this type of waste. To overcome this, various studies on recycling diaper waste are a top priority. Some researchers recycle diaper waste as building components, especially concrete. The results of those studies have a positive impact on waste recycling innovation. In this paper, we will try to approach the formula for calculating the recycling of diapers as concrete. In addition, a simple simulation is used to determine the capacity of diaper waste that can be recycled as building material. The study revealed that parameters to calculate waste capacity for disposable diapers is age population of 0-4 years, frequency of changing diaper and mass of diaper. Thus, parameters to be considered for replacement of concrete component by waste materials, especially disposable diapers is density of materials, mass of materials and percentage of replacement Based on the calculation simulation results, the use of diaper waste as a concrete material for road pavement is better to reduce waste production capacity than for building components.

1. Introduction

Population growth in the world currently reaches 7 billion and is predicted to increase to 10 billion by 2040 [1]. This population growth causes several environmental problems due to the consumption of natural resources to meet human needs such as land use, water, and energy [1]. In the term of environmental waste management, the population growth in the world leads to the increase of disposable diaper usage in the term of baby nursing. Disposable diapers become more popular since it was introduced in the 1960s due to the benefit of the circular economy within several versions of diapers and modified for a wider range of applications over the years [2]. While another benefit is in the social aspect, especially the convenience for parents by offer performances and affordability.

Beside it gives beneficial aspect of economy and social, in the term of waste management system, it also increases the difficulty of disposing and recycling once the diapers have been used and turned into waste. Currently disposable diapers contribute up to 4% from the total of solid waste and this amount has been increasing over the years [3]. Moreover, approximately 20 billion disposable diapers are thrown into landfills every year or more than 3.5 million tonnes of waste [4]. Based on a survey of household preferences for disposing of disposable diapers, around 5-40% of this waste is disposed of in landfills without further processing [5]. It leads to another problem of health.

Disposable diapers contain dioxins and dyes, as cancer agents that if released into the environment it can be accumulated in human and animals. Landfill causes serious environmental problems such as methane emissions, water pollution, land use, and odors [6-7]. Disposable diapers are also not degraded very well, it takes 500 years for nature to decompose them [8]. Furthermore, people often dispose of it wrapped in plastic [9-10]

In the term of infrastructure development, the material innovation that involved nondegradable waste material also become main focus in several research, especially to recycle disposable diapers waste as concrete components. The research proof that for mechanical properties and microorganism content, the disposable diaper concrete has no significant distinction with normal concrete [8] [11-12]. In addition, comparing to other waste management such as incineration and co-firing, by recycling disposable diapers as concrete components has more benefit in term of carbon emission and eco-costs [13]

However, the problem comes up when calculating the exact waste capacity of disposable diapers that may solved by recycling as concrete components. This paper is explained the calculation by using case study of urban city in Bandung, Indonesia. The study is conducted by collecting data of disposable diapers usage in Bandung, then using calculation approach of concrete mix design to assume the replacement component capacity of concrete that filled by disposable diapers waste. The final calculation is to assume the capacity of disposable diapers concrete that benefit to use as infrastructure materials, such as building components and road pavement.

2. Methods

2.1. Parameters Approach

To assume the disposable diapers waste capacity, some parameters are used. Some literatures state that the use of disposable diapers in household depends on background of parent such as age and sex [20][22][24], education [5][10][22], marital status [20], job and income [5][20][22-24]. While the main parameter is background of children such as age and sex [5][19-23], breast feeding [25-29] and toilet training [5][22-23][25-26][29-30]. These parameters have a relation to the usage of disposable diapers daily.

Table 1. Disposable diapers use frequency in daily household.

Age of baby	Diapers Use Frequency	Source
0 - 12 months	2 - 6 times per day	[31]
0 - 4 years	11 - 15 times per week = 2 - 3 per day	[5]
0 - 3 years	2 - 10 times per day	[30]
0 - 6 months	1- 6 per day	[26]
0 - 1 month	3 - 9 times per day	[27]
0 - 1 month	3 - 8 times per day	[28]
0 – 14 month	5 – 9 times per day	[32]

Based on the equation (1), then the weight of diapers can be decided 30.06 gr per piece. This parameter then marked as *md* (mass of diapers).

After all parameters decided, then to calculate the waste capacity of diapers per day is determined by population of age 0-4 years (Pb), frequency of diapers usage daily, monthly, or annual (Fd) and mass of diapers per piece (md) as formulized in the equation:

Based on table 1, the range frequency of children from 0 – 4 years age is 1 to 10 times of changing diapers a day. In this study the frequency is assumed based on simple equation:

$$Average = \frac{\frac{\sum n1+n2}{2}}{N} \quad (1)$$

Where *n1* and *n2* are the lowest and the highest value of the data. *N* is total data. Based on the equation, the average of children changing the diapers is 5 times or piece a day. This parameter then marked as *Fd* (frequency of diapers).

The other parameter to set in this calculation is weight of disposable diapers that show in table 2

Table 2. weight of disposable diapers per piece

Quantity	Weight	Source
12 pc	318.2 - 429.6 gr = 26.5 – 35.8 gr	[33][34]
1 pc	26.5 - 27.4 gr	[32][34]
1 pc	28.04 - 36.1 gr	[34]

Based on the equation (1), then the weight of diapers can be decided 30.06 gr per piece. This parameter then marked as md (mass of diapers).

After all parameters decided, then to calculate the waste capacity of diapers per day is determined by population of age 0-4 years (Pb), frequency of diapers usage daily, monthly, or annual (Fd) and mass of diapers per piece (md) as formulized in the equation (2) where *Cd* is marked as capacity of waste diapers.

$$Cd = Pb \text{ x } Fd \text{ x } md \quad (2)$$

2.2. *Concrete Mix Design Approach*

Based on several studies, the disposable diapers in concrete act as a substitute for fine aggregate. However, the replacement of fine aggregate by disposable diapers is to be considered in term of density distinction. To meet apple to apple of weight ratio, this calculation involves the formula:

$$\rho = \frac{m}{V} \quad (3)$$

Where:
ρ = density of material (gr/cm^3)
m = mass or weight of material (gr)
V = volume or three-dimensional space (cm^3)

In this scenario, the ratio of fine aggregate (fa) and disposable diaper (dd) is mutual in the volumetric. So, the calculation is decided by:

$$V\ fa : V\ dd$$
$$\frac{m}{\rho}\ fa : \frac{m}{\rho}\ dd \quad (4)$$

Then the mass of disposable to replace fine aggregate in the concrete mix design is determined:

$$md = \%\, rep\ x \left(\frac{\rho d}{\rho fa}\right) x\ m\ fa \quad (5)$$

Where:
ρd = density of disposable diaper (gr/cm^3)
m*d* = mass or weight disposable diaper (gr)
ρfa = density of fine aggregate (gr/cm^3)
m*fa* = mass or weight of fine aggregate (gr)
% *rep* = replacement percentage of fine aggregate by disposable diaper (%)

For density of disposable diapers is 0.07 - 0.21 g/cm^3 [35]. Then, by incorporate the equation (2) to equation (5), the equation of diaper capacity that can be used for concrete components is obtained:

$$Cdc = \frac{(Pb \text{ x } Fd \text{ x } md)}{\%\, rep\ x \left(\frac{\rho d}{\rho fa}\right) x\ m\ fa} \quad (6)$$

Then Cdc is marked as capacity of disposable diapers concrete that can be produced.

3. Result and discussion

3.1. Case study preview

Due to the population growth become the main issue of waste production, this case study involved metropolitan city in Indonesia. The other criteria in determining case study is population age of 0-4 years old, because the waste production for disposable diapers derived from this population. Another one is waste management in the case study, how the government or local communities deal with the waste production. These criteria for selected city are compiled from secondary data based on Indonesian Municipal Central Bureau of Statistics (BPS).

Based on the criteria, Bandung is appropriate to be the selected city due to the available information that provide in table 3.

Table 3. Comparative selection criteria for case study

No.	Metropolitan Cities	Population	Population Age 0-4	Waste Production	Waste Treatment	Source
1	East Jakarta	3 037 100	238 497	not available	not available	[14]
2	Bekasi	3 003 900	not available	not available	not available	[15]
3	Surabaya	2 896 000	183 975	not available	not available	[16]
4	Bandung	2 507 900	182 116	66 472 340	Categorised	[17]
5	West Jakarta	2 434 500	184 826	67 461 490	Not Identify	[18]

Table 4. Waste Type Separation in Bandung [17]

Garbage Type	(M3/Day)	Percentage %
Food waste and leaves	772.69	44.51%
Woods and twigs	69.09	3.98%
Paper	227.76	13.12%
Plastic	324.28	18.68%
Metal	15.62	0.90%
Cloths	82.46	4.75%
Rubber and leather	41.32	2.38%
Hazardous waste	131.42	7.57%
Others	71.35	4.11%
Total	1 735.99	100.00%

Based on population criteria, Bandung is fourth rank the most populous city in Indonesia. Compared to other populous cities, the information waste management in Bandung also available. Table 2 shows the garbage type that currently separated in municipal solid waste of Bandung. For disposable diapers waste, there is still categories to other type of garbage. Unlike other categories such as plastic, cloth or metal which have value to be recycled, disposable diaper is still unvaluable. The waste will be ended up in the incineration system. The recycling idea for disposable diapers will be triggered the government or society to separate the waste and turn it into new valuable items.

3.2. Calculation approach based on case study

To calculate the volume of diaper waste that can be recycled into building components, a concrete mix design calculation approach is based on Indonesian National Standard (SNI) 7394:2008 The Procedure for Calculating the Unit Price of Concrete Work for Building and Housing Construction. It should be noted that the use of concrete mix design in this case study is only for simplifying calculations. For more valid mix design results, it is necessary to carry out further laboratory tests, such as tests for aggregates and other mixed materials for concrete.

In this study, the concrete was designed in high strength of K-300 with the composition for 1 m^3 concrete is 413 kg coarse aggregate, 1021 kg fine aggregate and 681 kg Portland cement. The

disposable diapers is replacement for fine aggregate. The optimal percentage of replacement is 5% to 10% addition to reach the optimal compressive strength [8][11-12].
According to equation (2) and (5), the capacity of waste disposable diapers in case study is determined:
Pb = 182116 people (pc)
Fd = 5 pc/day
md = *30.06* gr/pc
ρd = 1400 kg/m^3 (middle range)
ρfa = 2082 kg/m^3
mfa = 681 kg
% *rep* = 7.5% (middle range)

Then the capacity of waste disposable diaper in Bandung can be determined based on equation (7) as:

$$Cd = Pb \text{ x } Fd \text{ x } md \tag{7}$$

$$Cd = 27372 \text{ kg/day} = 19.56 \text{ m}^3\text{/day}$$

Then the capacity of recycled disposable diapers for 1 m^3 concrete component is based on equation (8)

$$md = \%\, rep\ x \left(\frac{\rho d}{\rho fa}\right) x\ m\ fa \tag{8}$$

$$md = 34.35 \text{ kg} = 0.025 \text{ m}^3$$

Then, capacity of disposable diapers concrete that can be produced in Bandung is determined by equation (6) is 796.99 m^3.

Next the author tries to simulate the calculation of this diaper concrete production with a design proposal for a type 36 housing with the standard structure shown in the following table 5.

The table shows the capacity of diaper concrete that can be used to build a housing of type 36, which is 48.15 m^3. By considering the daily production of diaper concrete, which is 796.99 m^3, approximately 16 units of type 36 houses can be obtained. In addition, the author tries to simulate the disposable diapers concrete as road pavement. The calculation by designing a 4 m wide pavement with a thickness of 20 cm and the result is 0.997 km of road pavement can be constructed. However, it is noted that this calculation is for daily diaper waste production, if it is calculated in monthly or yearly terms, there will be even more amount of diaper waste that must be added recycled into building components or road pavements

Table 5. Design simulation of housing type 36

Building component	Dimension (m)	Unit	*DDC (m^3)
Colom	0.3 x 0.3 x 3	9	2.43
Beam	0.3 x 0.4 x 3	12	4.32
Wall Panel	3 x 3 x 0.05	12	5.4
Floor	6 x 6	1	36
Total DDC per housing unit of type 36			48.15

DDC = disposable diapers concrete

Based on those simple simulation and calculation approach, this paper is recommended for recycling the disposable diapers as concrete component, it would be more effective as road pavement that housing component

4. Conclusion

- Parameters to calculate waste capacity for disposable diapers is age population of 0-4 years, frequency of changing diaper and mass of diaper.

- Parameters to be considered for replacement of concrete component by waste materials, especially disposable diapers is density of materials, mass of materials and percentage of replacement. For disposable diapers, the optimal percentage replacement is 5-10 % as concrete structure.
- The utilization of disposable diapers as concrete component would be more effective to reduce the waste capacity by be used for constructing road pavement.

Acknowledgments
The Doctoral Program is financially supported by the Ministry of Education Culture, Sports, Science and Technology (MEXT), Japan

References

[1] David P; Huang X; Cordova A. and Pimentel M 1997 Impact of a Growing Population on Natural Resources: The Challenge for Environmental Management *Frontiers: The Interdisciplinary Journal of Study Abroad* **3** 105
[2] EDANA 2008 Sustainability Report 2007-2008: Absorbent Hygiene Products. European Disposables and Nonwovens Association (EDANA) Brussels
[3] Mfon E N, Emem O M, Ekom N E and Sajjad H 2020 *American Journal of Medicine and Public Health 2020* **1 (2)** 1009
[4] Environment Protection Agency (EPA) 2008 An updated life cycle assessment study for disposable and reusable nappies Science Report. SC010018/SR2 Bristol
[5] Wambui K.E, Joseph M and Makindi S 2015 Soiled Diapers Disposal Practices Among Caregivers In Poor And Middle Income Urban Settings *International Journal of Scientific and Research Publications,* **5, (10)** 154–163
[6] Virgen Q A, Taboada G P and Ojeda B S, 2014. Analysis of the feasibility of the recovery of landfill gas: a case study of Mexico. J. Clean. Prod. 79, 53–60.doi: 10.1016/j.jclepro.2014.05.025
[7] Smith A, Brown K, Ogilvie S, Rushton K and Bates J 2001. Waste Management Options and Climate Change: Final Report. European Communities, Luxembourg. Doi: 10.1016/S1352-2310(01)00532-5
[8] Daud M, Salmia B, Siti N S, Nur L M K, Zarina I, Khodrulnadiah M and Wani M S 2017 Self-curing Concrete using Baby Diapers Polymer *Indian Journal of Science and Technology* **10(4)**
[9] Valdemar R M E, Turpin M S, Delfín A I, Vázquez M A, María E V R, Sylvie T M, Irma D A and Alethia V M 2011 Disposable diapers biodegradation by the fungus pleurotus ostreatus *Waste Manag* **31** 1683–1688
[10] Valdemar R M E, Sotelo N P X, Quecholac P X, Beltrán V M, Ojeda B S and Vázquez M A. 2014 Biological recycling of used baby diapers in a small-scale composting system *Resour. Conserv Recycl.* **87**, 153–157
[11] Daud M; Salmia B, Nazirul M Z, Mohd H Z, Agusril S, Norfaraziela R and Wani K M S. 2018. The Performance of Diapers Polymer In Concrete As Self-Curing Agent In Term Of Chemical Properties *AIP Conference Proceedings* 2030, 020186
[12] Salmia B, Daud M, Fadzli M N, Siti N S and Mohamed G E 2008 Properties of Self-curing High Strength Concrete by using Baby Polymer Diapers MATEC Web of Conferences **203**, 06022
[13] Siswanti Z, Bart J D, Romi B M 2021 An Environmental Consideration of Disposable Diaper Waste Treatment as Concrete Component *Journal of Asian Institute of Low Carbon Design* 153-158
[14] BPS-Statistic of Jakarta Timur Municipality 2021. Jakarta Timur Municipality in Figures 2021 ISSN: 0215-4120 https://jaktimkota.bps.go.id/
[15] BPS-Statistic of Bekasi Municipality 2021 Bekasi Municipality in Figures 2021 ISSN:

1907.3763 https://bekasikota.bps.go.id/

[16] BPS-Statistic of Surabaya Municipality 2021 Surabaya Municipality in Figures 2021. https://surabayakota.bps.go.id/

[17] BPS-Statistic of Bandung Municipality 202. Bandung Municipality in Figures 2021 ISSN: 0215.2320 https://bandungkota.bps.go.id/

[18] BPS-Statistic of Jakarta Barat Municipality 2021 Jakarta Barat Municipality in Figures 2021 ISSN: 0215-4145 https://jakbarkota.bps.go.id/

[19] Robert J O, Veronica S, Y. Wang; Roger G, Donald L N, Mary B and Andrew N. C. 2019 Evaluation of the Impact of 2 Disposable Diapers in the "Natural" Diaper Category on Diapered Skin Condition. Clinical Pediatrics 2019 **58(7)** 806–815

[20] Amir S, William F, William J and Corinne D 1990. Diapers: What do Parents Choose and Why? Can.Fam.Physician **36**:1705-1708.

[21] Rahat S H, Sarkar A T, Rafie S A A and Hossain S 2014 Prospects of Diaper Disposal and Its Environmental Impacts on Populated Urban Area like Dhaka City *2nd International Conference on Advances in Civil Engineering 2014* (ICACE-2014) CUET, Chittagong, Bangladesh

[22] Lita Ns 2017 Gambaran Pemakaian Diapers Sekali Pakai Pada Anak Usia Prasekolah *Jurnal Photon* **7** 2

[23] Warner P and Kelly P 2006. *Mengajari Anak Pergi Ke Toilet* (Jakarta: Arcan)

[24] Widyawati U and Desy A. 2018. Hubungan Antara Pengetahuan Dan Perilaku Ibu Dalam Pemakaian Disposable Diapers Pada Batita Dengankejadian Ruam Popok *Jurnal Kedokteran Diponegoro* **7** 2

[25] Laurie A N-R, Meredith J H R C and Kathryn G D 2008 Newborn Wet and Soiled Diaper Counts and Timing of Onset of Lactation as Indicators of Breastfeeding Inadequacy, *February 2008 Journal of Human Lactation* **24(1)** 27-33.

[26] Protocol #3 Signs of Effective Breastfeeding, Breastfeeding Protocols for Health Care Providers Toronto Public Health

[27] Breastfed Baby Diaper Count https://www.healthy.arkansas.gov/breastfeeding

[28] Breasteeding Log , Northwestern Memorial Hospital, Patient Education care and treatment

[29] Ratih S and Dwi A F. 2019 Analysis of factors related to the continuity flow of breast milk in normal postpartum mother. Seminar Nasional Hasil Penelitian dan Pengabdian Pada Masyarakat IV Tahun 2019 "Pengembangan Sumberdaya menuju Masyarakat Madani Berkearifan Lokal" LPPM - Universitas Muhammadiyah Purwokerto

[30] Nining F 2013. Pengaruh PolaAsuh Orang Tua dan Intensitas Penggunaan Diapers terhadap Tingkat Kesiapan Toilet Training pada Anak Usia Toddler di Little Care Stikes Surya Global Yogyakarta. Publikasi Thesis. Surakarta: Program PascaKedokteran Fakultas Kedokteran Universitas Sebelas Maret.

[31] Anita D M n.d. Mother Perception in Disposable Baby Diapers Waste and The Management https://osf.io/preprints/inarxiv/kpgmc/ 10.31227/osf.io/kpgmc

[32] Diaper Size and Weight Chart Guide. 2019. https://www.pampers.com/en-us/about-us/diapers-and-wipes/article/diaper-size-and-weight-chart

[33] Mollyn V B 1971 Weighing Wet Diapers A New Approach to Measuring Infant Output College of Nursing. University of Utah 1971

[34] How much does a dry diaper weigh? https://earthmamadoulas.com/baby/how-much-does-a-dry-diaper-weigh.html. Access 11 May 2021

[35] Yahya B, Ayoub G; Fouad D, Ahmed B, Mohammed B 2020 A new approach for assessing the absorption of disposable baby diapers and superabsorbent polymers: A comparative study. *Results in Materials* **8**, 100156

Kinetics and thermodynamics study of organic waste combustion using thermogravimetric analysis

S Sukarni[1,2], A Prasetiyo[1], L Fidiah[1], A A Permanasari[1,2] and P Puspitasari[1,2]

[1]Center for Renewable and Sustainable Energy Engineering (CRSEE), Department of Mechanical Engineering, Universitas Negeri Malang, Jl.Semarang No 5, Malang, 65145, Indonesia
[2]Centre of Advanced Materials for Renewable Energy, Universitas Negeri Malang, Jl.Semarang No 5, Malang, 65145, Indonesia

sukarni.ft@um.ac.id

Abstract. Urban waste in Malang becomes a significant problem due to its increasing volume and negative impact on the environment. An incineration process offers a solution to reduce waste. The thermochemical method is one of the routes to reduce waste volume effectively. This paper discussed the combustion of organic waste with thermogravimetric analysis in a non-isothermal mode in which temperature escalated from 25 to 1000°C by a heating ramp of 20°C/min. The process of burning organic waste is divided into three steps of decomposition, namely the step of moisture loss, decomposition and combustion of volatile matter and fixed carbon, and the final step of ash decomposition from the residual matters. The kinetic parameters calculated using the Coats-Redfern method at the main combustion step, resulting in activation energy of 99.37 kJ/mol and a pre-exponential factor of 6.01E+08/s.

1. Introduction

The increasing population and living standard impact household needs and consumption, which increases the municipal waste production rate [1–5]. It influences the environment since municipal solid waste not only holds in storage and landfill areas but also produces disturbing odors and gas emissions. Continuous stockpiling impacts the contamination of water sources in the ground [5,6]. With a population of around 894,653, Malang can generate significant amounts of waste, roughly 660 tons daily, consisting of household and industrial waste [7]. Municipal waste includes organic and synthetic waste, where synthetic waste is difficult to degrade by the environment [8]. The proper incineration process can prevent the accumulation of environmental waste by converting municipal waste into energy through thermochemical processes [9–11]. Reducing waste mass and volume by 70-90% can be reached by means of incineration [2]. Therefore, converting waste into valuable energy through the combustion process is a great challenge that must be in-depth studied with a deliberate effort to generate significant energy and decrease the environmental issue.

Combustion technology can be well perceived by identifying the kinetics parameters. A popular technique for studying combustion kinetics is thermogravimetric analysis (TGA) [8]. Azam *et al.* [12] examined the kinetic combustion of solid waste by applying four isoconversional methods (Friedman, Kissinger-Akahira-Sunose (KAS), Vyazovkin, and Flynn-Wall-Ozawa (FWO)), and the results were compared with low-rank coals. They showed that for solid wastes, altogether, isoconversional methods have almost the same trends of activation energy in the selected conversion zones (0.1-0.6 and 0.7-0.9). As for coal, the Friedman model showed a lower and inconsistent activation energy value compared to the other selected isoconversional methods. Phasee & Areeprasert [13] have used hydrothermal treatment (HTT) to produce solid fuel from municipal solid waste (MSW), then it was analyzed by thermogravimetric technique. It was reported that the mass loss of raw-MSW by 56% occurred at the combustion temperature of 200-380°C, which was associated with organic and paper waste. In addition, the kinetic parameter analysis reported that HTT in the municipal waste combustion (MSW) process could increase the value of activation energy. The findings of this study can open a perspective of conversing waste into environmentally friendly energy and provide an alternative option for waste management policies.

This research intended to determine the decomposition of organic waste combustion with thermogravimetric analysis, evaluate its combustion kinetics using the Coats-Redfern method as recommended by [14] and analyze the thermodynamic parameters.

2. Materials and method

2.1. Material preparation

Organic waste was accumulated collectively from five landfill sites in the city of Malang, including Brantas, Comboran, Merjosari, Tanjung, and Tidar. Ten types of organic waste included coconut shells, beans, dry leaves, straw, corn husks, peanut shells, banana peels, wood husks, coconut fibers, and corn cobs. They were sun-dried for two weeks and were cut into small particles by scissors and mashed. In order to equalize the size, the samples were filtered with a mesh size of 40. Then, the waste was mixed with a mass ratio of 1. The organic waste sample was stored later in a desiccator.

2.2. Thermogravimetric analysis

The characteristics of organic waste combustion were studied with a thermogravimetric analyzer (TG/DSC1 of Mettler Toledo Instrument). For each test, 10 mg of organic waste sample was poured into the Al_2O_3 crucible and then put into the TG furnace. In this combustion experiment, the atmospheric air was used with a flow rate of 100 ml/min. The heating escalation in the furnace was set at a rate of 20°C/min, and it was initiated from room temperature (25°C) and was completed at 1000°C. A thermogravimetric (TG) and derivative thermogravimetric (DTG) graphs were achieved during the testing process. Any mass losses were recorded based on time and temperature.

2.3. Kinetics method

The material experienced thermal degradation during the combustion process, and its degraded fraction presented as conversion (α), is defined as:

$$\alpha = \frac{m_o - m_t}{m_o - m_f} \tag{1}$$

where m_0 is the initial mass of material; m_t is the mass at a time (t) and m_f is the final mass of the material. The rate of conversion denoted as $\frac{d\alpha}{dt}$, expressed as:

$$\frac{d\alpha}{dt} = kf(\alpha) = \mathrm{A}\, e^{\frac{-E}{RT}} f(\alpha) \tag{2}$$

where k is a temperature-dependent rate constant, and $f(\alpha)$ is the temperature-independent reaction model. In the non-isothermal studies with heating rate *($\beta = dT / dt$)*, Equation (2) becomes:

$$\frac{d\alpha}{dT} = \frac{A}{\beta} e^{\frac{-E}{RT}} f(\alpha) \tag{3}$$

where E is the activation energy, A is the pre-exponential factor, R is the ideal gas constant, and T is the absolute temperature. By considering the n-order reaction, $f(\alpha) = (1-\alpha)^n$, thus Equation (3) converted to be:

$$\frac{d\alpha}{(1-\alpha)^n} = \frac{A}{\beta} e^{\frac{-E}{RT}} dT \tag{4}$$

Equation (4) is the primary reference in calculating the kinetic parameters based on thermogravimetric (TG) graph data.

2.4. Thermodynamic parameters

The thermodynamic parameters could be determined based on the kinetic parameters obtained. The respective change in Gibbs free energy (ΔG), the change in entropy (ΔS), and the change in enthalpy (ΔH) were calculated based on the following equation:

$$\Delta G = E + R.Tm.\ln\left(\frac{k_B.Tm}{h.A}\right) \quad (5)$$

$$\Delta H = E\alpha - R.T \quad (6)$$

$$\Delta S = \frac{\Delta H - \Delta G}{Tm} \quad (7)$$

where k_B is the Boltzmann constant (1.381 × 10^{-23} J/K), h is the Plank constant (6,626 × 10^{-34} J/s), and Tm is the peak temperature on the DTG graph.

3. Results and discussion

3.1. Thermal characteristics

The thermal characteristics of organic waste combustion with a constant heating rate at 20°C/min in atmospheric air were depicted in the form of TG and DTG graphs, as shown in Figure 1. The graph of mass loss rate versus time (DTG) clearly showed three steps of decomposition in organic waste combustion. In step I, which occurred in the temperature range of 25-139°C, indicated the initial step of the combustion process was associated with moisture loss. Step II (139-601°C) was the main step of the decomposition and combustion of volatile matter and fixed carbon from organic waste, characterized by a fast mass degradation rate [14], with a significant mass loss of 63%. Step II was then divided into two zones. Zone I was the decomposition of organic waste characterized by a peak of DTG at a temperature of 294°C [13]. Zone II was the decomposition of coconut shell waste components characterized by the highest peak of DTG at a temperature of 339°C [15]. Step III (601-998°C) was the final step in the decomposition process of ash during the combustion process, accompanied by the oxidizing of the small amount of the remaining char from the previous step. Thus, it reduced the weight of biomass.

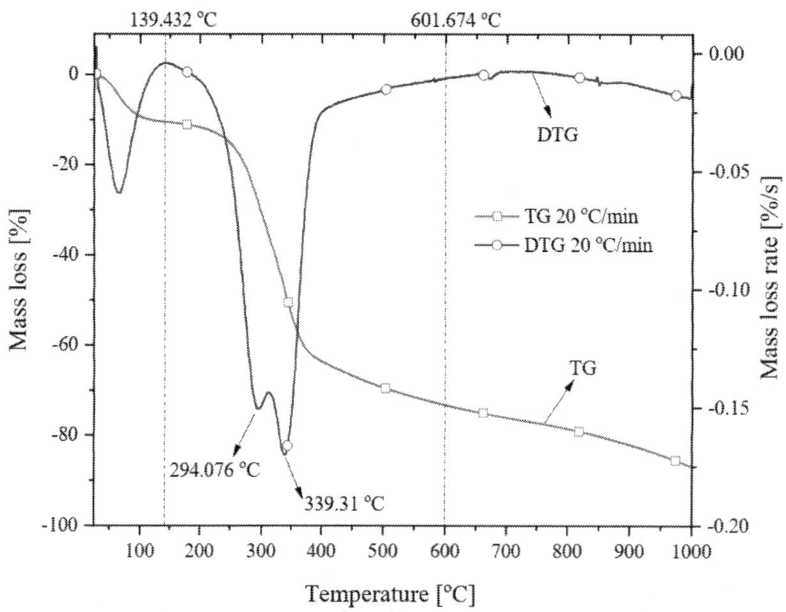

Figure 1. TG and DTG combustions of organic waste

3.2. Evaluation of kinetic parameters

Table 1 showed the temperature characteristics of organic waste combustion. In the step of main decomposition (Step II), the kinetic parameters were evaluated through the Coats-Redfern fitting model recommended by Acikallin [16].

Based on the integral of Equation (4), and in order to find an exact analytical solution, the equation is made simpler by bearing in mind that 2RT/E≪1 so that it becomes Equation (8):

$$ln\ g(\alpha) = -\frac{E}{RT} + ln\left(\frac{AR}{\beta E}\right) \tag{8}$$

where, if value;
$n = 1$, then $g(\alpha) = -(\ln(1-\alpha)/T^2)$,
$n \neq 1$, then $g(\alpha) = \{1-(1-\alpha).(1-n)\}/\{(1-n)T^2\}$.

Table 1. Characteristics of the temperature in organic waste combustion

Step	Characteristics of temperature				
	T_i (°C)	T_{max} (°C)	T_f (°C)	M_{max} (%/s)	Mass loss (%)
I	25	66.392	139.432	-0.059322	-10.39
II	139.432	339.31	601.674	-0.170196	-73.29
III	601.674	672.745	998.341	-0.010625	-86.84

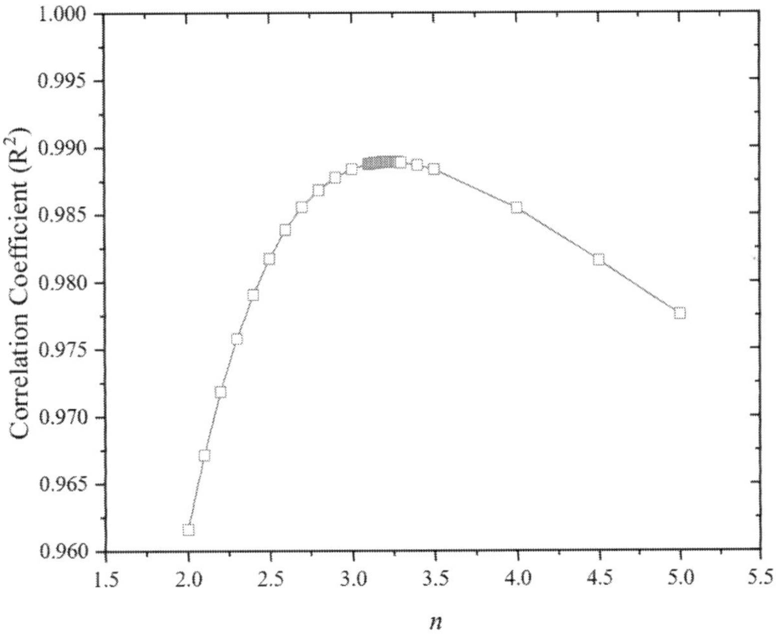

Figure 2. R^2-n combustion of organic waste in step II with the Coats-Redfern method

Figure 2 showed the R^2-*n* plot in finding the highest correlation coefficient (R^2) that is correlated with the most proper *n* value. The highest R^2 value of 0.989 was found at the selected *n* value of 3.23. This *n* value was then used to obtain the plot of ln g (α) vs. (1/T) to obtain the slope (-E/R) and intercept [ln ((AR)/(βE)}] as shown by Figure 3, in the conversion range of alpha (α) values which were calculated between 0.05-0.95.

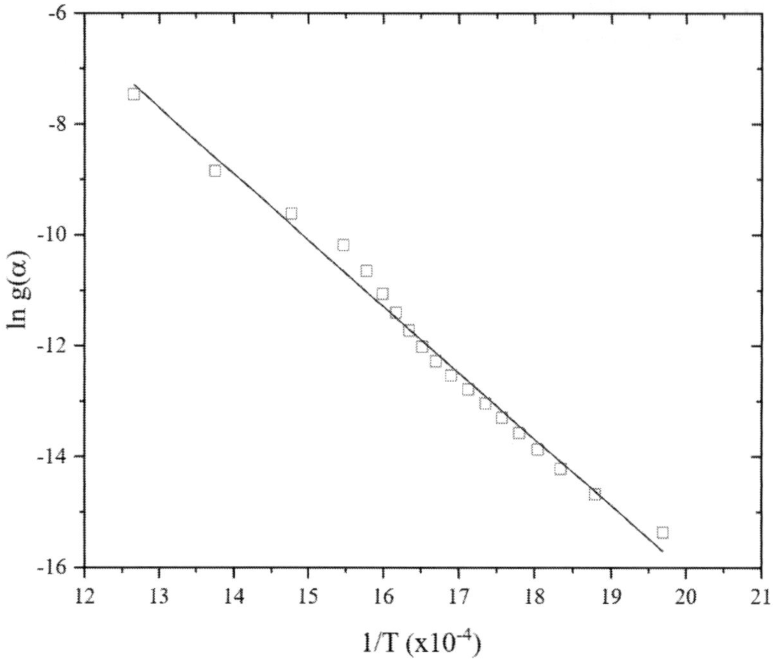

Figure 3. Slope and intercept of organic waste combustion in step II with the Coats-Redfern method

As presented in Table 3, the average activation energy value of organic waste combustion at a heating rate of 20°C/min, which was analyzed with the Coats-Redfern method in the main decomposition step, was 99.37 kJ/mol. The associated pre-exponential factor value was 6.01E+08/s. The results indicated that the activation energy of organic waste was greater than a mixture of synthetic waste and microalgae *Spirulina platensis* by 57.77 kJ/mol [17] and coal by 34 kJ/mol [18]; however, this value was smaller than corn silk biomass by 207.37 kJ/mol [19]. Sukarni et al. [20] stated that the smaller the value of activation energy is, the less energy needed by a particle to start reacting will be. Thus, this results in the activation energy of organic waste becoming the essential data for proper designing the power generation system.

Table 2. The kinetic parameters according to the Coats-Redfern method

Step	Trendline equation	R^2	Kinetic parameters		
			Ea (kJ/mol)	A (1/s)	*n*
II	*y= -11952x + 7.8297*	0.989	99.37	6.01E+08	3.23

3.3. Thermodynamic parameters

Table 3 showed the thermodynamic parameters of organic waste combustion in the main decomposition step at the heating rate of 20°C/min. The values of ΔH, ΔG, ΔS were 96.85; 150.10 and -0.087 kJ/mol, respectively. The ΔG showed the magnitude of energy change in the system. Besides, ΔS showed the change in the degree of irregularity in the system, and ΔH represented the difference in the energy of the reagent and the activated complex for the dissociation of chemical bonds in the reagent.

Table 3. Thermodynamic parameters

Step	Ea (kJ/mol)	Log A (1/min)	ΔH (kJ/mol)	ΔG (kJ/mol)	ΔS (kJ/mol)
II	99.37	8.778	96.58	150.10	-0.087

4. Conclusion

The characteristics of organic waste combustion have been analyzed to find the kinetics and thermodynamic parameters. Thermal decomposition of organic waste combustion is divided into three steps: the step of moisture loss, followed by decomposition of volatile matter and fixed carbon. The final step of combustion indicates the ash decomposition during the combustion. The organic waste combustion at the heating rate of 20°C/min showed the activation energy value at the main step of decomposition (in the range of 139-601°C) by 99.37 kJ/mol. Overall, the organic waste in Malang has the potential to be converted into valuable energy by combustion processes.

Acknowledgments

The financial support from the Ministry of Research, Technology and Higher Education, Republic of Indonesia, for 3 fiscal years (26.3.35/UN32.14/LT/2018, 188/SP2H/LT/DRPM/2019, and 188/SP2H/AMD/LT/DRPM/2020) was acknowledged.

References

[1] Xu D, Zhang Y, Zhou H, Meng Y and Wang S 2016 Characterization of adhesive penetration in wood bond by means of scanning thermal microscopy (SThM) *Holzforschung* **70** 323–30
[2] Kumar A and Samadder S R 2017 A review on technological options of waste to energy for effective management of municipal solid waste *Waste Manag.* **69** 407–22
[3] Huang J, Qiao Y, Wei X, Zhou J, Yu Y and Xu M 2019 Effect of torrefaction on steam gasification of starchy food waste *Fuel* **253** 1556–64
[4] Xu Z X, Shan Y Q, Zhang Z, Deng X Q, Yang Y, Luque R and Duan P G 2020 Hydrothermal carbonization of sewage sludge: Effect of inorganic salts on hydrochar's physicochemical properties *Green Chem.* **22** 7010–22
[5] Li Q, Faramarzi A, Zhang S, Wang Y, Hu X and Gholizadeh M 2020 Progress in catalytic pyrolysis of municipal solid waste *Energy Convers. Manag.* **226** 113525
[6] Sukarni S 2020 Thermogravimetric analysis of the combustion of marine microalgae Spirulina platensis and its blend with synthetic waste *Heliyon* **6** e04902
[7] Sukarni S 2016 Exploring the potential of municipal solid waste (MSW) as solid fuel for energy generation: Case study in the Malang City, Indonesia *AIP Conference Proceedings* vol 1778 p 020003
[8] Zhang D, Lin X, Zhang Q, Ren X, Yu W and Cai H 2020 Catalytic pyrolysis of wood-plastic composite waste over activated carbon catalyst for aromatics production: Effect of preparation process of activated carbon *Energy* **212** 118983
[9] Peng Y, Chen J, Lu S, Huang J, Zhang M, Buekens A, Li X and Yan J 2016 Chlorophenols in Municipal Solid Waste Incineration: A review *Chem. Eng. J.* **292** 398–414
[10] Istrate I R, García-Gusano D, Iribarren D and Dufour J 2019 Long-term opportunities for electricity production through municipal solid waste incineration when internalising external costs *J. Clean. Prod.* **215** 870–7
[11] Lu J W, Zhang S, Hai J and Lei M 2017 Status and perspectives of municipal solid waste incineration in China: A comparison with developed regions *Waste Manag.* **69** 170–86
[12] Azam M, Ashraf A, Jahromy S S, Raza W, Khalid H, Raza N and Winter F 2020 Isoconversional nonisothermal kinetic analysis of municipal solid waste, refuse-derived fuel, and coal *Energy Sci. Eng.* 1–12
[13] Phasee P and Areeprasert C 2017 Thermal Decomposition Behavior during Combustion of Hydrothermally Treated MSW by Thermogravimetric Analysis *Energy Procedia* **138** 616–21
[14] Sait H H, Hussain A, Salema A A and Ani F N 2012 Pyrolysis and combustion kinetics of date palm biomass using thermogravimetric analysis *Bioresour. Technol.* **118** 382–9
[15] Liyanage C D and Pieris M 2015 A Physico-Chemical Analysis of Coconut Shell Powder *Procedia Chem.* **16** 222–8
[16] Acikalin K 2012 Pyrolytic characteristics and kinetics of pistachio shell by thermogravimetric analysis *J. Therm. Anal. Calorim.* **109** 227–35

[17] Sukarni S, Prasetiyo A, Sumarli S, Nauri I M and Permanasari A A 2018 Kinetic analysis of co-combustion of microalgae spirulina platensis and synthetic waste through the fitting model *MATEC Web of Conferences* vol 204

[18] Tahmasebi A, Kassim M A, Yu J and Bhattacharya S 2013 Thermogravimetric study of the combustion of Tetraselmis suecica microalgae and its blend with a Victorian brown coal in O2/N2 and O2/CO2 atmospheres *Bioresour. Technol.* **150** 15–27

[19] Sumarli S, Bandhana Kusuma Himawan C U S, Wulandari R and Sukarni S 2019 Thermogravimetric analysis and the fitting model kinetic evaluation of corn silk thermal decomposition under an inert atmosphere *IOP Conference Series: Materials Science and Engineering* vol 494 (Institute of Physics Publishing)

[20] Sukarni, Sudjito, Hamidi N, Yanuhar U and Wardana I N G 2015 Thermogravimetric kinetic analysis of Nannochloropsis oculata combustion in air atmosphere *Front. Energy* **9** 125–33

Thermogravimetric and kinetic analyses of the *Skeletonema costatum* microalgae combustion using the fitting method

S Sukarni[1,2], A Prasetiyo[1], R Zulfambudy[1], A A Permanasari[1,2] and P Puspitasari[1,2]

[1]Center for Renewable and Sustainable Energy Engineering (CRSEE), Department of Mechanical Engineering, Universitas Negeri Malang, Jl.Semarang No 5, Malang, 65145, Indonesia
[2]Centre of Advanced Materials for Renewable Energy, Universitas Negeri Malang, Jl.Semarang No 5, Malang, 65145, Indonesia

sukarni.ft@um.ac.id

Abstract. This work was intended to understand the potential of *Skeletonema costatum* (*S. costatum)* microalgae being the solid biofuel feedstock by studying its combustion behavior under a thermogravimetric analyzer (TGA). The following parameters were used during the experiment: about 5 mg of weighed sample was put into a ceramic crucible; the atmospheric air of 100 ml/min was flowed to the chamber to ensure sufficient combustion air; heating was escalated with a rate of 10 °C/min, and combustion was performed at a temperature range from 30 °C to 1200 °C. The results showed that the combustion process of *S. costatum* microalgae could be divided into four stages of decomposition, including the evaporation stage of moisture content from a temperature of 31.68 °C to 150.93 °C, the devolatilization stage from a 150.93 °C to 505.053 °C, the stage of char combustion from a 505.053 °C to 1018.15 °C and the final stage of ash decomposition from 1018.15 °C to 1200 °C. The kinetic parameters were calculated in the 1st, 2nd, and 3rd stages based on the fitting model of the Coats-Redfern method by selecting reaction mechanisms. The activation energy values for 1st, 2nd, and 3rd stages were 71.18; 39.49, and 56.92 kJ /mol, respectively, and correlation coefficients (R^2) of 0.959; 0.988, and 0.978, respectively in the Chemical Reaction Order models.

1. Introduction

Microalgae are photosynthetic organisms living and growing rapidly in aquatic environments [1], with simple unicellular and multicellular shapes, sizes, and structures [2]. There are more than 50,000 species of microalgae worldwide [3], but only about 30,000 species are currently being analyzed and studied. Microalgae, with its potential, is a promising choice as a raw material for renewable fuels, but it still requires comprehensive exploration and understanding [4].

Various types of thermal conversion of solid renewable fuels can be obtained from microalgae, as well as the thermal pyrolysis process [5,6], combustion [7–9], converting to be biofuel [10], biodiesel [11], bioethanol [12], and biohydrogen [13]. The research on the potential of *S. costatum* microalgae into biodiesel has been presented in the paper of Gao *et al.* [14]. Kalhor *et al.*, [15], reporting the fatty acid profile of *S. costatum* microalgae as an initial step for biodiesel production, showed the existence of both palmitic acid and stearic acid in algal oil, where both contents were associated with a high cetane number of biodiesel produced. A higher biodiesel cetane number would yield more ignition delay in the diesel engine combustion chamber. However, in so far as the literature studied, the thermal conversion of *S. costatum* microalgae through the combustion process has not been found.

The thermogravimetric approach is a valuable method for studying the thermal behavior of microalgae combustion and its components [16]. The technique provides thermal degradation information of mass based on time and temperature. This knowledge then becomes the basis for determining the kinetics of the volatilization of biomass during a thermochemical reaction and being critical parameters for proper designing and modeling a combustion reactor system [17,18]. The combustion product is heavily reliant on the type of raw material, the heating rate, and the temperature setting used; therefore, a specific kinetics model is needed for a suitable conversion technique.

This research was on purpose to analyze the thermal behavior of *S. costatum* microalgae to understand its potential as raw material for solid biofuel by the route of a direct combustion process. The combustion kinetics parameters were analyzed as well, involve the value of activation energies and pre-exponential factors. The various kinetic models proposed by Naqvi et al. [19] are being used on the basis of the fitting technique of the Coats-Redfern.

2. Material and Method

2.1. Sample preparation

S. costatum have been cultured in the pond of the Center for Development of Brackish Water Aquaculture (BBPBAP) of Jepara, Indonesia. Microalgae in the pond were deposited with 150 ppm NaOH and resulted in 3:1 (mass ratio) of freshwater to microalgae. The precipitate was then filtered to remove the moisture content before being dried under the sun for 6 hours. To ensure an even drying, the microalgae precipitation was then re-dried in an oven at 400 °C for 3 hours. The microalgae were then mashed and filtered to uniform the size.

2.2. TG experiments

The combustion profile of the algal biomass was analyzed using thermogravimetric analysis by Mettler Toledo TG/DSC1. A sample mass of 5 mg was put into a crucible. The heating rate used was 10 °C/min, and 100 ml/min of an atmospheric air was flowed in the combustion chamber during the experiment that performed at range of 30 °C to 1200 °C. The mass loss presented as TGA graph and mass loss rate presented as differential thermogravimetric (DTG) graph related to both time and temperature were recorded continuously to determine the effect of thermal to the mass degradation and rate.

2.3. Kinetic analysis

The kinetic analysis of microalgae *S. costatum* combustion was easily defined by Arrhenius law, providing a numerical measure of the reaction rate. The conversion rate is basically expressed as the following equation:

$$\frac{d\alpha}{dt} = k(T)f(\alpha) \tag{1}$$

where,

$$\alpha = \frac{m_o - m_i}{m_0 - m_f} \tag{2}$$

where, m_o is the initial mass, m_i, and m_f are the mass at a certain time and the final mass of each end of the stage, respectively. The rate constant is a parameter that depends on temperature, which is representing the frequency of successful collisions, expressed as:

$$k(T) = A\,exp\left(-\frac{E_a}{RT}\right) \tag{3}$$

where the respective T, R, E_a, and A are sign of the absolute reaction temperature (K), the universal gas constant (0.008314 kJ/mol.K), the activation energy (kJ/mol), and the pre-exponential factor (1/min). The equations (1) and (3) are rewritten as a compound form becomes:

$$\frac{d\alpha}{dT} = \frac{A}{\beta} exp\left(-\frac{E_a}{RT}\right) f(\alpha) \tag{4}$$

where β is the heating rate ($\beta = dT/dt$). The reaction model $f(\alpha)$ then was integrated and its integral form is written as $g(\alpha)$. The analytical solution approach model is written into the following equations

$$g(\alpha) = \int_0^{\alpha} \frac{d\alpha}{f(\alpha)} = \frac{A}{\beta} \int_{T_0}^{T} exp\left(-\frac{E_a}{RT}\right) dT \tag{5}$$

2.4. The Coats-Redfern method

The Coats-Redfern selected model can be used in predicting reaction order, activation energy, and pre-exponential factors based on the variation of the selected $g(\alpha)$. The fundamental equation for the Coats-Redfern method expressed as:

$$ln\left[\frac{g(\alpha)}{T^2}\right] = ln\frac{AR}{\beta E_a}\left(1 - \frac{2RT}{E_a}\right) - \frac{E_a}{RT} \tag{6}$$

Plotting between $1/T$ and ln $[g(\alpha)/T^2]$ is resulted in a straight line, from which the slope and intercept values are found and in turn both are used for determining the activation energies and pre-exponential factors, respectively. Most of the thermal degradation reactions of solid materials categorized in five models are listed in Table 1.

Table 1. Model of reaction applied for kinetic parameters calculation

Reaction model	Symbol	Differential form $f(\alpha)$	Integral form $g(\alpha)$
Reaction order			
Chemical Reaction Order 1	F1	$1-\alpha$	$-\ln(1-\alpha)$
Chemical Reaction Order 1.5	F1.5	$(1-\alpha)]^{3/2}$	$2(1-\alpha)]^{3/2}-1]$
Diffusion			
Parabolic law	D1	$½\alpha$	α^2
Valensi equation	D2	$-[\ln(1-\alpha)]^{-1}$	$\alpha+(1-\alpha)\ln(1-\alpha)$
Ginstling-Broushtein equation	D3	$3/2[(1-\alpha)^{1/3}-1]^{-1}$	$(1-2/3\alpha)-(1-\alpha)^{2/3}$

3. Result and Discussion

3.1. Combustion characteristics of S. costatum

Figure 1 shows a graph of the combustion of microalgae *S. costatum* included TG and DTG, from room temperature up to 1200 °C.

Based on Figure 1, the combustion characteristics of the *S. costatum* microalgae were divided into four decomposition stages. The first stage, from the temperature of 30°C to 150.93°C, indicates the process of releasing the moisture content of *S. costatum*, which is indicated on the DTG graph as a very deep basin with a short temperature range and time, having a peak of mass loss rate at 49.13 °C and the mass is lost by 12.785%. Furthermore, the second stage, which occurred from 150.93 °C to 505.053 °C, characterized by the presence of the large size of the basin on the DTG curve at a longer time and temperature range, is associated to the devolatilization and combustion stage of volatile matter and char, as the similar profile was also reported by Chen *et al.*, [20]. The third stage, which taken place from 505.053 °C to 1018.15 °C, indicated by the TG graph having a gentle decline, is related to the burning of remaining fixed carbon and organic components. The fourth stage, from the termination of the third stage to 1201 °C, is related to the small loss of mass from the decomposition of the ash as the residual matter. The classification of the combustion characteristics of *S. costatum* is detailed in Table 2.

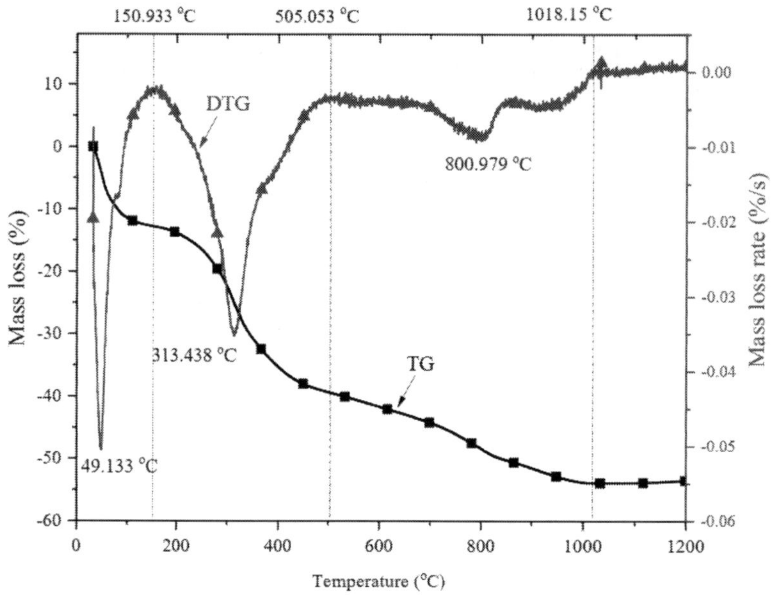

Figure 1. TG and DTG curves of *S. costatum* combustion at 10 °C/min

Table 2. Classification of temperature characteristics of *S. costatum* microalgae combustion

Stage	Temperature characteristics				
	T_i (°C)	T_{max} (°C)	T_f (°C)	Mass loss (%)	M_{max} (%/s)
I	31.68	49.133	150.933	-12.785	-0.05054
II	150.933	313.438	505.053	-39.438	-0.03529
III	505.053	800.979	1018.15	-53.865	-0.00911
IV	1018.15	1031.8	1201.69	-53.508	-0.00164

3.2. Kinetic analysis of combustion

Kinetic analysis was carried out at each stage of decomposition, including the first stage of releasing moisture content, the second stage related to the main devolatilization, and the third stage related to the burning of fixed carbon and organic components. The importance of knowing the value of the kinetics parameters at each stage is for estimating the amount of energy required for each stage and for determining the proper design and shape of the combustion reactor. Based on the reaction mechanisms on $f(\alpha)$ and $g(\alpha)$ selected in alpha (α) 0.05-0.95, then the evaluation of the kinetic parameters of *S. costatum* microalgae combustion with the Coats-Redfern method reaction mechanisms is shown in Table 3.

Table 3. The kinetic parameters of the *S. costatum* combustion at various selected models

Model name	Stage I			Stage II			Stage III		
	Ea (kJ/mol)	R^2	Log A (1/min)	Ea (kJ/mol)	R^2	Log A (1/min)	Ea (kJ/mol)	R^2	Log A (1/min)
F1	37.21	0.793	5.21	39.49	0.978	2.40	56.92	0.988	1.52
F1.5	71.18	0.959	11.65	71.87	0.976	6.35	102.62	0.929	4.96
D1	52.49	0.651	7.18	61.36	0.903	3.94	91.01	0.942	2.90
D2	59.62	0.701	8.15	68.62	0.933	4.44	101.57	0.965	3.29
D3	62.78	0.723	8.06	71.75	0.945	4.12	106.08	0.973	2.93

In the Chemical Reaction Order 1.5 (F1.5) model, Stage I shows the highest activation energy value of 71.18 kJ/mol with R^2 0.959, then it is followed by stage III at 56.92 kJ/mol with R^2 0.988, and the lowest one is stage II of which value of activation energy is 39.49 kJ/mol with R^2 0.978. The mechanism model with linear regression of R^2 close to 0.99 is considered as a suitable model. The Chemical Reaction Order (F1 and F1.5) model shows a simple model used in homogeneous kinetics, as the reaction rate is directly proportional to the conversion value [21].

4. Conclusion

This study has investigated the combustion characteristics of 5 mg pulverized biomass of *S. costatum* at a thermogravimetric analyzer with a 10 °C/min of heating ramp. Four prominent stages were noticeable throughout the sample combustion process, among which three peaks were strongly recognized. The first stage (I) was correlated to the release of moisture content, and it was followed by the devolatilization stage (II) of the char and volatile matter. The third stage (III) was related to the burning of remaining fixed carbon and organic components, and the last stage (IV) was the decomposition of the ash as the residual matter. The activation energies were analyzed using the Coats-Redfern model in stages I, II, and III, respectively, those are 71.18; 39.49, and 56.92 kJ/mol in Chemical Reaction Orders (F1 and F1.5) models.

Acknowledgments

The authors would like to acknowledge the Universitas Negeri Malang for supporting this research.

References

[1] Wen Z and Jhonson M B 2009 Microalgae as a feedstock for biofuel production N. Biotechnol. 25 S278–9
[2] Wolkers H, Barbosa M, Kleinegris D M . and Rouke 2011 Microalgae: the green gold of the future? Large-scale Wageningen UR
[3] Helliwell K E, Lawrence A D, Holzer A, Kudahl U J, Sasso S, Kräutler B, Scanlan D J, Warren M J and Smith A G 2016 Cyanobacteria and Eukaryotic Algae Use Different Chemical Variants of Vitamin B12 Curr. Biol. 26 999–1008
[4] Phukan M M, Chutia R S, Konwar B K and Kataki R 2011 Microalgae Chlorella as a potential bio-energy feedstock Appl. Energy 88 3307–12
[5] Widiono A E, Sukarni S, Wulandari R, Prasetiyo A, Suryanto H and Yanuhar U 2020 Pyrolytic thermal decomposition behavior and kinetic parameters of Tetraselmis chuii microalgae AIP Conf. Proc. 2255
[6] Sukarni S, Widiono A E, Wulandari R, Prasetiyo A and Puspitasari P 2020 Thermogravimetric study on the thermal characteristics of tetraselmis chuii microalgae pyrolysis in the presence of titanium dioxide Key Eng. Mater. 851 KEM 156–63
[7] Prasetiyo A, Sukarni S, Wulandari R, Suryanto H and Yanuhar U 2019 Investigation on kinetic parameters during combustion of Tetraselmis chuii microalgae under thermogravimetric analyzer
[8] Prasetiyo A, Sukarni S, Wulandari R and Puspitasari P 2020 A Kinetic Study on Tetraselmis chuii Combustion: The Catalytic Impact of Nanoparticle Titanium Dioxide (TiO2) Additive J. Adv. Res. Fluid Mech. Therm. Sci. 71 39–49
[9] Sukarni S, Ardianto P, Retno W, Aloon E W and Poppy P 2020 Thermal characteristic of tetraselmis chuii combustion influenced by titanium dioxide (Tio2) nanoparticle Key Eng. Mater. 851 KEM 149–55
[10] Raheem A, Prinsen P, Vuppaladadiyam A K, Zhao M and Luque R 2018 A review on sustainable microalgae based biofuel and bioenergy production: recent developments
[11] Yusuf C 2007 Biodiesel from microalgae Biotechnol. Adv. 25 294–306
[12] de Farias Silva C E and Bertucco A 2016 Bioethanol from microalgae and cyanobacteria: A review and technological outlook Process Biochem. 51 1833–42
[13] Levin D B, Pitt L and Love M 2004 Biohydrogen production: Prospects and limitations to practical application Int. J. Hydrogen Energy 29 173–85
[14] Gao G, Wu M, Fu Q, Li X and Xu J 2019 A two-stage model with nitrogen and silicon limitation enhances lipid productivity and biodiesel features of the marine bloom-forming diatom

Skeletonema costatum Bioresour. Technol. 289 121717
[15] Kalhor A X, Movafeghi A, Mohammadi-Nassab A D, Abedi E and Bahrami A 2017 Potential of the green alga Chlorella vulgaris for biodegradation of crude oil hydrocarbons Mar. Pollut. Bull. 123 286–90
[16] Vamvuka D, Tsamourgeli V and Galetakis M 2013 Study on Catalytic Combustion of Biomass Mixtures with Poor Coals Combust. Sci. Technol.
[17] Sukarni S 2020 Thermogravimetric analysis of the combustion of marine microalgae Spirulina platensis and its blend with synthetic waste Heliyon 6 e04902
[18] Vamvuka D and Sfakiotakis S 2011 Combustion behaviour of biomass fuels and their blends with lignite Thermochim. Acta 526 192–9
[19] Naqvi S R, Tariq R, Hameed Z, Ali I, Chen W, Ceylan S, Rashid H, Ahmad J and Taqvi S A 2018 Pyrolysis of high ash sewage sludge: kinetics and thermodynamic analysis using Coats-Redfern method Renew. Energy
[20] Chen C, Ma X and Liu K 2011 Thermogravimetric analysis of microalgae combustion under different oxygen supply concentrations Appl. Energy 88 3189–96
[21] Fedunik-Hofman L, Bayon A and Donne S W 2019 Kinetics of Solid-Gas Reactions and Their Application to Carbonate Looping Systems

IOP Conf. Series: Earth and Environmental Science **847** (2021) 012017 doi:10.1088/1755-1315/847/1/012017

Physicochemical properties and porosity of coconut chell waste (CSW) biomass

A Prasetiyo[1], S Sukarni[1,2], A Irawan[1], A A Permanasari[1,2] and P Puspitasari[1,2]

[1]Center for Renewable and Sustainable Energy Engineering (CRSEE), Department of Mechanical Engineering, Universitas Negeri Malang, Jl.Semarang No 5, Malang, 65145, Indonesia

[2]Centre of Advanced Materials for Renewable Energy, Universitas Negeri Malang, Jl.Semarang No 5, Malang, 65145, Indonesia

sukarni.ft@um.ac.id

Abstract. Utilizing waste biomass as a reserve energy source can play an essential role in reducing expanding fossil fuel's environmental impact. This paper presented an analysis of physical, chemical, morphological, calorific values and porosity of biomass from coconut shell waste (CSW) as an alternative for biofuel feedstock. The physical properties of CSW biomass were dominated by volatile matter and fixed carbon content of 73.8 and 22.7 (wt.%, DB), respectively. The calorific value is 20.39 MJ/kg, giving a significant energy effect when the biomass is burned. The chemical properties of CSW biomass were dominated by C and O content of 41.04 and 57.47 (wt.%), respectively, contributing most of the oxygen to the thermal process. However, most elements of chemical properties contributed to the formation of ash in the combustor, including minor and major elements (S, K, Si, Al, P, Cl, Ca, Mg, and Fe). Besides, they had a negative effect in the form of slagging in the combustion residue. The porosity analysis of CSW biomass was 67.30 – 72.92%, which was able to increase the size of char particles during the thermal process. The analysis of physicochemical properties and porosity of CSW biomass confirmed that this material could be beneficial as an alternative energy reserve in the future.

1. Introduction

The world's energy demand is increasing gradually to meet the needs of industrial and commercial fields. Meanwhile, it is projected that fossil fuel sources face dwindling reserve sources [1]. The use of renewable energy becomes an alternative way to substitute fossil fuels by converting them into bioenergy. This solution is a promising alternative, can be renewed continuously, is environmentally friendly, and is estimated to contribute >10% of world energy needs; from biomass origin, the agricultural sector contributes a vast biomass resource [2,3].

Indonesia is one of the largest coconut producers in the world. It has a total annual production of up to 3 million tons and is estimated to produce 360 thousand tons CSW. The production of CSW in large quantities certainly requires proper management to maintain the quality of biomass [4]. As primary raw material requirements for biofuel feedstock, it is necessary to pay attention to material quality and characteristics, such as calorific value. However, other factors determine the biomass conversion process between the two extremes of dry and wet. High humidity in the Indonesian environment greatly affects biomass quality related to the absorbed water in the material feedstock [3,5]. The amount of water content in the biomass affects the energy level required for evaporating. It also significantly determine its combustion behavior in the combustor [6,7].

In order to ensure the thermal process continuity in the reactor, it is critical to know the fundamental characteristics of biomass [8]. Therefore, it is crucial to characterize the biomass material content to find out the various aspects involved in it. Identifying the physical properties content makes it possible to predict feedstock's quality and give the essential data to determine the proper handling mode during thermal processes [9]. Likewise, the chemical properties of biomass become one of the parameters which need to be considered since they determine the amount of energy contained in the biomass, and their inorganic components would affect the formation of ash and slag deposits in the boiler components [10]. Studies that hitherto discuss the thermal characteristics of CSW pyrolysis, such as those of Ali et al. [11], have been reported. However, it is also necessary to know in-depth the characteristics of CSW's physical, chemical, morphological properties. Moreover, the fundamental research correlated the material porosity with its potential as a fuel feedstock has not been found in the literature.

In this study, the essential characteristics in terms of physical, chemical, morphological, calorific value, and porosity in coconut shell waste (CSW) were analyzed to understand their potential as future and long-term energy reserves.

2. Material and Method

2.1. Material

CSW biomass was obtained from Kalipare traditional market, Malang Regency, Indonesia. It was separated from the outer coconut fiber and then air-dried; after that, it was mashed by means of ground. The powder was then filtered with a standard size of 60 mesh. The powder material was preserved and stored in an airtight bottle and then characterized.

2.2. Proximate and Gross Calorie Value

The composition of the CSW sample's physical properties was analyzed with ASTM D 3173-17, ASTM D 3174-12, ASTM D 3175-17, ASTM D 3172-13, ASTM D 4239-14e1, and ASTM D 5865-13 test methods, respectively to evaluate contents of moisture, ash, volatile matter, fixed carbon, sulphuric and gross calorific values.

2.3. Chemical composition

Evaluating the chemical composition of CSW samples was determined by Energy-dispersive X-ray (EDX) spectrometry. In order to obtain the image, an FEI Inspect S50 (AMETEX EDAXTSL) x-ray micro-analysis was used through a scanning electron microscope at three magnifications. Based on the scanned image of each magnification, the material porosity could then be determined.

3. Result and Discussion

3.1. Physical properties of CSW

Table 1 showed the physical properties of CSW biomass, including contents of moisture, ash, volatile matter, fixed carbon, sulphuric, and gross calorie value (GCV) in the air-dried analysis or as received (ar) condition and dry basis (db). It was clearly observed that the CSW biomass was dominated by volatile matter 73.8 (wt%, db), followed by fixed carbon 22.7 (wt%, db), moisture content 4.0 (wt.%, ar), ash 3.5 (wt%, db), and sulfur 0.10 (wt%, db).

The large percentage of the volatile matter showed the ease level of biomass material for combustion at low temperatures, which impacted high levels of reactivity [6]. It made it easier for the biomass to be more inflammable and reactive for combustion during the thermal process [12]. Volatile matter is the main constituent of biomass matter, including carbon monoxide, hydrogen, methane, oxygen, and other hydrocarbons formed chemically and generated during thermal processes [13].

The second-largest content of CSW biomass was fixed carbon of 22.7 (wt%, db). A high percentage of fixed carbon indicated a large amount of rest material which could be decomposed after the release of volatile matter through gas-phase oxidation so that the large VM/FC ratio affected the reactivity level of solid fuels [13,14]. It is known that the fixed carbon content of CSW biomass is lower than that of volatile matter, indicating that CSW biomass is more reactive and easier to ignite in the combustion system.

Table 1. Physical Properties of CSW

Parameters	Unit	*ar	**db	Test Method
Moisture content	wt%	4.0	---	ASTM D 3173-17
Ash	wt%	3.4	3.5	ASTM D 3174-12
Volatile Matter	wt%	70.8	73.8	ASTM D 3175-17
Fixed Carbon	wt%	21.8	22.7	ASTM D 3172-13
Sulfur	wt%	0.10	0.10	ASTM D 4239-14e1
Gross Calorific Value	kCal/kg	4680	4875	ASTM D 5865-13

*ar: as received; **db:dry basis

A moisture content level of 4.0 (wt%, ar) indicated relatively little water contained in CSW biomass. The importance of knowing the moisture content in biomass is that it affected the design of the combustor furnace, system efficiency, and combustion behavior. Lower water content was considered more profitable since it required less energy for the evaporation process during the thermal process, and in turn, it increases the combustion efficiency [15]. On the other hand, the higher water content in the biomass could increase the energy level of evaporation; a bigger size of the furnace is required due to a longer residence time of the material in the reactor and more greenhouse gas production release in the atmosphere [16].

Finally, ash and sulfur were present in small levels that are 3.4 and 0.1 (wt.%, ar), respectively. The amount of ash content was influenced by inorganic elements remaining after burning combustible substances [17]. A small percentage of ash content was desirable since it affected overall operating costs (transportation, storage, piping maintenance). The low ash content of CSW biomass was preferred because it was more efficient in disposal costs and the cleaning of slag deposits at the boiler's bottom [16,18].

Table 1 showed the calorific value of CSW biomass by 4875 kCal/kg or the equivalent of 20.39 MJ/kg. This calorific value was equivalent to the coconut waste studied by Borel et al. [8] at 18.2 MJ/kg and corn silk at 19.50 MJ/kg [19], greater than *N. Oculata* at 16.80 MJ/kg [6], sewage sludge at 16.71 MJ/kg [20] and lower than bituminous coal at 25.44 MJ/kg [21]. The calorific value indicated the measure of the energy amount produced by burned biomass [13]. This calorific value resulted from this research confirms that CSW is strongly possible to be an alternative solid fuel for power generation.

3.2. Chemical properties of CSW

The chemical content of CSW biomass was analyzed with SEM and EDX spectrograms. Figure 1 showed the SEM photo of CSW biomass with three magnification variations at 200, 500, and 1000 times. Figure 2 showed the average value of three measurements in CSW biomass chemical elements, including carbon (C), oxygen (O), sulfur (S), potassium (K), silicon (Si), aluminum (Al), phosphorus (P), chlorine (Cl), calcium (Ca), magnesium (Mg) and iron (Fe).

The chemical content of both major elements with a composition >1% (S, K, Si, Al, P, Cl, Ca, and Fe) and minor ones with a composition of 1-0.1% (Mg) are relevant as ash fusing components. During combustion processes, these elements lead to generate adverse effects related to the formation of sediment, fly ash, aerosol emissions, and ash melting. The increase in the melting point behavior of ash is influenced by Mg and Ca elements, but the K element can decrease it [18]. Along with aerosol emissions, the elemental chlorine and sulfur biomass of CSW impacts corrosion in the combustor. However, chlorine (Cl) significantly reduces the melting point of ash. As the melting materials of the ash change due to the decreasing temperature, it causes slag in the combustor and the formation of hard deposits on the heat exchanger tube [6,18].

The concentrations of chemical elements C and O of CSW biomass were 41.04 and 57.47 (wt.%), respectively. The C element contributed to the generation of calorific value, and the O element served

IOP Conf. Series: Earth and Environmental Science **847** (2021) 012017 doi:10.1088/1755-1315/847/1/012017

to supply the oxygen needed for thermal processes [6]. The ratio of C to O elements in the biomass contributed to the reactivity of the feedstock during thermal conversion.

(a) (b) (c)

Figure 1. SEM Images of CSW with (a) 200x, (b) 500x and (c) 1000x magnifications

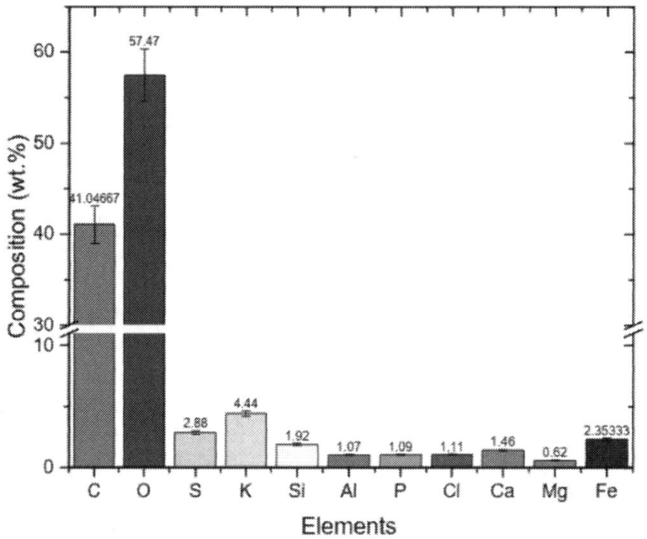

Figure 2. Chemical element composition of CSW biomass, where the error bar showed the error in the mean of three replications

3.3. Porosity of CSW

Figure 3 showed CSW biomass porosity analysis results from the SEM photograph with magnifications of 200, 500, and 1000 times. The CSW biomass material at 200-time magnification showed the percentage ratio of porous volume and solid one is 67.30%. A 500-time magnification showed a pore percentage of 72.92%. The analysis of the 1000-time magnification showed that a pore percentage of 69.81%. This qualitative measurement confirmed that CSW biomass microstructure with a large pore percentage increased the char particles product during the conversion process when an endothermic reaction occurred.

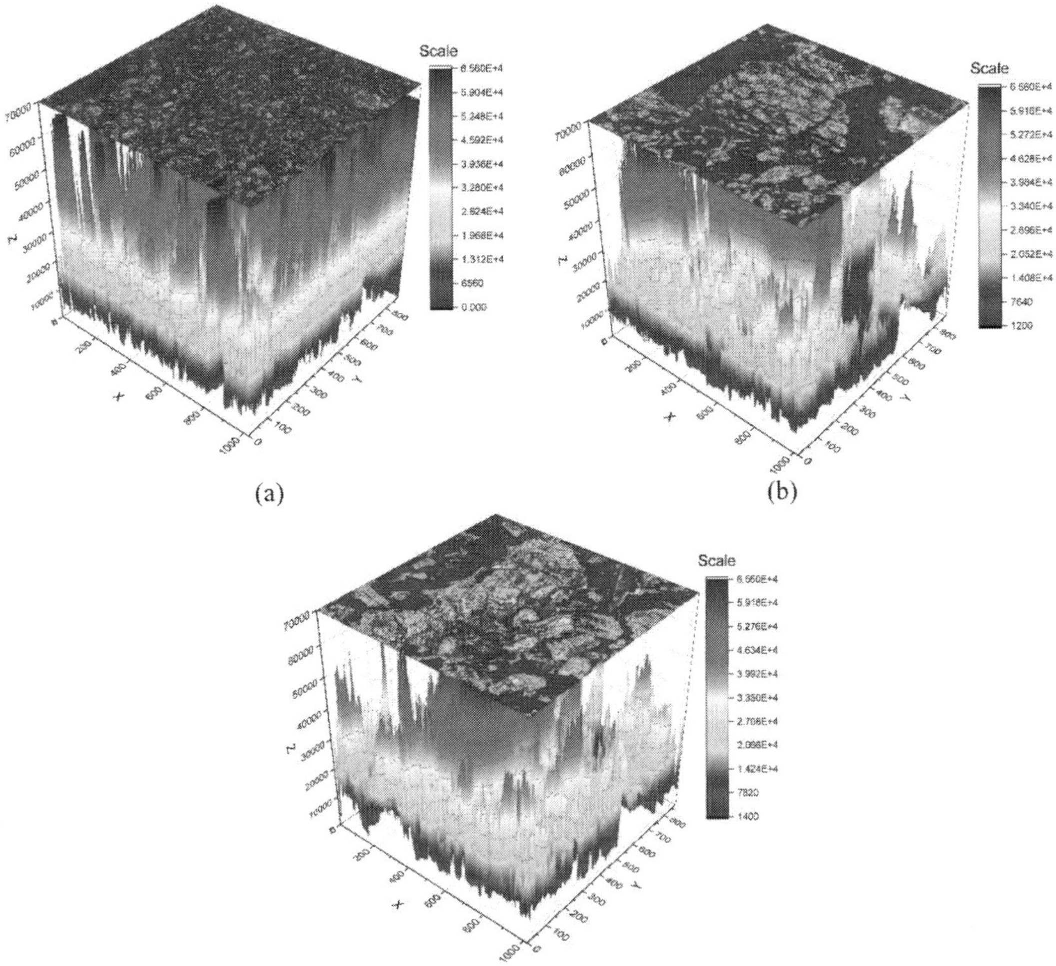

(a) (b)

(c)

Figure 3. CSW biomass porosity from SEM photos with (a) 200x, (b) 500x and (c) 1000x magnifications

4. Conclusion

The physical, chemical, and porosity properties of CSW biomass have been studied. According to the proximate analysis results, the physical properties of CSW biomass are dominated by volatile matter and fixed carbon, which are 73.8 and 22.7 (wt%, db), respectively, which shows that the biomass has high reactivity. The CSW calorific value is equivalent to coal. The analysis of chemical elements indicated that CSW is dominated by C and O, followed by a small portion of inorganics elements of ash formation. These investigative data show the potential of CSW biomass as an alternative fuel. In addition, this work analyzes the porosity percentage of the porous material with a value range of 67.30-72.92%, which supports the increase of char particles product during thermal processes.

Acknowledgments

The authors would like to thank for the financial support from Universitas Negeri Malang to this research.

References

[1] Inayat M, Sulaiman S A and Naz M Y 2018 Thermochemical Characterization of Oil Palm Fronds, Coconut Shells, and Wood as A Fuel for Heat and Power Generation MATEC Web Conf. 225

[2] Inayat M and Sulaiman S A 2018 Effect of Blending Ratio on Quality of Producer Gas from Co-Gasification of Wood and Coconut Residual MATEC Web Conf. 225 1–6

[3] McKendry P 2002 Energy production from biomass (part 1): overview of biomass Bioresour. Technol. 83 37–46

[4] Irawan A, Latifah Upe S and Meity Dwi I P 2017 Effect of torrefaction process on the coconut shell energy content for solid fuel AIP Conf. Proc. 1826

[5] Chen W H, Peng J and Bi X T 2015 A state-of-the-art review of biomass torrefaction, densification and applications Renew. Sustain. Energy Rev. 44 847–66

[6] Sukarni, Sudjito, Hamidi N, Yanuhar U and Wardana I N G 2014 Potential and properties of marine microalgae Nannochloropsis oculata as biomass fuel feedstock Int. J. Energy Environ. Eng. 5 279–90

[7] Sukarni S, Sumarli S, Nauri I M, Purnami P and Yanuhar U 2018 Exploring the prospect of marine microalgae Isochrysis galbana as sustainable solid biofuel feedstock J. Appl. Res. Technol. 53–66

[8] Borel L D M S, de Lira T S, Ataíde C H and de Souza Barrozo M A 2021 Thermochemical conversion of coconut waste: material characterization and identification of pyrolysis products J. Therm. Anal. Calorim. 143 637–46

[9] Sequeiros A and Labidi J 2017 Characterization and determination of the S/G ratio via Py-GC/MS of agricultural and industrial residues Ind. Crops Prod. 97 469–76

[10] Kopejan S van loo and jaap 2008 the hand book of biomass combustion and co-firing

[11] Ali I, Bahaitham H and Naibulharam R 2017 A comprehensive kinetics study of coconut shell waste pyrolysis Bioresour. Technol.

[12] Sukarni S, Zakaria Y, Sumarli S and Wulandari R 2019 Physical and Chemical Properties of Water Hyacinth (Eichhornia crassipes) as a Sustainable Biofuel Feedstock 515 1–7

[13] Miller B G and Tillman D A 2008 Coal Characteristics Combust. Eng. Issues Solid Fuel Syst. 33–81

[14] Sukarni S 2016 Exploring the potential of municipal solid waste (MSW) as solid fuel for energy generation: Case study in the Malang City, Indonesia AIP Conference Proceedings vol 1778 (American Institute of Physics Inc.)

[15] Yazdani M G, Hamizan M and Shukur M N 2012 Investigation of the fuel value and the environmental impact of selected wood samples gathered from Brunei Darussalam Renew. Sustain. Energy Rev. 16 4965–9

[16] Lubwama M, Yiga V A, Muhairwe F and Kihedu J 2019 Physical and combustion properties of agricultural residue bio-char bio-composite briquettes as sustainable domestic energy sources Renew. Energy

[17] McKendry P 2002 Energy production from biomass (part 1): Overview of biomass Bioresour. Technol. 83 37–46

[18] Obernberger I, Brunner T and Bärnthaler G 2006 Chemical properties of solid biofuels-significance and impact Biomass and Bioenergy 30 973–82

[19] Sumarli S, Upendra C, Bandhana S and Himawan K 2019 Physicochemical Characteristics of Corn Silk as Biomass Fuel Feedstock 515 1–5

[20] Wang Z, Hong C, Xing Y, Li Y, Feng L and Jia M 2018 Combustion behaviors and kinetics of sewage sludge blended with pulverized coal: With and without catalysts Waste Manag. 74 288–96

[21] Wu Z, Wang S, Zhao J, Chen L and Meng H 2014 Synergistic effect on thermal behavior during co-pyrolysis of lignocellulosic biomass model components blend with bituminous coal Bioresour. Technol. 169 220–8

Characteristics and modeling kinetics of oil palm frond petiole pyrolysis using thermogravimetric analyzer

A Prasetiyo[1], S Sukarni[1,2], A A Wibowo[1], A A Permanasari[1,2] and P Puspitasari[1,2]

[1]Center for Renewable and Sustainable Energy Engineering (CRSEE), Department of Mechanical Engineering, Universitas Negeri Malang, Jl.Semarang No 5, Malang, 65145, Indonesia

[2]Centre of Advanced Materials for Renewable Energy, Universitas Negeri Malang, Jl.Semarang No 5, Malang, 65145, Indonesia

sukarni.ft@um.ac.id

Abstract. As one of the largest palm oil-producing countries globally, Indonesia produces abundant oil palm biomass; indeed, it brings large oil palm frond petiole (OPFP). OPFP is one of the oil palm biomass that can be converted into an energy source. This research focuses on analyzing thermal decomposition characteristics and pyrolysis kinetics of OPFP using thermogravimetric analysis (TGA). The results showed that the thermal decomposition of pyrolysis was divided into three steps, including the evaporation stage of moisture content, the releasing of volatile compounds and organic matter, and the final stage of lignin and tar decomposition. Both the activation energy and the pre-exponential factors were analyzed with the Coats-Redfern method in step II active pyrolysis zone are 80.56 kJ/mol and 7.21/min, respectively, as shown in the Chemical Reaction Order F1.5 model with the accuracy of a correlation coefficient of 0.99. These results provided the fundamental data useful for properly designing the reactor system for industrial purposes.

1. Introduction

Currently, both the domestic and industrial sectors of electricity are still dominated by fossil energy, in which coal is the primary fuel supply. On the other hand, the transportation sector is still heavily reliant on the use of oil fuel. The projection of renewable energy supply is undergoing rapid growth, up to 6.5% per year, including biofuel, wind, geothermal, biogas, biomass, and municipal waste. The potential for renewable energy development in Indonesia is relatively high, but it has not been optimally utilized [1]. The use of renewable energy has positive impacts since it is able to reduce CO_2 emissions and the greenhouse gas effect [2,3].

Indonesia and Malaysia are the most significant oil palm-producing countries globally, in which both Kalimantan and Sumatra are the largest producing regions for palm oil in Indonesia [4,5]. Indonesia's largest total palm oil exports in 2017 reached USD 18.7 billion [6]. Furthermore, the demand for palm oil supply increased from 8 million tonnes in 2015 to 15 million tonnes in 2018. The surge in the amount of palm oil production definitely caused a considerable increment in palm biomass waste, including oil palm shell (OPS), empty fruit bunch (EFB), and oil palm fiber (OPF) [7] as well as OPFP. The biomass waste has the great potential to be used as an energy source due to its inexhaustible renewal source. Today, along with technological developments, an oil palm tree can be utilized for many applications [8]. Fresh fruit bunches (FFB) can be converted into crude palm oil (CPO), and palm oil waste can be converted into chemical compounds, fertilizers, biomaterials, and alternative fuel sources [5]. However, inappropriate disposal of oil palm fronds and stems trigger serious air pollution problems [9]. Therefore, utilization of oil palm waste is a critical challenge in overcoming its environmental issues and providing renewable energy reserves.

The combustion is a thermochemical conversion producing heat, in which the materials are oxidized directly without being converted into other forms of fuels [10]. In contrast, pyrolysis is a thermal conversion producing the final products in the form of char, tar, and gaseous coproducts [11]. One of the promising techniques for the analysis of thermochemical characteristics during biomass pyrolysis is thermogravimetric analysis. Nudri et al. [12] studied the pyrolysis of oil palm trunk (OPT) for producing biochar, and the results showed that thermal degradation of OPT biomass pyrolysis was divided into three decomposition steps with the first stage was between temperatures of 70 and 110 °C and the second stage, with a maximum mass loss at 360 °C, was related to lignin degradation. Sakulkit et al. [13] reported that OPT pyrolysis observed through TGA analysis occurred in 4 steps, the first stage was related to moisture evaporation, the second stage was light volatile evaporation, the third stage was the evaporation of cellulose and hemicellulose organic matter, and the fourth stage was related to lignin decomposition. However, concerning energy applications, the previous study has not reported an analysis related to the utilization of oil palm frond petiole (OPFP) waste for energy applications through the pyrolysis process to evaluate its thermal characteristics and kinetics.

Based on the presented facts, this study characterized the thermal behavior of OPFP throughout the pyrolysis process and evaluated its kinetic parameters. The analyzed kinetic parameters involve the value of the activation energy, the pre-exponential factor, and the suitable kinetic model.

2. Material and Method

2.1. Sample preparation
The OPFP material was obtained from oil palm smallholder plantations in Harapan Jaya Village, Muara Enim, South Sumatra, Indonesia. The OPFP sample was dried in an oven for 4 hours at a temperature of 110 °C. The sample was then powdered and being filtered using 40 mesh to equalize the size. After that, the OPFP sample powder was stored in an airtight bottle and maintained in a desiccator, and finally, the OPFP sample was ready to be tested.

2.2. TG experiments
The pyrolysis thermal decomposition behavior of OPFP samples was observed through the Mettler toledo TG/DSC1 thermogravimetric analyzer (TGA). An amount of 10 mg of OPFP was put into the crucible. The sample was heated at a temperature of 25 - 1000 °C in inert air gas of nitrogen. The heating rate of 20 °C/min was used, and nitrogen was flow at a rate of 100 ml/min.

2.3. Kinetics modeling
The pyrolysis kinetics of OPFP was basically determined by Arrhenius law. The basic equation of the conversion rate is written as.

$$\frac{d\alpha}{dt} = k(T)f(\alpha) \tag{1}$$

where α is a mass fraction of degraded material and expressed as:

$$\alpha = \frac{m_o - m_i}{m_0 - m_f} \tag{2}$$

where, m_o is the initial mass, m_i, and m_f are the mass at a certain time and the final mass of each end of the stage, respectively.

$$k(T) = A\exp\left(-\frac{E_a}{RT}\right) \tag{3}$$

where T is the reaction temperature (K), R is the universal gas constant (0.008314 kJ/mol.K), Ea is the activation energy (kJ/mol), and A is the pre-exponential factor (/min). Based on equations (1) and (3), for a non-isothermal experiment ($dT/dt = \beta$), it is rewritten as a compound form becoming

$$\frac{d\alpha}{dT} = \frac{A}{\beta}\exp\left(-\frac{E_a}{RT}\right) f(\alpha) \tag{4}$$

The integral form of the reaction model is written as $g(\alpha)$. The analytical solution approach model is written into the following equations:

$$g(\alpha) = \int_0^\alpha \frac{d\alpha}{f(\alpha)} = \frac{A}{\beta}\int_{T_0}^{T} exp\left(-\frac{E_a}{RT}\right) dT \tag{5}$$

2.4. Coats-Redfern method

The Coats-Redfern method can be applied to the various selected model for predicting reaction order, activation energy, and pre-exponential factors. The basic equation for the Coats-Redfern method expressed as:

$$ln\left[\frac{g(\alpha)}{T^2}\right] = ln\frac{AR}{\beta E_a}\left(1 - \frac{2RT}{E_a}\right) - \frac{E_a}{RT} \tag{6}$$

where R is the universal constant (0.008314 kJ/mol K), β is the heating rate and $g(\alpha)$ is the kinetic function of the integration of $f(\alpha)$. Plotting a graph between 1/T and ln $[g(\alpha)/T^2]$ is able to form a straight line to get the slope and intercept values to determine the activation energy and pre-exponential factor, respectively. Most of the thermal degradation reactions mechanism of solid materials are presented in the five categories such as listed in Table 1.

Table 1. Reaction mechanisms model fitting and model names with their $f(\alpha)$ and $g(\alpha)$

Reaction mechanism Symbol	 Model name	Differential form $f(\alpha)$	Integral form $g(\alpha)$
Reaction order			
F1	Chemical Reaction Order 1	$1-\alpha$	$-\ln(1-\alpha)$
F1.5	Chemical Reaction Order 1.5	$(1-\alpha)]^{3/2}$	$2(1-\alpha)]^{3/2}-1]$
Diffusion			
D1	Parabolic law	$½\alpha$	α^2
D2	Valensi equation	$-[\ln(1-\alpha)]^{-1}$	$\alpha+(1-\alpha)\ln(1-\alpha)$
D3	Ginstling-Broushtein equation	$3/2[(1-\alpha)^{1/3}-1]^{-1}$	$\left(1-\frac{2}{3}\alpha\right)-(1-\alpha)^{2/3}$
Nucleation and growth			
N1.5	Avrami-Erofeev equation of Nucleation and growth	$3(1-\alpha)[-\ln(1-\alpha)]^{2/3}$	$[-\ln(1-\alpha)]^{2/3}$
N2	Avrami-Erofeev equation of Nucleation and growth	$2(1-\alpha)[-\ln(1-\alpha)]^{1/2}$	$[-\ln(1-\alpha)]^{1/2}$
Phase of interfacial reaction			
S1	Shrinkage geometrical column	$2(1-\alpha)^{1/2}$	$1-(1-\alpha)^{1/2}$
S2	Shrinkage geometrical Spherical	$2(1-\alpha)^{2/3}$	$1-(1-\alpha)^{1/3}$
Power law			
P	Power law	1	α

3. Result and Discussion

3.1. OPFP pyrolysis characteristics

The thermal behavior of the OPFP sample was observed through TGA and DTG profiles, as shown in Figure 1. These graphs show that there are three steps of OPFP pyrolysis decomposition.

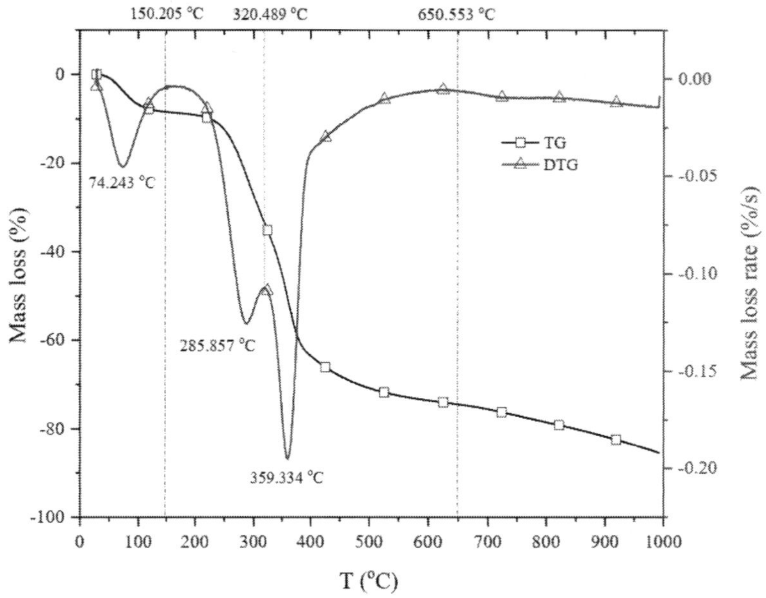

Figure 1. TG and DTG graph

The first stage occurs at 30-150.2 °C temperatures range, where the sample undergoes a mass loss of up to 8.5% due to evaporation of the moisture content. The second stage occurs at 150.2-650.5 °C temperatures range, with a mass loss of about 74.4%, related to the evaporation of both volatile compounds and volatile organic materials, including cellulose and hemicellulose [14,15]. The third stage occurs above 650.5 °C, with mass loss of almost 85.4%. The thermal degradation of OPFP at this stage is relatively sloping due to the decomposition of lignin and the remaining char. The temperature characteristics of the OPFP pyrolysis are shown in Table 2.

Table 2. OPFP pyrolysis temperature characteristics

Steps	Temperature characteristics				
	T_i (°C)	T_{max} (°C)	T_f (°C)	Mass loss (%)	M_{max} (%/s)
I	25	74.243	150.205	-8.451	-0.04526
II	150.205	359.334	650.553	-74.474	-0.19526
III	650.553	711.659	993.28	-85.485	-0.00931

3.2. Kinetic analysis of OPFP biomass pyrolysis

Table 3 shows the OPFP kinetics parameters calculated using the Coats-Redfern method proposed by Naqvi et al. [16]. The values of activation energy, R^2, and Log A are estimated at intervals of α from 0.05 to 0.95. The R^2 (correlation coefficient) value close to 1 is considered to have good accuracy.

The Chemical Reaction Order F1.5 model shows the highest R^2 value of 0.99, in which the activation energy of OPFP pyrolysis is 80.56 kJ/mol and Log A 7.21/min. This activation energy value is comparable to the microalgae *Spirulina platensis* and *Cassava stalk,* 84.61 and 91.19 kJ/mol, respectively [17,18], and it is higher than that of *water hyacinth* and palm date biomass, which are 60.74 and 22 - 43.6 kJ/mol, respectively [19,20], but it is lower than *Cornsilk, T.chuii,* and *N.cculata* which are 207.37; 101.05, and 251 kJ/mol, respectively [21–23]. The Chemical Reaction Order F1.5 model shows a homogeneous kinetics model, and the thermal conversion value directly proportional to the reaction rate [24].

Table 3. The kinetic parameters of OPFP biomass pyrolysis

Models Name	Stage II		
	Ea (kJ/mol)	R^2	Log A (1/min)
F1	42.77	0.903	2.87
F1.5	80.56	0.990	7.21
D1	64.62	0.796	4.34
D2	72.79	0.838	4.89
D3	76.37	0.856	4.61
N1.5	25.05	0.873	1.20
N2	16.20	0.829	0.29
S1	34.00	0.822	1.62
S2	36.70	0.852	1.74
P	27.12	0.726	1.16

4. Conclusion

OPFP is abundant in oil palm biomass, and it can be used as an alternative fuel for energy generation. The characteristics of OPFP pyrolysis show that it is divided into three steps of decomposition, including the stage of the evaporation of moisture, followed by thermal degradation of cellulose and the hemicellulose stage, and the final stage indicating the decomposition of lignin and remaining char. The activation energy value refers to the F1.5 Chemical Reaction Order model is 80.56 kJ/mol, with the accuracy of the model of R^2 is 0.99 and Log A 7.21/min. Based on this thermal characterization, OPFP is profoundly feasible as an environmentally friendly future renewable energy alternative.

Acknowledgments

The authors would like to acknowledge the Universitas Negeri Malang for supporting this research.

References

[1] BPPT Indonesia Energy Outlook 2020 Indonesia Energy Outlook 2020 - Special Edition Dampak Pandemi COVID-19 terhadap Sektor Energi di Indonesia Diterbitkan

[2] Prasetiyo A, Sukarni S, Wulandari R, Widiono A E, Suryanto H and Yanuhar U 2020 Investigation on kinetic parameters during combustion of Tetraselmis chuii microalgae under thermogravimetric analyzer AIP Conference Proceedings vol 2255

[3] Sanchez-Silva L, Lopez-Gonzalez D, Garcia-Minguillan A M and Valverde J . 2013 Bioresource Technology Pyrolysis , combustion and gasification characteristics of Nannochloropsis gaditana microalgae Bioresour. Technol. 130 321–31

[4] Arafat Hossain M, Ganesan P, Jewaratnam J and Chinna K 2017 Optimization of process parameters for microwave pyrolysis of oil palm fiber (OPF) for hydrogen and biochar production Energy Convers. Manag. 133 349–62

[5] Hambali E and Rivai M 2017 The Potential of Palm Oil Waste Biomass in Indonesia in 2020 and 2030 IOP Conf. Ser. Earth Environ. Sci. 65

[6] Voora V, Larrea C, Bermudez S and Baliño S 2019 Global Market Report : Palm Oil Int. Inst. Sustain. Dev. 16

[7] Ismail W M S W, Mohd Thaim T and Abdul Rasid R 2019 Biomass gasification of oil palm fronds (OPF) and Koompassia malaccensis (Kempas) in an entrained flow gasifier: A performance study Biomass and Bioenergy 124 83–7

[8] Sulaiman S A, Basmufleh H S, Tamili S N A M I and Naz M Y 2016 Characterization of date palm fronds as a fuel energy production Chem. Soc. Ethiop. 30 465–72

[9] Nudri N A, Wan Abdul Karim Ghani W A, Thomas Bachmann R, Baharudin B T H T, Ng D K S and Md Said M S 2021 Co-combustion of oil palm trunk biocoal/sub-bituminous coal fuel blends Energy Convers. Manag. X 100072

[10] Sukarni S, Ardianto P, Retno W, Aloon E W and Poppy P 2020 Thermal characteristic of

tetraselmis chuii combustion influenced by titanium dioxide (Tio2) nanoparticle Key Eng. Mater. 851 KEM 149–55

[11] Sukarni S, Widiono A E, Wulandari R, Prasetiyo A and Puspitasari P 2020 Thermogravimetric study on the thermal characteristics of tetraselmis chuii microalgae pyrolysis in the presence of titanium dioxide Key Eng. Mater. 851 KEM 156–63

[12] Nudri N A, Bachmann R T, Ghani W A W A K, Sum D N K and Azni A A 2020 Characterization of oil palm trunk biocoal and its suitability for solid fuel applications Biomass Convers. Biorefinery 10 45–55

[13] Sakulkit P, Palamanit A, Dejchanchaiwong R and Reubroycharoen P 2020 Characteristics of pyrolysis products from pyrolysis and co-pyrolysis of rubber wood and oil palm trunk biomass for biofuel and value-added applications J. Environ. Chem. Eng. 8 104561

[14] Palamanit A, Khongphakdi P, Tirawanichakul Y and Phusunti N 2019 Investigation of yields and qualities of pyrolysis products obtained from oil palm biomass using an agitated bed pyrolysis reactor Biofuel Res. J. 6 1065–79

[15] Khongphakdi P, Palamanit A, Phusunti N, Tirawanichakul Y and Shrivastava P 2020 Evaluation of oil palm biomass potential for bio-oil production via pyrolysis processes Int. J. Integr. Eng. 12 226–33

[16] Naqvi S R, Tariq R, Hameed Z, Ali I, Chen W, Ceylan S, Rashid H, Ahmad J and Taqvi S A 2018 Pyrolysis of high ash sewage sludge: kinetics and thermodynamic analysis using Coats-Redfern method Renew. Energy

[17] Sukarni S, Sumarli S, Nauri I M, Prasetiyo A and Puspitasari P 2019 THERMOGRAVIMETRIC ANALYSIS ON COMBUSTION BEHAVIOR OF MARINE MICROALGAE Spirulina platensis INDUCED BY MgCO3 AND Al2O3 ADDITIVES Int. J. Technol. 10 1174–83

[18] Sukarni Sukarni and Ramadhan M R 2020 Pyrolytic characteristics and kinetic parameters evaluation of cassava stalks using thermogravimetric analyzer Key Eng. Mater. 851 KEM 137–41

[19] Sukarni S, Widiono A E, Sumarli S, Wulandari R, Nauri I M and Permanasari A A 2018 Thermal decomposition behavior of water hyacinth (eichhornia crassipes) under an inert atmosphere MATEC Web of Conferences vol 204, ed P Puspitasari, A H Suryanto, T Lenggo Ginta, I Srikun, A Ayu Permanasari, M Niaz Akhtar, M Sharifi Jebeli, Y Rohmat Aji Pradana, D Zakariya Lubis, A Larasati and A Dwi Astuti (EDP Sciences) p 00010

[20] Sait H H, Hussain A, Salema A A and Ani F N 2012 Pyrolysis and combustion kinetics of date palm biomass using thermogravimetric analysis Bioresour. Technol. 118 382–9

[21] Sumarli S, Bandhana Kusuma Himawan C U S, Wulandari R and Sukarni S 2019 Thermogravimetric analysis and the fitting model kinetic evaluation of corn silk thermal decomposition under an inert atmosphere IOP Conf. Ser. Mater. Sci. Eng. 494 1–8

[22] Prasetiyo A, Sukarni S, Wulandari R and Puspitasari P 2020 A kinetic study on Tetraselmis chuii combustion: The catalytic impact of nanoparticle titanium dioxide (TiO2) additive J. Adv. Res. Fluid Mech. Therm. Sci. 71 39–49

[23] Sukarni, Sudjito, Hamidi N, Yanuhar U and Wardana I N G 2015 Thermogravimetric kinetic analysis of Nannochloropsis oculata combustion in air atmosphere Front. Energy 9 125–33

[24] Fedunik-Hofman L, Bayon A and Donne S W 2019 Kinetics of Solid-Gas Reactions and Their Application to Carbonate Looping Systems

IOP Conf. Series: Earth and Environmental Science **847** (2021) 012019 doi:10.1088/1755-1315/847/1/012019

Determining of black spot location in Purbalingga Regency using road geometric approach

G Sugiyanto[1], Y A Prasetyo[1], E W Indriyati[1], Y Yanto[1] and M Y Santi[2]

[1]Jenderal Soedirman University, Department of Civil Engineering, Purbalingga, Central Java, Indonesia
[2]Health Polytechnic Ministry of Health of Yogyakarta, Mangkuyudan Street MJ III/304, Yogyakarta, Indonesia

gito.sugiyanto@unsoed.ac.id

Abstract. Traffic accidents are still one of the serious problems faced by the Indonesian Government. Based on traffic accident data from Traffic Corps Republic of Indonesia the number of traffic accidents in 2020 is 100,028 incidents and the number of deaths due to accidents is 23,529 people. Traffic accidents are caused by several factors such as vehicles, human error of road users, road conditions and the environment. Road geometry is one factor that contributes to the high value of the risk of accidents. One effort to reduce the number of traffic accidents is identification and handling the black spot location. The purpose of this research was to determine the black spot locations using road geometric facilities. Thus, the black spot location can be predicted before the incidence of a traffic accident occurs through the weighting parameter road geometric. Research was conducted on arterial and collector roads in Purbalingga. Six road geometric parameters are used to determine black spot locations, namely vehicle speed, road lane width, road shoulder width, road pavement width, road median width, and shoulder drop-off (difference in elevation between road shoulder and the pavement edge). Weighting intervals for road geometrics are 1 to 5 with interval 1. As a result of this study, there are differences in the ranking of black spot locations based on the equivalent accident number method and the road geometric parameter approach. The further research is conducted to determine the coefficient value of each road geometric parameter and combined with road equipment facilities.

1. Introduction

According to data released by the United Nations about 1.3 million people every year or about 3,000 people every day die due to traffic accidents. About 90% of deaths due to road accidents occur in developing countries between the ages of 5-44 years. If there are no effective efforts to reduce the number of accidents, deaths from traffic accidents will be the fifth leading cause of death in the world with an estimated number of around 2.4 million deaths every year. According to the Traffic Corps Republic of Indonesia in 2011-2018, the accident rate of 34.48% occurred in the morning and 24.14% in the afternoon. Based on the type of vehicle that had an accident, 52.5% were motorcycles; private cars 20%, trucks 17.5% and buses 10%. Accidents that occur can involve various ages who use the road as an area for activities. On average 20% of people who die in traffic accidents in developing countries are under the age of 15 [1]. The number of traffic accidents in Indonesia in 2020 as many as 100,028 incidents [1]. The same thing happened in Purbalingga Regency, Central Java, Indonesia where the number of traffic accidents increased from 2010 to 2019 [2].

Four factors that caused traffic accidents are vehicles, human error, road conditions and environment [3-4]. From the four factors, human error is a factor that often results in traffic accidents [5]. To minimize the road infrastructure deficiencies, there are three aspects, namely a forgiving road, a self-explaining road, and regulating road [6]. Mobility is one of the components that contribute to traffic accidents [7-9]. Therefore, the road safety model focuses mainly on risk factors [10]. The high number of traffic accidents indicates that there are various factors that cause accidents.

One of the factors that cause traffic accidents is the road, which includes the geometric conditions of the road [11]. Traffic accidents tend to be concentrated on certain roads. Several parameters used to analyze the causes of road accidents are road geometry, driving behavior, weather conditions, and speed limits [12]. This shows that in addition to driver error, road characteristics are also an important factor in the occurrence of traffic accidents. Road safety audit was introduced to handle the black spot location and reduce the severity of crashes, the accident cost [13-14], driving behavior [15], the generalized cost [16-17], and increased efficiency in transportation cost [18-21].

One of the efforts that can be done to reduce the traffic accidents number is by identification and handling the location of black spots. Black spot locations are areas that have a high accident rate, high risk and potential for accidents on a road section [22]. One of the methods to determine the black spot location is using the equivalent accident number and the upper control limit method. However, this method has a weakness, namely that traffic accident-prone locations can be determined after waiting for data on accidents and traffic accident victims [13, 23]. For this reason, a study using another approach is needed. Another method in determining the black sport locations without having to wait for accidents and victims is based on the road geometry facilities approach. The purpose of this research was to determine the black spot locations using road geometric facilities approach. Thus, the black spot location can be predicted before the incidence of a traffic accident occurs through the weighting of parameter road geometric facilities.

2. Method

2.1. *Study location*

The study was conducted on 10 (ten) arterial and collector roads in Purbalingga Regency, Central Java, Indonesia namely: Serang Village road, Selanegara Village road, Gembong Village road, Mangunegara Village road, Karanganyar Village road, Mayjend Sungkono Street Kalimanah Wetan, Bajong Village road, Bojongsari Village road, Padamara Village road and Tlahab Lor Village road.

2.2. *Analysis approach*

2.2.1 Equivalent accident number and upper control limit method

Accident data that has been grouped by fatality level then weighted using the equivalent accident number (EAN). The EAN weighting using a weighting formula from Road Research and Development Center Ministry of Public Work Republic of Indonesia [24]. The equivalent accident number with fatal: seriously injured: slightly injured: property damage only = 12:3:3:1 [24]. After being weighted for each level of victim fatality, then the accident equivalent number is total for each road segment.

After obtaining the EAN value for each road section, then calculate the upper control limit (UCL) value for each road section to determine which road sections are categorized as black spot locations. A road segment is declared a black spot location if the EAN value is greater than UCL (EAN > UCL).

2.2.2 Road geometric approach

There are six road geometric parameters used to determine the black spot locations, namely vehicle speed, road lane width, road shoulder width, road pavement width, road median width, and shoulder drop-off. Design speed is the speed chosen for planning purposes for each section of the highway such as horizontal alignment, road slopes, visibility, stopping sight distance and others [25]. The standards for the six road geometric parameters used refers to Urban Road Geometric RSNI T-14-2004 [26], AASHTO [27], Government regulations Number 34, 2006 [28], and Law of the Republic of Indonesia Number 22 of 2009 concerning Road Traffic and Transportation [29].

Weighting intervals for road geometrics are 1 to 5 with interval 1. In this case, 1 is the best weight where the geometric conditions of the road exist and the size is in accordance with a predetermined standard. While the weight of 5 is the value with the worst condition where the geometric parameter of the road does not exist or the value exceeds the predetermined limit. For example, the vehicle speed, the higher the vehicle speed and the greater the difference with the design speed, the greater the effect on the rate of traffic accidents. This is related to the greater potential for traffic accidents for higher speeds. Mohammed in 2013 examined the effect of road geometric design characteristics on traffic accidents, mentioning several road geometric parameters that can affect traffic safety. The eight geometric parameters that influence the road are speed, lane width, number of lanes, shoulder width and type, median width and type, climbing lane length, access density, and median barrier. A reduction of 1.6 km/h from the average speed will reduce the incidence of injury by about 5%. Wider paths are theoretically associated with higher operating speeds and thus increased security. The more the number of lanes, the higher the traffic accident rate [30].

3. Result and discussion

3.1. *Road geometric parameter weighting*

The first road geometric parameter was vehicle speed. The vehicle speed used for weighting was based on the 85th percentile speed of all types of vehicles in the field. Five weighting intervals are set based on vehicle speed with a weighting number from 1 to 5, with interval 1. Vehicle speed weighting intervals for arterial and collector roads can be seen in Table 1.

Table 1. Vehicle speed weighting interval for arterial and collector roads

Interval of vehicle speed for arterial roads (km/h)	Weight	Interval of vehicle speed for collector roads (km/h)
≤ 60.9	1	≤ 50.9
61-66.9	2	51-55.9
67-72.9	3	56-60.9
73-78.9	4	61-65.9
≥79	5	≥ 66

The second road geometric parameter was the road lane width. The weighting interval was determined based on the difference in the width of the road lane in every 0.20 m multiple of the standard size. Weight 1 is the best weight where the road lane width is in accordance with the regulation standards. The road lane width weighting intervals for arterial and collector roads are presented in Table 2.

Table 2. Road lane width weighting interval for arterial and collector roads

Interval of road lane width for arterial roads (m)	Weight	Interval of road lane width for collector roads (m)
≥ 3.50	1	≥ 3.00
3.30-3.49	2	2.80-2.99
3.10-3.29	3	2.60-2.79
2.90-3.09	4	2.40-2.59
< 2.90	5	≤ 2.39

The third geometric parameter of the road was the shoulder width. The weighting interval was determined based on the difference in the width of the road shoulder compared to the standard shoulder width. Weight 1 is the best weight where the road shoulder width is in accordance with the standard. The shoulder width weighting interval is divided into two conditions, namely for roads with sidewalks and roads without sidewalks for both arterial and collector roads as presented in Table 3.

Table 3. Shoulder width weighting intervals for arterial and collector roads with sidewalks and without sidewalks

Interval of shoulder width weighting intervals for arterial and collector roads with sidewalks (m)	Weight	Interval of shoulder width weighting intervals for arterial and collector roads without sidewalks (m)
≥ 0.25	1	≥ 2.00
0.15-0.249	2	1.20-1.999
0.10-0.149	3	0.80-1.199
0.05-0.099	4	0.40-0.799
< 0.049	5	< 0.399

The fourth road geometric parameter was the width of the sidewalk. The weighting interval was determined based on the difference between the width of the sidewalk and the width of the standard sidewalk. The weight of 1 is the best weight where the pavement condition is in accordance with the standard rules, while the weight that is worth 5 is the worst weight where the road segment does not have a sidewalk or sidewalk width is < 0.299 m. The weighting intervals for the sidewalk width for arterial and collector roads can be seen in Table 4.

Table 4. Sidewalk width weighting interval for arterial and collector roads

Interval of sidewalk width for arterial roads (m)	Weight	Interval of sidewalk width for collector roads (m)
≥ 1.50	1	≥ 1.50
0.90-1.499	2	0.90-1.499
0.60-0.899	3	0.60-0.899
0.30-0.599	4	0.30-0.599
< 0.299	5	< 0.299

The fifth road geometric parameter was the median width. The weighting interval was determined based on the difference in the road median width compared to the standard median width. Weight 1 is the best weight where the median width of the road is in accordance with the standards. The median width weighting intervals for arterial and collector roads can be seen in Table 5.

Table 5. Median width weighting interval for arterial and collector roads

Interval of median width for arterial roads (m)	Weight	Interval of median width for collector roads (m)
≥ 2.50	1	≥ 1.50
1.50-2.49	2	0.90-1.49
1.00-1.49	3	0.60-0.89
0.50-0.99	4	0.30-0.59
< 0.499	5	< 0.299

The sixth road geometric parameter was the shoulder drop-off (elevation difference between road shoulder and pavement edge). The weighting interval was determined based on the difference in elevation difference for each increment of 4 cm. Weight 1 is the best weight where the difference in elevation between the road shoulder and the edge of the pavement is < 1.99 cm. The elevation difference weighting intervals between the shoulder and the pavement edge for arterial and collector roads are presented in Table 6.

Table 6. Shoulder drop-off weighting interval for arterial and collector roads

Interval of drop-off (elevation difference between road shoulder and pavement edge) for arterial roads (cm)	Weight	Interval of drop-off (elevation difference between road shoulder and pavement edge) for collector roads (cm)
< 1.99	1	< 1.99
2.0-5.99	2	2.0-5.99
6.0-9.99	3	6.0-9.99
10-13.0	4	10-13.0
> 13.0	5	> 13.0

3.2. *Black spot location based on EAN-UCL*

The analysis results of the equivalent accident number and upper control limit values for ten roads in Purbalingga Regency, Central Java which are as black spot locations are shown in Table 7.

Table 7. Equivalent accident number and upper control limit values for ten roads

Name of road	Equivalent Accident Number	Upper Control Limit
Tlahab Lor Village Road, Karangreja	168	40.549
Bojongsari Village Road, Bojongsari	126	37.813
Serang Village Road, Karangreja	96	34.806
Mangunegara Village Road, Mrebet	87	34.374
Mayjend Sungkono Street Kalimanah Wetan	75	33.169
Gembong Village Road, Bojongsari	75	33.169
Selanegara Village Road, Kaligondang	75	32.749
Karanganyar Village Road, Karanganyar	72	32.853
Padamara Village Road, Padamara	69	32.531
Bajong Village Road, Bukateja	63	31.447

3.3. *Black spot location based on road geometric approach*

The weighting of the road geometric parameters is carried out by comparing the data of the geometric parameters of the roads measured in the field with the weighting intervals. The weighting of road geometric parameters for 10 roads can be seen in Table 8.

The results of the research, the ranking of black spot locations based on the weighting of the equivalent accident number-upper control limit method and the road geometric parameter method was different. It can be seen that the order of black spot locations tends to be not much different between two methods. For comparison, Tlahab Lor Village Road ranks first based on the EAN-UCL method with a total weight value of 168. Meanwhile, based on the road geometric parameter method, also the first ranks with a total weight value of 22. The maximum value for all parameters is 30. Meanwhile, Bajong Village Road which ranks 10th in the EAN-UCL method, changes to the 9th in the road geometric parameter.

There are several other factors that have a major influence on the occurrence of accidents on several roads which cause inconsistency of ranking results on several roads with both methods. The severity of traffic accident victims is also influenced by natural factors. The result of this study is relevant with the study of Shaheed et al that the driver's visibility, road surface conditions [31]. Natural factors, for example when it rains, will also affect the amount of vehicle speed chosen by the driver and the vehicle speed limit [32]. Operating speed has a significant contribution to the total number of crashes [33]. Tarko in 2009 examined the impact of speed on driving behavior [15]. The effect of speed on travel time by Nilsson in 1991 [5] and the determination of the maximum speed value on urban roads [34]. The road pavement type affects the road roughness value [35]. Road safety strategies potentially are improved by applying the system theory and safety models [7].

Table 8. The weighting of road geometric parameters for ten roads

Name of road	Road function	Vehicle speed (km/h)	Lane width (m)	Shoulder width (m)	Sidewalk width (m)	Median width (m)	Drop-off (cm)	Total weight
Tlahab Lor Village Road, Karangreja	Collector road	46.00	3.40	0	0	0	44.00	
Weight		1	1	5	5	5	5	22
Bojongsari Village Road, Bojongsari	Collector road	51.00	4.00	1.00	0	0	13.00	
Weight		2	1	3	5	5	4	20
Serang Village Road, Karangreja	Collector road	56.00	1.99	1.50	0	0	0	
Weight		3	5	2	5	5	1	21
Mangunegara Village Road, Mrebet	Collector road	52.00	4.00	1.00	0	0	8.00	
Weight		2	1	3	5	5	3	19
Mayjend Sungkono (to Purbalingga)	Secondary arterial	50.00	2.70	0.55	0	1.50	6.00	
Weight		1	5	4	5	2	2	19
Gembong Village Road, Bojongsari	Collector road	57.00	4.00	1.30	0	0	11.00	
Weight		3	1	2	5	5	3	19
Selanegara Village Road, Kaligondang	Collector road	53.00	3.30	0.85	0.91	0	1.00	
Weight		2	1	1	2	5	1	12
Karanganyar Village Road, Karanganyar	Collector road	56.00	2.20	1.80	0	0	0.00	
Weight		3	5	2	5	5	1	21
Padamara Village Road, Padamara	Collector road	49.00	3.00	0.50	0	0	7.00	
Weight		1	1	4	5	5	2	18
Bajong Village Road, Bukateja	Arterial road	58.00	3.50	2.00	0	0	4.00	
Weight		1	1	1	5	5	2	15

The result of the study is similar to Aljanahi et al. stated that road crashes are influenced by many factors such as road geometric design, traffic volume, vehicles speed, weather, driver's physical and mental conditions [36]. For two-lane rural highways, there is a relationship between geometric design consistency and road accidents in homogeneous traffic conditions [37]. The factors that affect the safety performance of the horizontal curve include: traffic volume, geometric features of the curves, cross section, roadside hazards, stopping sight distance, coordination between horizontal and vertical alignment, distance between curves, pavement friction and traffic control devices [38]. Super elevation, degree of curve, shoulder width and average daily traffic also contributes to road crashes [38-39]. Accident rate in horizontal curves increases monotonically and at an accelerating rate as curve radius is reduced [40]. According to AASHTO, factors that influence the safest speed depends on design features, conditions of road, traffic volumes, cross-traffic volumes, weather conditions, roadside development, spacing of intersecting roads and other factors [41-42].

In the other study, arterial roads with heavier traffic volumes and higher speed limits tended to have more severe crashes. Medians were helpful in reducing the severity of crashes [43] and reducing crash rate up to 15% [44-45]. The lower roadside hazard rate decreases the risk of more severe crashes

[46]. The horizontal curve and its degree of curvature must be considered when predicting the total crash frequency on horizontal curves [47]. In the other study, the shoulder width had a significant effect on actual speed and on perceived safe driving speed, but only when a guardrail was present [48].

4. Conclusion and further research

Based on the results of study determining the black spot locations using road geometric facilities approach, the following conclusion can be drawn:

a. Six road geometric parameters are used to determine the black spot locations, namely vehicle speed, road lane width, road shoulder width, road pavement width, road median width, and shoulder drop-off (the difference in elevation between road shoulder and the pavement edge).
b. There are differences in the ranking of black spot locations based on the equivalent accident number method and the road geometric parameter approach.

In the further research is conducted to determine the coefficient value of each road geometric parameter and combined with road equipment facilities.

Acknowledgments

"The authors fully acknowledged the Ministry of Education, Culture, Research and Technology Republic of Indonesia for the approved fund the Basic Research in 2021 which makes this important research viable and effective".

References

[1] Traffic Corps (*Korlantas Polri*) 2021 *Kecelakaan Lalu Lintas dalam Angka 2020* (Jakarta: Traffic Corps.

[2] *Satuan Lalu Lintas* (Satlantas) Polres Purbalingga 2020 *Data Kecelakaan Lalu Lintas Tahun 2010-2019* (Purbalingga: *Kepolisian Resor* Purbalingga).

[3] Treat J R, Tumbas N S, McDonald S T, Shinar D, Hume R D and Meyer R E 1977 *Tri-level Study of the Causes of Traffic Accidents* (Washington: NTHSA).

[4] Sugiyanto G, Fadli A and Santi M Y 2017 Identification of black spot and equivalent accident number using upper control limit method *ARPN J. Eng. Appl. Sci.* **12**(2), 528-35.

[5] Nilsson G 1991 Speed limits, enforcement and other factors influencing speed. In Koornstra M J and Christensen J. (Eds.), *Enforcement and Rewarding: Strategies and Effects* pp. 46-50 (Netherlands: SWOV Leidschendam).

[6] Directorate General of Highway 2006 *Kajian Kebutuhan Pelaksanaan Keselamatan Jalan di Indonesia* (Jakarta: Ministry of Public Work Republic of Indonesia).

[7] Hughes B P, Anund A and Falkmer T 2015 System theory and safety models in Swedish, UK, Dutch and Australian road safety strategies *Accid. Anal. Prev.* **74**, 271-78. https://doi.org/10.1016/j.aap.2014.07.017.

[8] Wang X, Yuan J, Schultz G G and Fang S 2018 Investigating the safety impact of roadway network features of suburban arterials in Shanghai *Accid. Anal. Prev.* **113**, 137-48. https://doi.org/10.1016/j.aap.2018.01.029.

[9] Sugiyanto G, Jajang, and Santi M Y 2019 The impact of lowering speed limit on mobility and the environment. *AIP Conference Proceedings* **2094**(1) 020019, pp. 1-8. https://doi.org/10.1063/1.5097488.

[10] Schepers P, Hagenzieker M, Methorst R, Wee B V and Wegman F 2014 A conceptual framework for road safety and mobility applied to cycling safety *Accid. Anal. Prev.* **62**, 331-40. http://dx.doi.org/10.1016/j.aap.2013.03.032.

[11] Sugiyanto G, Suciningtyas R, Jajang and Santi M Y 2020 Strategies for handling black spot area to increased road safety *ARPN J. Eng. Appl. Sci.* **15**(9), 1101-110.

[12] Atashafrazeh M and Yadollahi M M 2013 Geometric design consistency model based on speed and safety in rural highways in some European countries: A review *Inter. J. Structural Civil Eng. Res.* **2**(2), 129-36.

http://www.ijscer.com/uploadfile/2015/0427/20150427034549611.pdf.
[13] Sugiyanto G 2017 The cost of traffic accident and equivalent accident number in developing countries (case study in Indonesia) *ARPN J. Eng. Appl. Sci.* **12**(2), 389-97.
[14] Sugiyanto G and Santi M Y 2017 Road traffic accident cost using human capital method (case study in Purbalingga, Central Java, Indonesia) *J. Teknol.* **79**(2), 107-16. https://doi.org/10.11113/jt.v79.5375.
[15] Tarko A P 2009 Modeling drivers' speed selection as a trade-off behavior *Accid. Anal. Prev.* **41**, 608-16. https://doi.org/10.1016/j.aap.2009.02.008.
[16] Sugiyanto G 2016 The impact of congestion pricing scheme on the generalized cost and speed of motorcycle to the city of Yogyakarta, Indonesia *J. Eng. Appl. Sci.* **11**(8), 1740-746.
[17] Sugiyanto G 2018 The effect of congestion pricing scheme on the generalized cost and speed of motorcycle *Walailak J. Sci. & Tech. (WJST)* **15**(1), 95-106. http://wjst.wu.ac.th/index.php/wjst/article/view/2347.
[18] Sugiyanto G, Santosa P B, Wibowo A and Santi M Y 2016 Hub and spoke airport networks in Indonesia based on Herfindahl-Hirschmann Index (HHI) *J. Eng. Appl. Sci.* **11**(8), 1804-10.
[19] Sugiyanto G, Santosa P B, Jajang, Fadli A and Santi M Y 2018 Evaluation of hub-spoke airport networks in Sumatra Island, Indonesia to increase efficiency of air transportation *MATEC Web of Conference* **195**(04009), pp. 1-9. https://doi.org/10.1051/matecconf/201819504009.
[20] Sugiyanto G, Santosa P B, Wibowo A and Santi M Y 2015 Analysis of hub-and-spoke airport networks in Java Island, Indonesia based on cargo volume and freight ratio *Procedia Eng.* **125**, 556-63. http://dx.doi.org/10.1016/j.proeng.2015.11.061.
[21] Sugiyanto G, Santosa P B and Santi M Y 2020 Hub and spoke airport networks in Sulawesi Island, Indonesia based on freight ratio *ARPN J. Eng. Appl. Sci.* **15**(10), 1201-209.
[22] Directorate General of Land Transportation 2007 *Pedoman Operasi* ABIU/UPK (Accident Black spots Investigation Unit/*Unit Penelitian Kecelakaan*) (Jakarta: Directorate General of Land Transportation, Ministry of Transportation Republic of Indonesia).
[23] Sugiyanto G, Fadli A, Suciningtyas R, Indriyati E W and Santi M Y 2019 Road safety audit at black spot area (Case study in Tlahab Lor, Karangreja, Purbalingga). AIP Conference Proceedings **2094**(1) 020020, pp. 1-10. https://doi.org/10.1063/1.5097489.
[24] Pusat Penelitian dan Pengembangan Prasarana Transportasi 2005 *Penanganan Lokasi Rawan Kecelakaan Lalu Lintas: Pd.T-09-2004-B* (Jakarta: Departemen Permukiman dan Prasarana Wilayah, Ministry of Public Work Republic of Indonesia).
[25] Sukirman S 1994 *Dasar-Dasar Perencanaan Geometrik Jalan* (Bandung: Nova).
[26] Badan Standardisasi Nasional (BSN) 2004 *Geometri Jalan Perkotaan* RSNI T-14-2004 (Jakarta: Direktorat Bina Teknik).
[27] AASHTO 2001 A Policy on Geometric Design of Highways and Streets, The American Association of State Highway and Transportation Officials (Washington DC).
[28] Ministry of State Secretariat of the Republic of Indonesia 2006 *Peraturan Pemerintah Republik Indonesia Nomor 34 Tahun 2006 tentang Jalan*.
[29] Ministry of State Secretariat of the Republic of Indonesia 2009 *Undang-Undang Republik Indonesia Nomor 22 Tahun 2009 tentang Lalu Lintas dan Angkutan Jalan* (Jakarta: Ministry of State Secretariat of the Republic of Indonesia).
[30] Mohammed H A 2013 The influence of road geometric design element of highway safety *Inter. J. Civil Eng. Technol.* **4**(4), 146-62.
[31] Shaheed M S, Gkritza K, Carriquiry A L and Hallmark S L 2016 Analysis of occupant injury severity in winter weather crashes: A fully Bayesian multivariate approach *Anal. Methods Accid. Res.* **11**, 33-47. https://doi.org/10.1016/j.amar.2016.06.002.
[32] Soriguera F, Martinez I, Sala M and Menendez M 2017 Effects of low speed limits on freeway traffic flow *Transp. Res. Part C: Emerg. Technol.* **77**, 257-74. https://doi.org/10.1016/j.trc.2017.01.024.
[33] Vayalamkuzhi P and Amirthalingam V 2016 Influence of geometric design characteristics on

safety under heterogeneous traffic flow *J. Traffic Transp. Eng. (Engl. Ed.)* **3**(6), 559-70. http://dx.doi.org/10.1016/j.jtte.2016.05.006.

[34] Sugiyanto G and Malkhamah S 2018 Determining the maximum speed limit in urban road to increase traffic safety *J. Teknol.* **80**(5), 67-77. https://doi.org/10.11113/jt.v80.10489.

[35] Sugiyanto G 2017 Characterization of asphalt concrete produced from scrapped tire rubber *Engineering J.* **21**(4), 193-206. https://doi.org/10.4186/ej.2017.21.4.193.

[36] Aljanahi A A M, Rhodes A H, and Metcalfe A V 1999 Speed, speed limits and road traffic accident under free-flow conditions *Accid. Anal. Prev.* **31**(1/2), 161-68. https://doi.org/10.1016/S0001-4575(98)00058-X.

[37] Cafiso S, Graziano A D and Silvestro G D 2010 Development of comprehensive accident models for two-lane rural highways using exposure, geometry, consistency and context variables *Accid. Anal. Prev.* **42**(4), 1072-79. https://doi.org/10.1016/j.aap.2009.12.015.

[38] Aram A 2010 Effective safety factors on horizontal curves of two-lane highways *J. Appl. Sci.* **10**(22), 2814-22. https://doi.org/10.3923/jas.2010.2814.2822.

[39] Dadashova B, Ramírez B A, McWilliams J M and Izquierdo F A 2016 The identification of patterns of interurban road accident frequency and severity using road geometry and traffic indicators *Transp. Res. Proc.* **14**, 4122-29. https://doi.org/10.1016/j.trpro.2016.05.383.

[40] Elvik R 2013 International transferability of accident modification functions for horizontal curves *Accid. Anal. Prev.* **59**, 487-96. https://doi.org/10.1016/j.aap.2013.07.010.

[41] AASHTO 2011 A Policy on Geometric Design of Highways and Streets 6th Edition, The American Association of State Highway and Transportation Officials (Washington DC).

[42] AASHTO 2018 A Policy on Geometric Design of Highways and Streets 7th Edition, The American Association of State Highway and Transportation Officials (Washington DC).

[43] Ma M, Yan X, and Abdel-Aty M 2010 Safety analysis of urban arterials under mix-traffic patterns in Beijing *Transp. Res. Rec.* **2193**, 105-15. https://doi.org/10.3141/2193-13.

[44] Frawley W E and Eisele W L 2005 Investigation of Access Point Density and Raised Medians: 27 Crash Analysis and Micro-simulation. FHWA/TX-05/04221-P1 (Texas: Texas A&M Transportation Institute).

[45] Sawalha Z and Sayed T 2003 Statistical issues in traffic accident modeling. The 82nd Transportation Research Board Annual Meeting (Washington DC).

[46] Haghighi N, Liu X C, Zhang G and Porter R J 2018 Impact of roadway geometric features on crash severity on rural two-lane highways *Accid. Anal. Prev.* **111**, 34-42. https://doi.org/10.1016/j.aap.2017.11.014.

[47] Gooch J P, Gayah V V and Donnell E T 2016 Quantifying the safety effects of horizontal curves on two-way, two-lane rural roads *Accid. Anal. Prev.* **92**, 71-81. https://doi.org/10.1016/j.aap.2016.03.024.

[48] Ben-Bassat T and Shinar D 2011 Effect of shoulder width, guardrail and roadway geometry on driver perception and behavior *Accid. Anal. Prev.* **43**(6), 2142-52. https://doi.org/10.1016/j.aap.2011.06.004.

Assessment on biochemical oxygen demand (BOD) removal using treatment train system for agricultural run-off

A A A Razak[1], H Takaijudin[1], M Osman[1], K W Yusof[1], W Ali[1] and S Ishak[2,3]

[1]Department of Civil & Environmental Engineering, Faculty of Engineering, Universiti Teknologi PETRONAS, 32610 Seri Iskandar, Perak, Malaysia
[2]HICoE-Centre for Biofuel and Biochemical Research, Institute of Self-Sustainable Building, Universiti Teknologi PETRONAS, 32610 Seri Iskandar, Perak, Malaysia
[3]Chemical Engineering Department, Universiti Teknologi PETRONAS, 32610 Seri Iskandar, Perak, Malaysia

husna_takaijudin@utp.edu.my

Abstract. Agricultural runoff is non-point source pollution that is unpredictable and difficult to control. Best Management Practices (BMPs) and Low Impact Development (LID) has been utilized as a method to manage agricultural runoff. Treatment train systems are LID-BMPs arranged in a series that aims to treat and manage agricultural runoff. This study focuses on the use of treatment train systems to reduce the concentration of BOD in agricultural runoff. A treatment train system was set up with three varying configurations to evaluate the efficiency of BOD removal. The three configurations consisted of no vegetation set which acts as the control, vegetated set, and vegetated set with the saturated zone. All three configurations showed good final removal rates of BOD at 56% for the control and 75% and 85% for both vegetated sets without and with the saturated zone. The control is classified classifying it as Class III according to Water Quality Standards, while both vegetated set without and with saturated zone had classified both sets as Class II and I, respectively. The performance of the treatment train system was compared to the use of a single bioretention system.

1. Introduction

Agricultural runoff is a non-point source of water pollution where water flows from agricultural land use and ultimately reaches bodies of water such as streams and lakes or be absorbed into the Earth. The source of contamination stems from fertilizers, sediments, pesticides, and waste coming from croplands and live-stock operations [1]. The use of inputs such as fertilizers and phosphorus not only help crops flourish, but once it is washed off by the rain, the agricultural runoff carries the input as well and when it reaches water bodies such as rivers and lakes, it acts as a catalyst to the growth of algae which leads to a phenomenon called an algal bloom. Unfortunately, the funding required to treat the occurrence of algal bloom is very costly [2]. The occurrence of algal bloom creates areas in the water body where there is no oxygen present due to all the oxygen content in the water being consumed by algae and releasing carbon dioxide. The lack of oxygen leads organisms such as bacteria and fish to suffocate and die. The algae also block sunlight from reaching the bottom of the water body which leads to vegetation not being able to perform photosynthesis.

According to Wiens [1], non-point sources of water pollution are becoming more of a concern as point sources are easier to control. Being a non-point source, control over where and when it occurs becomes more difficult because of its random and unpredictable nature. However, the consequences it brings is not something to dismiss as the impact it brings towards 2 the ecosystem may be devastating if not controlled. Wiens [1] mentions that agricultural land use and water systems are linked due to the hydrologic cycle that occurs, where both land use and water systems are affected by each other. This leads to water bodies such as rivers containing relatively high levels of bacterial contamination from the agricultural watersheds. For example, Bryan's 1976 study (as cited in [1]) stated that runoff coming from dairy farms containing manure had shown an increase in fecal coliform levels, which in turn would lead to high concentrations of BOD.

The Department of Irrigation and Drainage of Malaysia (DID) has prepared an Urban Stormwater Management Manual that includes standards and practices that should be followed for all matters related to water management. One of the indicators that can be used to characterize pollutants for runoff is biochemical oxygen demand (BOD). A high BOD value would indicate a high amount of organic pollution in a body of water which has the potential to disrupt the local ecosystem. Agricultural runoff from plantations contains all sorts of pollutants such as pesticides and fertilizers that have seeped into the soil. Treatment train systems are introduced to act as a filter to remove BOD in the runoff. This study aims to evaluate the performance of BOD removal using a treatment train system for BOD removal for agricultural runoff treatment.

2. Materials and method

2.1. Site sampling

Samples were obtained from a nearby palm oil plantation located at Jalan Felcra Nasaruddin for characterization and use as a benchmark. The BOD concentrations in the samples were determined accordingly. Two types of samples were collected, which were from the palm oil water channel and the river where the water channels flow into. Samples were taken from three different locations along the water channel and three more locations along the river as shown in figure 1. Based on the finding from the characterization study, the pollutant source can be used as a benchmark to determine the performance of the treatment train system in reducing the BOD value of the synthetic runoff used.

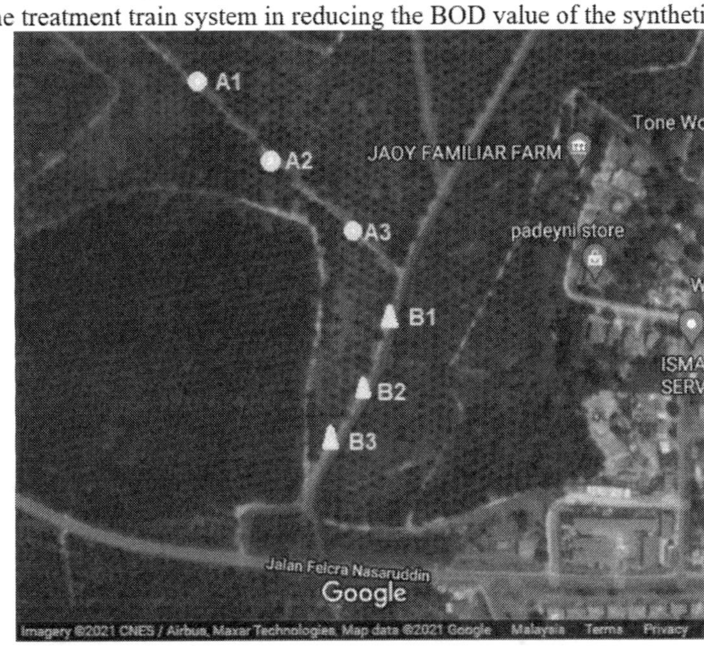

Figure 1. Six (6) sampling points from palm oil plantation to the river

2.2. Treatment train setup

Figures 2 and 3 demonstrate how a series of three bioretention columns were set up. Three alternative column arrangements were employed, each with its own set of configurations, as follows:

a.) No vegetation with 300mm soil depth (control)
b.) Vegetation with 300mm soil depth
c.) Vegetation with 300mm soil depth and saturated zone.

The bioretention column consisted of a ponding zone, soil media, sand layer, and gravel layer. A Ti plant (Cordyline fruticosa) was planted in the ponding zone for columns with vegetation as a replacement to the Red Hot Chinese Hibiscus plants that were planted prior because it was found that the plants were wilting. The soil media was made up of a mixture of 60 percent sand, 30 percent topsoil, and 10 percent compost. While the soil media layer was 300mm deep, the sand layer was 95mm deep, and the gravel layer was 100mm deep. The water tank holds 150L of synthetic runoff acting as the influent. The system was regulated by valves, which allow the influent to flow via column 1, where it will eventually flow to columns 2 and 3. The outlets of each column simulate a free falling of the runoff which helps aerate the runoff to allow for more supply of oxygen to the runoff.

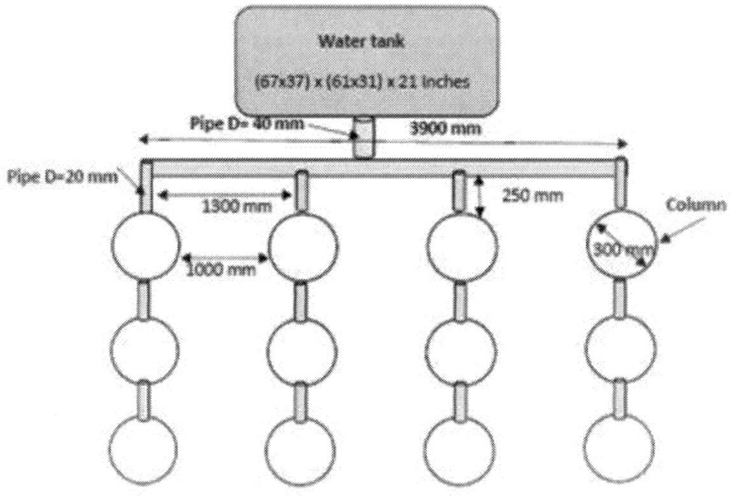

(a)

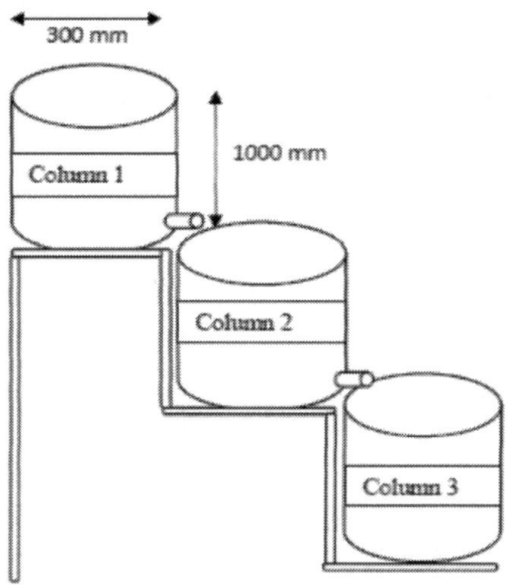

(b)

Figure 2. (a) Treatment train system setup: layout of treatment train; (b) Arrangement of bioretention columns in series

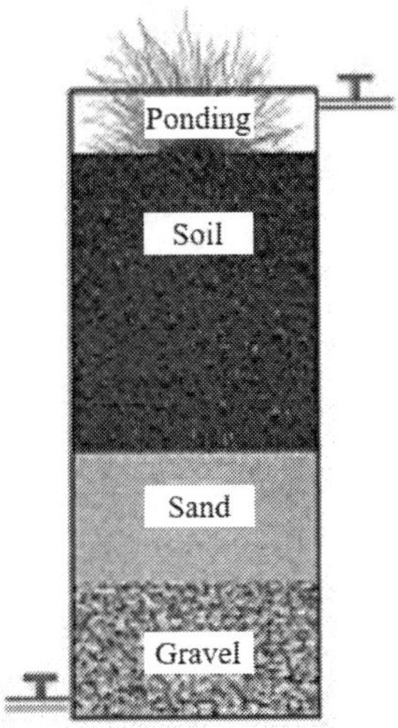

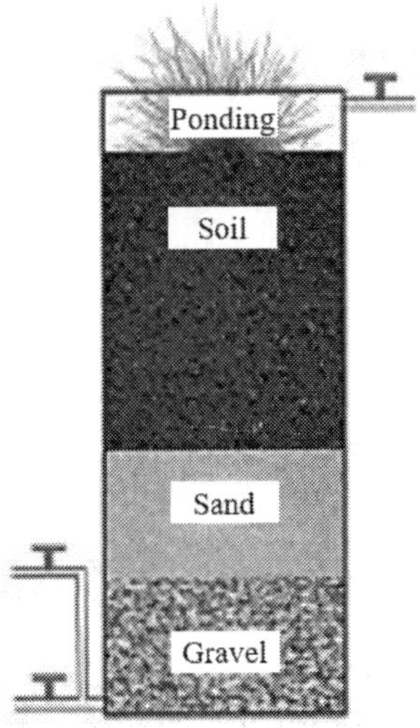

Figure 3. Column non-saturated zone (left) and column with the saturated zone (right)

2.3. Preparation of Synthetic Agricultural Runoff

In this work, the synthetic agricultural runoff was used to overcome the drawback of using actual agricultural runoff, which would require multiple trips to obtain large volumes of the agricultural runoff that would act as an influent on the system. The synthetic runoff was prepared with measurements according to table 1 which includes the measurements scaled according to the 60-gallon water tank (227 liters) that was used. The calculations were based on the characterization study and compared to previous research.

The synthetic runoff was prepared by measuring each chemical constituent using a weighing scale with a tolerance of 0.01g. The chemical constituents were then put into a beaker and mixed with tap water. The synthetic runoff was then carefully poured slowly into the water tank while being constantly stirred. The synthetic runoff was prepared prior for every run, as it begins to be contaminated with algae after 3 to 4 days.

Table 1. Composition chemical substances for the preparation of synthetic agricultural runoff

Synthetic Agricultural Runoff Constituents	Chemical substances	Amount (g)
Nitrate	Potassium nitrate (KNO_3)	11.824
Ammonium	Ammonium chloride (NH_4Cl)	2.753
Total Phosphorus	Potassium di-hydrogen phosphate (KH_2PO_3)	1.870
Organic Nitrogen	Urea (CH_4N_2O)	0.466
Glucose	Glucose ($C_6H_{12}O_6$)	22.720

2.4. Sampling and BOD testing

About 150 mL of each sample was filled into BOD bottles. The dissolved oxygen of each BOD bottle was measured with the calibrated DO meter and the results were recorded accordingly. A sufficient amount of distilled water was added up to half of the b BOD bottle's neck, and a stopper was inserted to prevent air bubbles in each of the prepared sample bottles. The BOD bottles were capped off, covered with aluminium foil, and labeled immediately to prevent evaporation. The BOD bottles were then kept in the incubator for five days at a temperature of 20°C. After 5 days of the incubation period, the remaining dissolved oxygen was measured and recorded, respectively. Since no bacterial seed was added throughout the research, the BOD was determined using equations (1), (2), and (3).

$$BOD_5=(DO_{initial}-DO_{final})\times \text{Dilution factor} \quad (1)$$

$$\text{Dilution factor}=\frac{\text{Bottle volume}}{\text{Sample Volume}} \tag{2}$$

The removal rate of BOD can be calculated as,

$$\text{Removal rate (\%)}=\frac{BOD_{influent}-BOD_{effluent}}{BOD_{influent}}\times 100\% \tag{3}$$

3. Results and discussion

3.1. Characterization of Agricultural Runoff

The BOD_5 values of water samples taken from the Felcra Nasaruddin palm oil plantation are displayed in figure 4. Samples A1, A2, and A3 were obtained from the palm oil water channel, while samples B1, B2, and B3 were collected from the river into which the water channels flow. It was found that the BOD_5 values for agricultural runoff from the palm oil water channel were determined to be in an average of 12.08mg/L. On the other hand, the BOD_5 values for the main river were in an average of 5.48mg/L, which is lower than the BOD_5 values for agricultural runoff from the palm oil water channel. This result was considered acceptable and in agreement with the finding reported by other studies [3, 4, 5, 6, and 7]. Moreover, the more it goes downstream showing lower in BOD_5 values since the constituents being diluted [3].

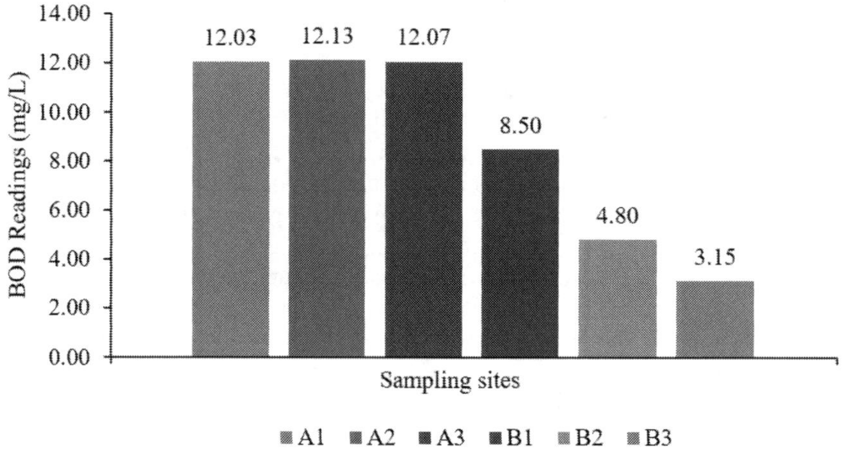

Figure 4. BOD concentration for agricultural runoff from palm oil plantation

3.2. Biochemical Oxygen Demand (BOD) Analysis

In this study, five runs of experiments were conducted with the three trial runs to get used to the process and improving the system. Each run takes five days from the start of the run to the measurement of BOD of the samples. Each run would necessitate the preparation of a new synthetic runoff due to the high possibility of algae accumulation in the water tank after two days, which would affect the results. The first three runs helped determine the appropriate volume of sample to be used, and the fourth run helped identify any problems in collecting samples from the vegetated set with the saturated zone. From the fourth run, it was found that running the synthetic runoff through the vegetated set with the saturated zone required a minimum of two hours and constant monitoring, as leaving the saturated zone inside the column configurations lead to the vegetation wilting. Despite the fact that five runs were carried out, only the fifth yielded promising and acceptable data that could be used for further investigation. Figure 5 shows the average BOD_5 measurements from the five runs. It can be seen that certain BOD readings cannot be accepted due to several issues such as the presence of bubbles in the bottle, a remaining DO reading of less than 1mg/L, a small difference between final DO and initial DO (excluding the final effluents), and unable to collect the sample due to time constraints.

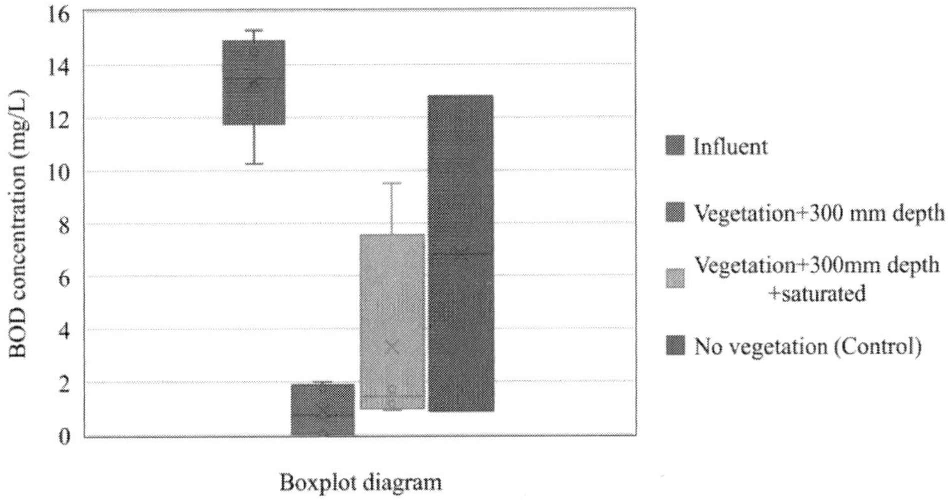

Figure 5. BOD concentration before and after treatment using treatment train.

Figure 6 exhibits the reduction of the BOD measurements when the agricultural runoff passes through each bioretention configuration. Moreover, each column configuration shows an initial BOD reduction of greater than 82%. It was found that the control set shows mean BOD removal rate of 52%, while the vegetated sets shows slightly better which mean BOD removal rates with 75% and 85% for Set A and B, respectively. Out of the three configurations, the vegetated set with saturated zone shows the best performance with an 88% initial removal rate and 98% final removal rate. Though the vegetated set with the saturated zone requires higher maintenance and monitoring, as leaving the saturated zone in the setup causes the vegetation to wilt. From the results, the control set can be classified as Class II according to Water Quality Standards, while both vegetated set without and with saturated zone can be classified as Class I. If each set was to consider the first column as a single bioretention system, all three configurations would be classified as Class IV, which shows that the arrangement of bioretention systems in a series contributes greatly in removing BOD of agricultural runoff. The reduction of the BOD measurements can be attributed to the soil media as it filters the pollutants present in the runoff. In addition to that, the cascading discharge of each column configuration assisted in aerating the runoff as it cascades into the next column.

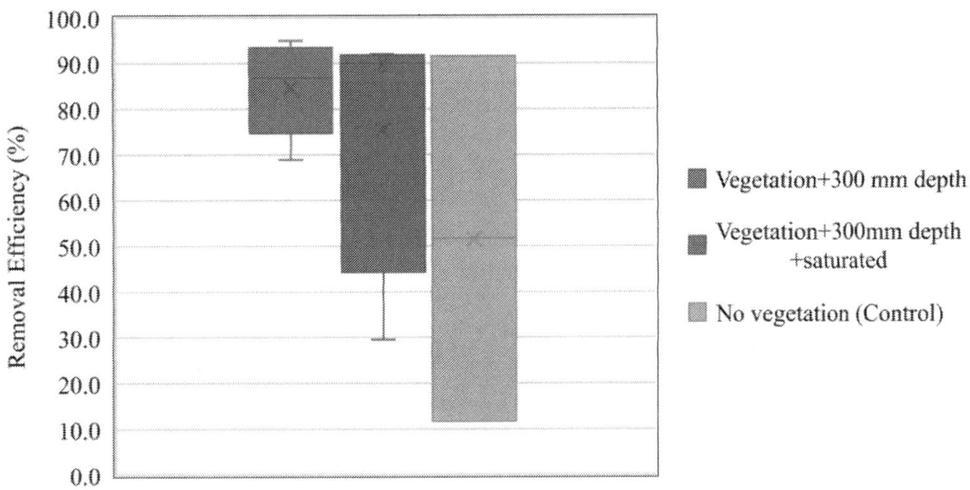

Figure 6. BOD reduction (%) using treatment train.

4. Conclusion

Agricultural runoff from a nearby palm oil plantation was characterized and used as a benchmark for the synthetic runoff and it was found that the samples collected from the palm oil water channel had an average BOD5 reading of 12.08mg/L and the main river had a lower average reading at 5.48mg/L. The three configurations showed high final BOD removal rates at 96% for the control, and 98% for both vegetated sets without and with saturated zone and the control set was classified as Class II according to Water Quality Standards, while the two vegetated sets without and with saturated zone were classified as Class I. Based on the results, it can be attributed to the cascading flow of discharge as well as soil media in filtering the pollutants. Lastly, the effectiveness of BOD removal in treatment train system was compared to a single bioretention system. When only the first bioretention cell was taken into consideration, the control set had the lowest BOD_5 removal rate of 52%, the vegetated set with 300 mm soil depth had a greater BOD_5 removal rate of 75% compared to the control set, while the highest removal rates is 85% with the presence of saturated zone. This would classify all three sets as Class IV, so it can be said that treatment train systems exhibit better performances in BOD removal compared to single bioretention systems.

Acknowledgement

The research is funded by Yayasan Universiti Teknologi PETRONAS Cost Centre (015LC0-141) entitled "Rehabilitation of Artificial Lakes by Integrating BIOECODS system for Flash Flood Management". The authors express their gratitude to Final Year Students, Amir al-Hafiz Abd Razak for his assistance completing this work. Special thanks to Universiti Teknologi PETRONAS for giving fully support of this project.

References

[1] Wiens J H 1980 *Can.Water Resour.* J. **5**(3),78–89
[2] Biello D 2014 *Deadly Algae Are Everywhere, Thanks to Agriculture*, Scientific American, New York, Springer Nature
[3] Schreiber J D and Neumaier E E 1987 *J. Environ. Qual.* **16**(1) 6-10
[4] Wang M Zhang D Dong Jand Tan S K 2018 *Hydrobiologia* **805**(1) 1-31
[5] Laurenson G Laurenson S Bolan N Beecham S and Clark I 2013 *The Role of Bioretention Systems in the Treatment of Stormwater*. In Advances in Agronom vol 120 Elsevier pp 223-274
[6] Revitt D M Ellis J B and Lundy L 2017 *Urban Water* J. **14**(8) 839–845
[7] Simpson M G and Roesner L A 2018 J. *Sustainable Water Built Environ.* **4**(2) 05018003.

Energy dissipation of graded concrete beams on maximum reinforcement ratio

Y Novitasari[1] and M M A Pratama[2]

[1]Laboratory of Structures, Faculty of Engineering, Universitas Negeri Malang, Indonesia
[2]Department of Civil Engineering, Faculty of Engineering, Universitas Negeri Malang, Indonesia

mirza.abdillah.ft@um.ac.id

Abstract. Energy dissipation is a parameter that contributes to the deformation process of structural elements. Factors affecting the value of energy dissipation include concrete compressive strength, reinforcement ratio, cross-sectional dimensions, and history of loading that occurs on the structure. In this study, the graded concrete (GC) beams with a maximum reinforcement design was tested under flexural test to analyse the resulting energy dissipation. The resulting data was then compared to the reference beams. The results showed that: 1) The use of cross-sectional dimension ratio of ½ in GC beams using maximum reinforcement can increase the energy dissipation by more than 2000%; 2) Design of reinforcement area should be based on the higher concrete strength to increase energy dissipation up to 100%; 3) The use of GC save on construction costs can increase energy dissipation by up to 400% with a reduction in load capacity of 25%; 4) The use of GC to improve the properties of structural elements can increase energy dissipation by more than 60% while increasing load performance by almost 50%.

1. Introduction

In terms of designing earthquake-resistant structural elements, structural engineers should ensure that the structure can perform in a ductile manner. Energy dissipation is a parameter that contributes to the deformation process of a structural element. The ductile deformation can be achieved through large energy dissipation. Energy dissipation is the ability of a structure to transmit energy through the process of yielding the reinforcement, so it determines the damage mechanism of the structures. Generally, the damage mechanism is exposed as structural cracks or yielding of reinforcement. Therefore, large dissipation energy is an ideal condition for structural elements. When the energy dissipation is large, plastic deformation can be performed before failure occurs. Park et al. [1] stated that energy dissipation is highly related to ductility. The resulting ductility is in a linear relationship to the energy dissipation. A structural element with a high ductility can withstand more loads due to large inelastic deformations before it reaches its failure.

Studies related to energy dissipation have been carried out by several researchers. Nao et al. [2] examined the effect of temperature on the resulting energy dissipation of concrete structures. The results show that an increase in temperature causes the resulting energy dissipation increase. In addition, the increase in energy dissipation is accompanied by an increase in the strain rate. Investigations related to the effect of lateral reinforcement in beam-column joints on energy dissipation have been carried out by Ugale et al. [3]. It was found that the energy dissipation of the beam-column connection increases when installed hoop reinforcement. The energy dissipation can be improved by adding restraints on a structural element. Priastiwi et al. [4] confirmed the application of confinement crossties can increase energy dissipation up to four times. On a structural level, Salazar [5] examined a portal with red brick walls. The red brick walls can increase the energy dissipation by 3.5 times compared to a portal without walls. Based on Hasan's research [6], an eccentrically braced steel frame system with the application of a wing-plate link and a hybrid link can increase the energy dissipation performance compared to the use of a built-up link. Furthermore, the energy dissipation is influenced by the compressive strength of concrete, the ratio of tensile fibre reinforcement, the ratio of compression fibre reinforcement, the cross-sectional dimensions and the history of loading that occurred in the structure.

In general, the energy dissipation can be determined from the definite integral of the load-deflection curve. Another method is to calculate the area under the load-deflection curve using a CAD-based program. This study aims to identify the structural characteristics of graded concrete (GC) beams with the use of maximum reinforcement focusing on the resulting energy dissipation. Material properties of graded concrete have been completely studied on previous studies [7–10]. The behaviour of reinforced concrete beam incorporating GC materials has been studied to understand the resulting load-deflection relationship, stress response, and the ductility [11–19].The research finding is then used as reference materials and improvements for future research.

2. Experimental programme

2.1. Concrete mix proportion and specimen design

This research was conducted experimentally in the Laboratory of Structures, Universitas Negeri Malang. The concrete mixture used in this study consisted of two types of concrete strength, namely 20 MPa and 40 MPa. The concrete used was supplied concrete batching plant of PT. Tasa Beton. The consideration for the selection of ready mix concrete is that researchers want to create uniform concrete quality for all specimens. Concrete making on different batch resulted in uncontrolled quality on testing specimens. The specimen specifications are presented in Table 1, while the proportions of the concrete mixture are presented in Table 2. The Indonesian code [20] stipulates that the reinforcing steel used in the concrete structural elements is of the deformed type. However, this study used the plain type for all diameters for ease in the installation of strain gauges. Researchers also want to eliminate factors related to bond strength on different rebar diameter that commonly affected by different rebar types used during experimental testing.

Table 1. Specimen details

Code	Beam dimension			Strength configuration (T-C side) (MPa)	Rebar ratio	Reinforcement details
	Span (mm)	Width (mm)	Depth (mm)			
G1/2 20	1,720	130	260	20-40	0.0159	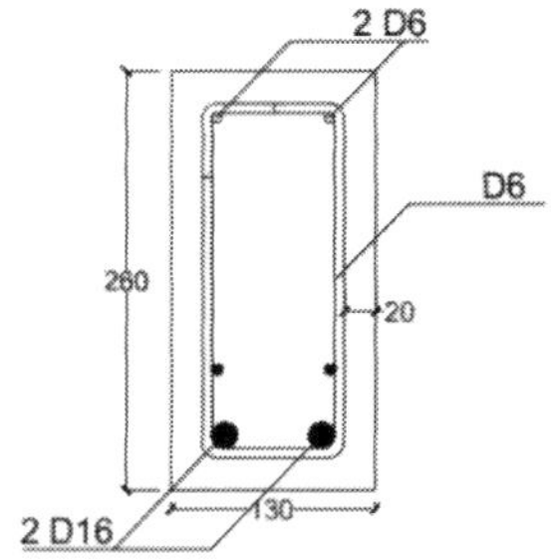

Table 1. Specimen details (contd.)

Code	Beam dimension			Strength configuration (T-C side) (MPa)	Rebar ratio	Reinforcement details
	Span (mm)	Width (mm)	Depth (mm)			
G1/2 40	1,720	130	260	20-40	0.0478	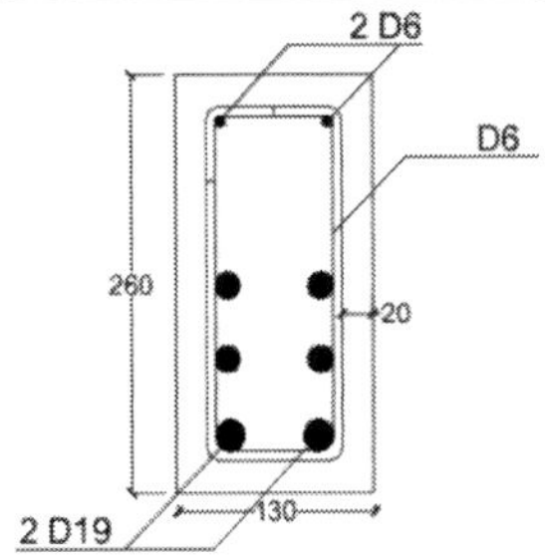
G2/3 20	1,400	130	195	20-40	0.0159	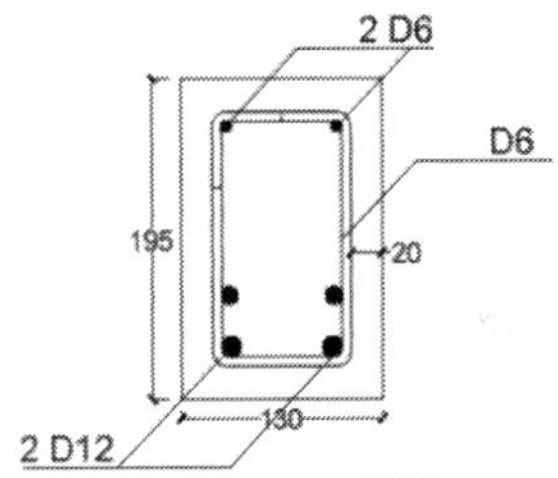
G2/3 40	1,400	130	195	20-40	0.0478	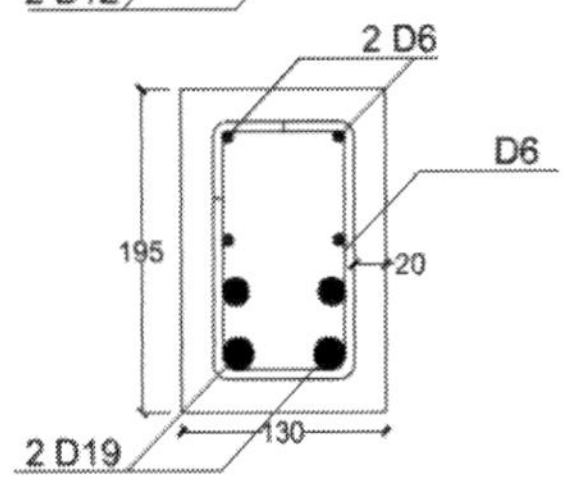
R1/2 20	1,720	130	260	20	0.0159	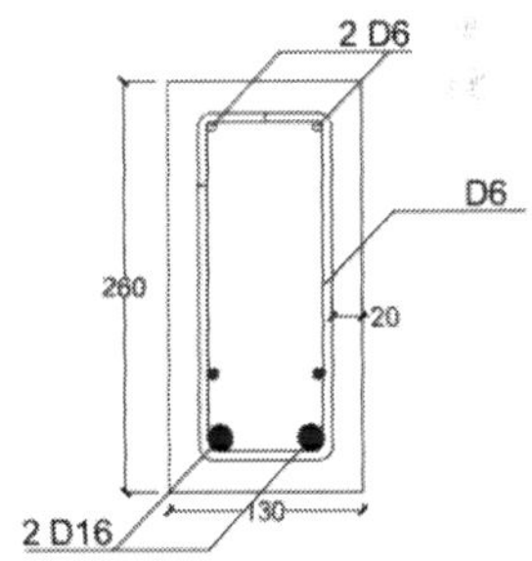
R2/3 20	1,720	130	260	20	0.0159	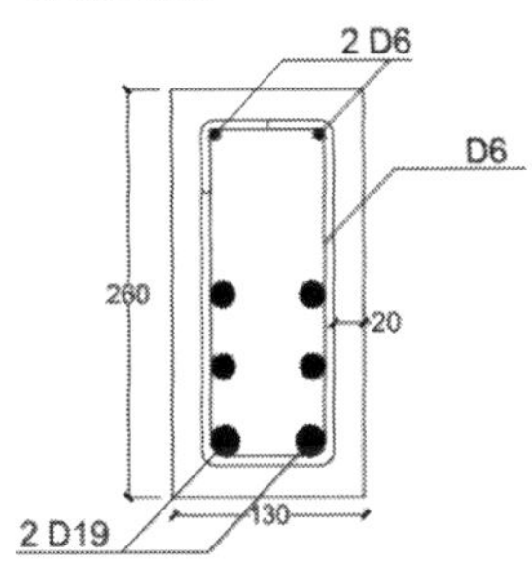

Table 1. Specimen details (contd.)

Code	Beam dimension			Strength configuration (T-C side) (MPa)	Rebar ratio	Reinforcement details
	Span (mm)	Width (mm)	Depth (mm)			
R1/2 40	1,400	130	195	40	0.0478	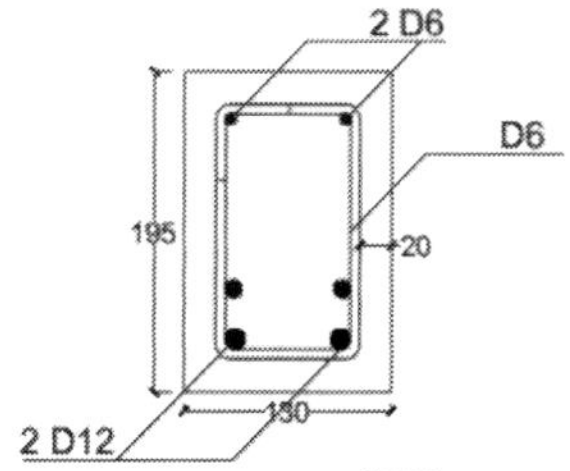
R2/3 40	1,400	130	195	40	0.0478	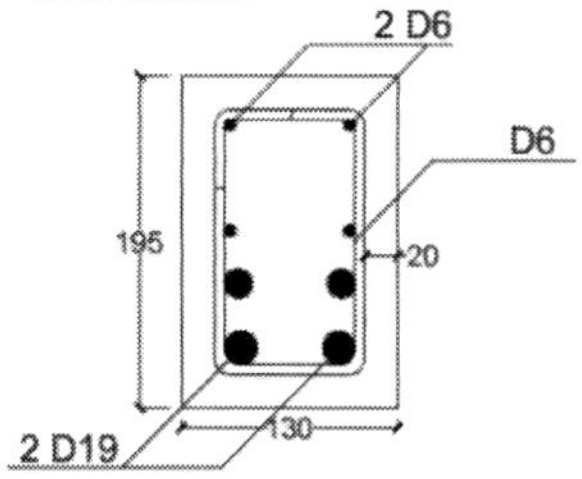

Table 2. Concrete mix design per-cubic meter

Concrete strength (MPa)	Water (kg)	PC (kg)	w/c (%)	Aggregate		Additive (kg)
				Coarse (kg)	Fine (kg)	
20	154	300	0.513	899	1030	1.04
40	140	760	0.210	854	681	2.64

2.2. Specimen casting

Before specimen moulding, the formwork was laid on a flat surface and protected from weather exposure. The assembled reinforcement was placed on the formwork and ensures that the concrete cover is provided as planned by putting concrete blocks on some points. For GC beams, concrete moulding began by pouring a 20 MPa concrete mixture on the bottom of the mould until it reaches half of the beam depth. Compaction was carried out with a rubber mallet to ensure that the mass of concrete descends to the bottom, filling the gaps between reinforcement while removing air voids trapped in the moulded concrete. A 40 MPa concrete mix is poured over the previous layer until full. Mechanical compaction was continuedly applied until the air voids rise to the surface.

2.3. Specimen curing

Specimen curing was carried out for 28 days starting from the dismantling of the beam formwork. Curing was carried out by covering all parts of the reference specimens and the GC beams with wet burlap sacks. The humidity of the burlap sacks was always maintained by routinely wetting them with water. The purpose of this concrete curing is to prevent over evaporation of freshly harden concrete specimen which causes microcracks that can reduce the performance of the beam.

2.4. Specimen testings

The air-dried beam specimens were then cleaned and painted white. This painting aims to provide a clean surface that eases the researchers to observe cracks occurring during the test. The beams were then mounted on the loading frame positioning a concrete layer of 20 MPa at the bottom and 40 MPa at the top. The test setup was equipped with several instruments that aim to obtain data readings, including load cells, dial gauges, and data loggers. Loading was given manually using a hydraulic jack to maintain a load increment of 1 kN.

3. Results and discussion

3.1. Energy dissipation of specimens

Energy dissipation is obtained by calculating the area under the load-deflection curve from the experimental test. The load-deflection data series in spreadsheet format is imported into the CAD program. The area can be obtained by plotting the load-deflection curve using the polyline command, block area using the hatch command, and read the area information in the properties window. The area reading from P-Δ curve is shown in Figure 1 and the recapitulation are shown in Table 3.

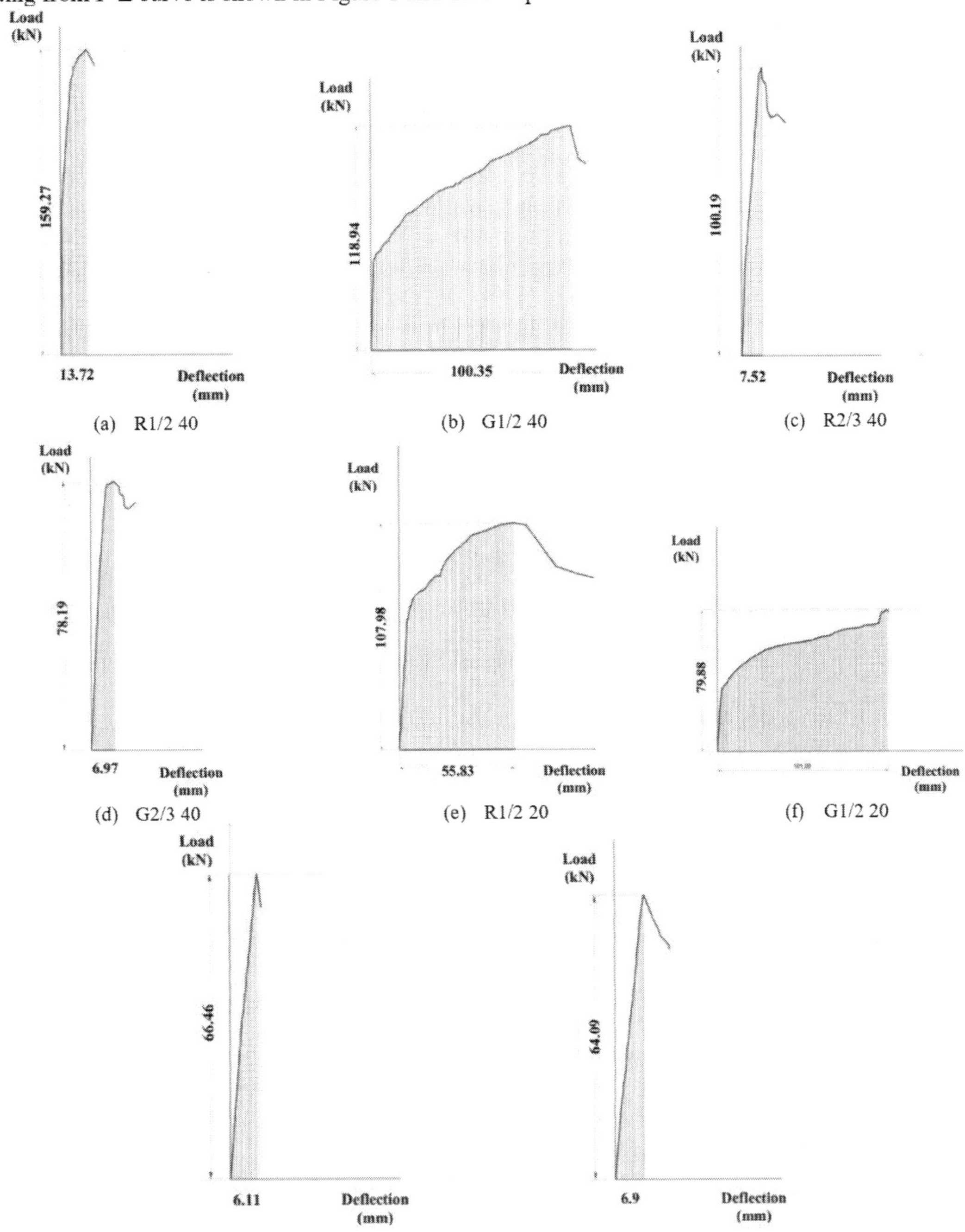

(a) R1/2 40 (b) G1/2 40 (c) R2/3 40
(d) G2/3 40 (e) R1/2 20 (f) G1/2 20
(g) R2/3 20 (h) G2/3 20

Figure 1. Calculation of energy dissipation

Table 3. Energy dissipation of testing specimens

Specimen types	Code	Energy dissipation (kN.mm)
GC beams	G1/2 40	9,880.58
	G2/3 40	382.28
	G1/2 20	4,950.73
	G2/3 20	220.73
Reference beams	R1/2 40	1,827.79
	R2/3 40	453.02
	R1/2 20	6,105.77
	R2/3 20	235.69

3.2. Effects of cross-sectional dimension ratio on the energy dissipation of GC beams

Based on Table 4, if the energy dissipation is analysed based on the ratio of the cross-sectional dimension ratio, the percentage of energy dissipation of G1/2 40 increased by 2,484.64% when compared to G2/3 40. For G1/2 20, the value of the energy dissipation produced exceeds G2/3 20 of 2,142.89%. For the reference specimens, when R1/2 40 is compared to R2/3 40, the energy dissipation increases by 303.47%. An increase in energy dissipation is also shown by R1/2 20, where there is an increase of 2,490.59% when compared to R2/3 20. Park et al. [1] stated that the factor affecting the energy dissipation is the resulting ductility. A structural element that has a high ductility value will experience large inelastic deformations to withstand additional loads acting on the structure so that the structural element deforms in large deformation before reaching its collapse. A change in the cross-sectional dimension ratio from 2/3 to 1/2 can increase the energy dissipation performance by more than 2,000%. The increase in dissipation energy is also due to the beam with a cross-sectional ratio of 1/2 having a deeper ratio so that the crack propagation occurring in the cross-section is longer than in the 2/3. Therefore, the GC beam with a cross-sectional dimension ratio of 1/2 has better performance than 2/3. The increase in the resulting energy dissipation indicates that the structural elements can withstand earthquakes better through a gradual failure mechanism. The structural elements that adopting GC technology can be applied in earthquake-prone areas.

Table 4. Effects of cross-sectional dimension ratio on the energy dissipation of GC beams

Specimen types	Code	Energy dissipation (kN.mm)	Significance (%)
GC beams	G2/3 40	382.28	+2,484.64
	G1/2 40	9,880.58	
	G2/3 20	220.73	+2,142.89
	G1/2 20	4,950.73	
Reference beams	R2/3 40	453.02	+303.47
	R1/2 40	1,827.79	
	R2/3 20	235.69	+2,490.59
	R1/2 20	6,105.77	

3.3. Effects of reinforcement area on the energy dissipation of GC beams

From the aspect of reinforcement area, G1/2 40 produces 99.58% greater energy dissipation than G1/2 40. The difference in energy dissipation of G2/3 40 to G2/3 20 is 73.19%. If the R1/2 40 is compared to R1/2 20, the resulting energy dissipation increases by 70.06%, whereas if R2/3 40 is compared to R2/3 20, the increase in energy dissipation is 87.12%. Research that focuses on reviewing the reinforcement ratio has been carried out by Sabelli et al. [21]. He showed that the addition of the reinforcement area causes greater resulting energy dissipation. A significant increase in energy dissipation is found in GC beams with a dimension ratio of 1/2. The use of higher strength concrete as a basis for calculating the reinforcement area increases energy dissipation by almost 100%. Analysis of energy dissipation due to varying reinforcement areas are shown in Table 5.

By using the same data provided in Table 5, it can be further analysed regarding the energy dissipation of the GC beam compared to the reference beam. The GC can be applied as a solution to reduce construction costs by replacing 50% of the volume of high-strength concrete in beams with a mixture of lower concrete strength. In this study, G1/2 40 is expected to replace R1/2 40 and G2/3 40 can replace R2/3 40. In beams with a dimension ratio of 1/2, the use of GC can increase the energy dissipation by more than 400% despite a decrease load capacity of 25%. By using the same rebar configuration as the R1/2 40, the G1/2 40 can dissipate more energy but at a lower cost of implementation. Different patterns were found in beams with a dimension ratio of 2/3, therefore it requires further evaluation. The analysis of energy dissipation for cost-saving purpose is shown in Table 6.

Table 5. Effects of reinforcement area on the energy dissipation of GC beams

Specimen types	Code	Energy dissipation (kN.mm)	Significance (%)
GC beams	G1/2 20	4,950.73	+99.58
	G1/2 40	9,880.58	
	G2/3 20	220.73	+73.19
	G2/3 40	382.28	
Reference beams	R1/2 20	6,105.77	-70.06
	R1/2 40	1,827.79	
	R2/3 20	235.69	+87.12
	R2/3 40	453.02	

Table 6. Analysis of energy dissipation for cost-saving purpose

Code	Load (kN)	Energy dissipation (kN.mm)	Significance in: (%)	
			Load	Energy dissipation
R1/2 40	159.2	1,827.7	-25.3	440.5
G1/2 40	118.9	9,880.5		
R2/3 40	100.1	453.0	-21.9	-15.6
G2/3 40	78.19	382.2		

In other cases, the GC can also be applied as an alternative to improve structural performance by replacing 50% of the volume of lower-strength concrete with the higher-strength concrete in the compression fibre and increasing the reinforcement area based on the higher-strength concrete. Improving the material properties of structural elements is an option for structural planners dealing with designing structural elements with strict and limited dimensions. For example, R1/2 20 is upgraded to G1/2 40 and R2/3 20 is upgraded to G2/3 40. The increase in beam properties allows the designer to maintain the same cross-sectional dimensions as the existing condition. This scenario can increase the energy dissipation in the range of 60% for both types of dimensional ratios. The most significant increase in capacity was found in the GC beams with a dimension ratio of 1/2, an increase in load was obtained by almost 50%. Further analysis can be seen in Table 7.

Table 7. Analysis of energy dissipation on properties improvement purpose

Code	Load (kN)	Energy dissipation (kN.mm)	Significance in: (%)	
			Load	Energy dissipation
R1/2 20	79.8	6,105.7	48.8	61.8
G1/2 40	118.9	9,880.5		
R2/3 20	64.0	235.6	22.0	62.1
G2/3 40	78.1	382.2		

4. Conclusion

- The use of cross-sectional dimension ratio of ½ in GC beams using maximum reinforcement can increase the energy dissipation by more than 2000%;
- Design of reinforcement area should be based on the higher concrete strength to increase energy dissipation up to 100%
- The use of GC save on construction costs can increase energy dissipation by up to 400% with a reduction in load capacity of 25%;
- The use of GC to improve the properties of structural elements can increase energy dissipation by more than 60% while increasing load performance by almost 50%

Acknowledgement
This work was funded by PNBP Universitas Negeri Malang 2021

References

[1] Park H and Eom T 2004 Energy Dissipation Capacity of Flexure-Dominated Reinforced Concrete Members *13th World Conf. Earthq. Eng. World Conf. Earthq. Eng.*

[2] Lv N, Wang H, Zong Q, Wang M and Cheng B 2020 Dynamic Tensile Properties and Energy Dissipation of High-Strength Concrete after Exposure to Elevated Temperatures *Materials (Basel).* **13** 5313

[3] Ugale A B and Khante S N 2020 Study of energy dissipation of reinforced concrete beam-column joint confined using varying types of lateral reinforcements *Mater. Today Proc.* **27** 1356–61

[4] Priastiwi Y A, Imran I, Nuroji and Hidayat A 2017 Confinement hoops of compression zone in beam under cyclic loading *AIP Conference Proceedings* p 020009

[5] Reyes-Salazar A, Bojórquez E, Bojorquez J, Valenzuela-Beltran F and Llanes-Tizoc M D 2018 Energy Dissipation and Local, Story, and Global Ductility Reduction Factors in Steel Frames under Vibrations Produced by Earthquakes *Shock Vib.* **2018** 1–19

[6] Hasan W M, El-Khatieb M and Al-Ani H 2011 Steel reinforcement ratio dependency of plastic rotational capacity of reinforced concrete beams *Jordan J. Civ. Eng.* **5**

[7] Gan B S, Aylie H and Pratama M M A 2015 The Behavior of Graded Concrete, an Experimental Study *Procedia Eng.* **125** 885–91

[8] Pratama M M A 2015 *An Experimental Study and Finite Element Approach to the Behavior of Graded Concrete* (Semarang: Universitas Diponegoro)

[9] Pratama M M A, Sthenly Gan B, Han Ay Lie H A L and Nur Rahma Putra A B 2019 A Numerical Analysis of The Modulus of Elasticity of The Graded Concrete *Proceedings of the 2nd International Conference on Vocational Education and Training (ICOVET 2018)* (Paris, France: Atlantis Press)

[10] Pratama M M A, Umniati B S, Arumsari Mutiara Wulandari B, Han A L, Sthenly Gan B and Zhabrinna Z 2018 Modulus elasticity of the graded concrete, a preliminary research ed P Hajek, A L Han, S Kristiawan, W T Chan, M b. Ismail, B S Gan, R Sriravindrarajah and B A Hidayat *MATEC Web Conf.* **195** 01005

[11] Prabowo A Y N 2018 *Kapasitas Lentur Balok Beton Gradasi* (Universitas Negeri Malang)

[12] Margaretha D I 2019 *Pengaruh Konfigurasi Mutu Beton Pada Perilaku Lentur Balok Beton Gradasi* (Universitas Negeri Malang)

[13] Pratama M M A, Abifurrizal S, Novitasari Y, Damayanthi W, Puspitasari P and Permanasari A A 2021 Potential use of graded concrete as structural elements of multi-storey buildings *J. Phys. Conf. Ser.* **1833** 012040

[14] Pratama M M A, Arifanda W, Karyadi K, Nindyawati N, Sulaksitaningrum R and Prayuda H 2019 Numerical and experimental investigation on the shear resistance of functionally graded concrete (FGC) beams *IOP Conf. Ser. Mater. Sci. Eng.* **669** 012055

[15] Pratama M M A, Umniati B S, Nindyawati N, Margaretha D I, Suhud R K, Puspitasari P and Permanasari A A 2021 Effect of concrete strength configuration on the structural behaviour of reinforced graded concrete beams *J. Phys. Conf. Ser.* **1808** 012017

[16] Pratama M M A, Vertian T, Umniati B S and Yoh W H 2019 Flexural behaviour of the functionally graded concrete beams using two-layers and three-layers configuration *IOP Conf. Ser. Mater. Sci. Eng.* **669** 012054

[17] Pratama M M A, Suhud R K, Puspitasari P, Kusuma F I and Putra A B N R 2019 Finite element analysis of the bending moment-curvature of the double-layered graded concrete beam *IOP Conf. Ser. Mater. Sci. Eng.* **494** 012064

[18] Pratama M M A 2021 ANALISIS DAKTILITAS BALOK BETON GRADASI DENGAN DISPARITAS MUTU BETON SERAT TEKAN DAN SERAT TARIK | Pratama | BANGUNAN *Bangunan*

[19] Prayuda H, Pratama M M A, Novitasari Y, Damayanthi W, Arifurrizal S, Puspitasari P and

Permanasari A A 2021 Seismic performance of low-rise building using graded concrete as flexural elements *IOP Conf. Ser. Mater. Sci. Eng.* **1144** 012014

[20] Badan Standardisasi Nasional 2019 SNI 2847-2019 *Persyaratan Bet. Strukt. untuk bangunan gedung dan penjelasan*

[21] Sabelli R, Sabol T and W. 2011 NEHRP Seismic Design Technical Brief No. 5 - Seismic Design of Composite Steel Deck and Concrete-filled Diaphragms: A Guide for Practicing Engineers

Model prediction and scenario of urban land use and land cover changes for sustainable spatial planning in Lhokseumawe, Aceh, Indonesia

A Achmad[1], N Fadhly[2], A Deli[3], I Ramli[4] and R Hadi[5]

[1]Architecture and Planning Department, Universitas Syiah Kuala, Banda Aceh, Indonesia
[2]Civil Engineering Department, Universitas Syiah Kuala, Banda Aceh, Indonesia
[3]Agribusiness Department, Universitas Syiah Kuala, Banda Aceh, Indonesia
[4]Agricultural Engineering Department, Universitas Syiah Kuala, Banda Aceh, Indonesia
[5]Aceh Regional Settlement Infrastructure Center, Banda Aceh, Indonesia

ashfa.achmad@unsyiah.ac.id

Abstract. The purpose of this research is to find the best-fit scenario to support spatial planning, which can optimize the ecological function of urban areas. This study was conducted in Lhokseumawe city, a city which is the center of national activities. The city is also known as an industrial city and special economic zone. Prediction of future growth for this city needs to be done to anticipate the environmental damage that will occur. This prediction uses 2018 land use/cover (LUC) map, to get a prediction of the LUC pattern until 2032. The first stage is to create a raster map for the four variables reviewed in this research, namely a) distance to activity center (DAC), b) distance to road (DRD), c) distance to river (DRV), and d) reserve area (RAR) which are made in the form of a raster layer. The next stage is a scenario model simulation. There are two prediction scenarios assessed in this research to choose the best-fit scenario for 2032, business as usual (BAU) and reserve area interest (RAI). The results showed that the scenario should be chosen is the (RAI) scenario.

1. Introduction

Urbanization is a complex process which can lead to changes in economic, social, and cultural aspects [1]. In addition, significant changes can also occur in ecological and environmental aspects [2]. The urban area is a place where the population is concentrated, as part of the urbanization process. More than 50% of the world's population lives in urban areas, and it is estimated that it will reach 69.6% in 2050 [3]. The occurrence of urbanization is also marked by changes in the regional and urban landscape, namely land use/cover (LUC), which in the long term and uncontrolled can result in detrimental to the ecological function. Recent studies show that changes in LUC also have an effect on the emergence of an urban heat island, local climate change, and reduce the value of ecosystem services [4–6].

Lhokseumawe City, a center of national activities dominated by settlements, offices and industry, was chosen as the object of the study because of its high urbanization and unique typology of industrial cities [7]. Also, this city is a Special Economic Area (SEZ), which was established by the government of the Republic of Indonesia. This research was conducted to obtain the scenario best-fit for spatial planning [8], in an effort to control urban growth sustainable.

Determination of scenarios and future LUC models can assist in the management of urban expansion, change and the development of alternative plans before the transformation of land into protected areas [9][10]. Therefore, simulating future LUC growth scenarios can assist policymakers in evaluating alternative development schemes, and form the basis for urban planning policy recommendations for sustainable urban development.

Cellular Automata (CA) which is integrated with the Analytic Hierarchy Process (AHP) is used as a simulator in this study [8], in an effort to gain researchers insights regarding changes in LUC in Lhokseumawe. Until now, no simulation based on future growth scenarios for Lhokseumawe has been reported, and this research is one of the few studies that uses driving and inhibiting factors to analyze and compare based on the simulation result scenario [10–12]. In this paper, we present a scenario-based urban simulation of Lhokseumawe. The aim is to understand and predict the dynamics of urban growth in Lhokseumawe with the six LUC categories, and to assess the best-fit scenario, for sustainable spatial planning.

2. Materials and methods

2.1. Study location

Located in Lhokseumawe City, which is also a Special Economic Zone. The city which has an area of 181 km^2 is located at 4°-5° North Latitude and 96°-97° East Longitude. The mean surface elevation is 13 m above sea level. The boundaries of Lhokseumawe City, to the north are bordered by the Malacca Strait, to the south with Kuta Makmur District (North Aceh), to the East with the Syamtalira Bayu District (North Aceh), and to the West with Dewantara District (North Aceh) (Figure 1).

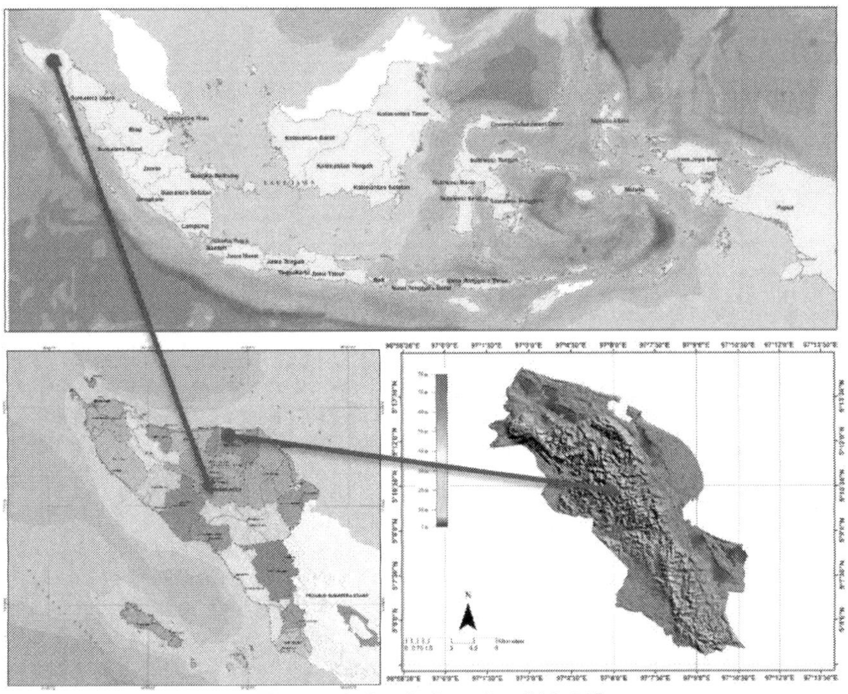

Figure 1. Study location [13-14]

The economy of Lhokseumawe city leads to the residential, lodging, restaurant and industrial sectors [15]. This sector is increasing from year to year. The level of demand for lodging in this city is also fairly high, because it is a transit city between Medan and Banda Aceh. In addition, public and private employees who work in Lhokseumawe City often look for lodging while on assignment, considering of they come from outside Lhokseumawe City. Based on the results of the Geological research, in this city area there are excavation type C such as limestone, landfills and sand/ gravel besides that, there are also natural resources in the form of natural gas. Gas processing result in the form of *condensate is* also utilized by the Aromatix Factory which was built in 1998 and other large companies such as fertilizer factories [7].

2.2. The data used
This study uses data for land use/cover 2018, which has been classified from the 2018 Lansat satellite imagery, obtained from [16] LUC is divided into six categories, namely 1. built-up area; 2. vegetation; 3. agriculture; 4. bare land; 5. Wetland, and 6. water body. Each LUC map has a kappa coefficient of 0.86 LUC maps for 2018 can be seen in Figure 2.

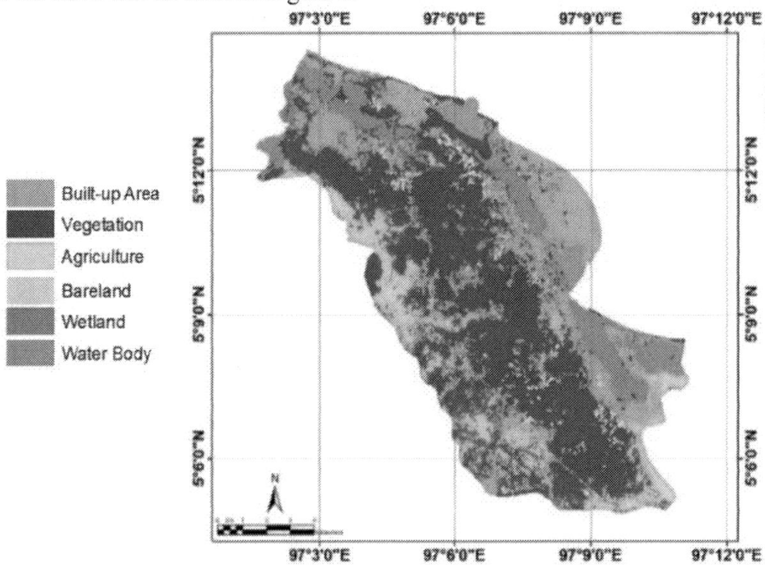

Figure. 2. LUC 2018 [14]

2.3. Methods
The simulation model for predicting land use change in Lhokseumawe City from 2018 to 2032 was carried out using software LanduseSim 2.3.1 based on cellular automata (CA) [8,18,19]. Simulations using this software were also carried out by previous researchers with good results [20]. The Predictions incorporate land use or land use Lhokseumawe in 2018 as the initial land use, factors (proximity to roads, proximity to the river, and the proximity to the center of city services) and inhibiting factors (reverse area plan), the fuzzy sets value of the growth potential of land, a neighborhood filter 3x3, and progression transition map.

The first stage is to create a raster map for the four variables reviewed in this research, namely a) distance to activity center (DAC), b) distance to road (DRD), c) distance to river (DRV), and d) reserve area (RAR). RAR is an inhibiting factor, while other variables are driving factors. DAC, DRD, and DRV variables were created using the Euclidean distance tool in ArcGIS 10.1, using data on field conditions. Activity center data, apart from referring to spatial plans, is also adjusted to existing conditions. The RAR raster variable is created by converting the protected area vector data that has been defined in the spatial plan into raster form.

The next stage is a scenario model simulation. Both scenarios are predictions of growth that will occur in 2032, which is the end of the validity period of the Lhokseumawe City spatial plan. Two scenarios are compared in this study. These two scenarios, apart from being based on previous studies, are also the results of in-depth interviews with several experts related to spatial planning, as follows:

a. Business as usual (BAU) [18]; This scenario is based solely on the trend of changes that occur in the coming years, where the 2018 LUC is used as the basis for determining future predictions, using only the driving variables of DAC, DRD, and DRD.
b. Reserve area of interest (RAI); This scenario is based on efforts to provide strong protection in areas that are dominated by vegetation. In addition to the driving variable, an inhibiting variable is also applied, namely RAR.

The driving and inhibiting factor accessibility raster maps that have been converted into ASCII format in the previous stage still use the ESRI system in the simulation stage in the software LanduseSim, so that the data is converted again into raster format (.tif) with the LanduseSim system, but the difference is that there is no real value yet on this raster. Weights are processed using analysis Analytical Hierarchy Process (AHP) to obtain weight with expert considerations, namely proximity to roads (0.40), proximity to rivers (0.20), and proximity to city service centers (0.40).

Next is to form the Neighborhood Filter (NF), which is a neighbor calculation process that works on a grid system. Basically, the neighborhood system that is most commonly used is the neighborhood system with a 3x3 filter, because the system is able to accommodate the changing ability of each grid so that it has the most optimal value.

3. Results and discussion

3.1. Results

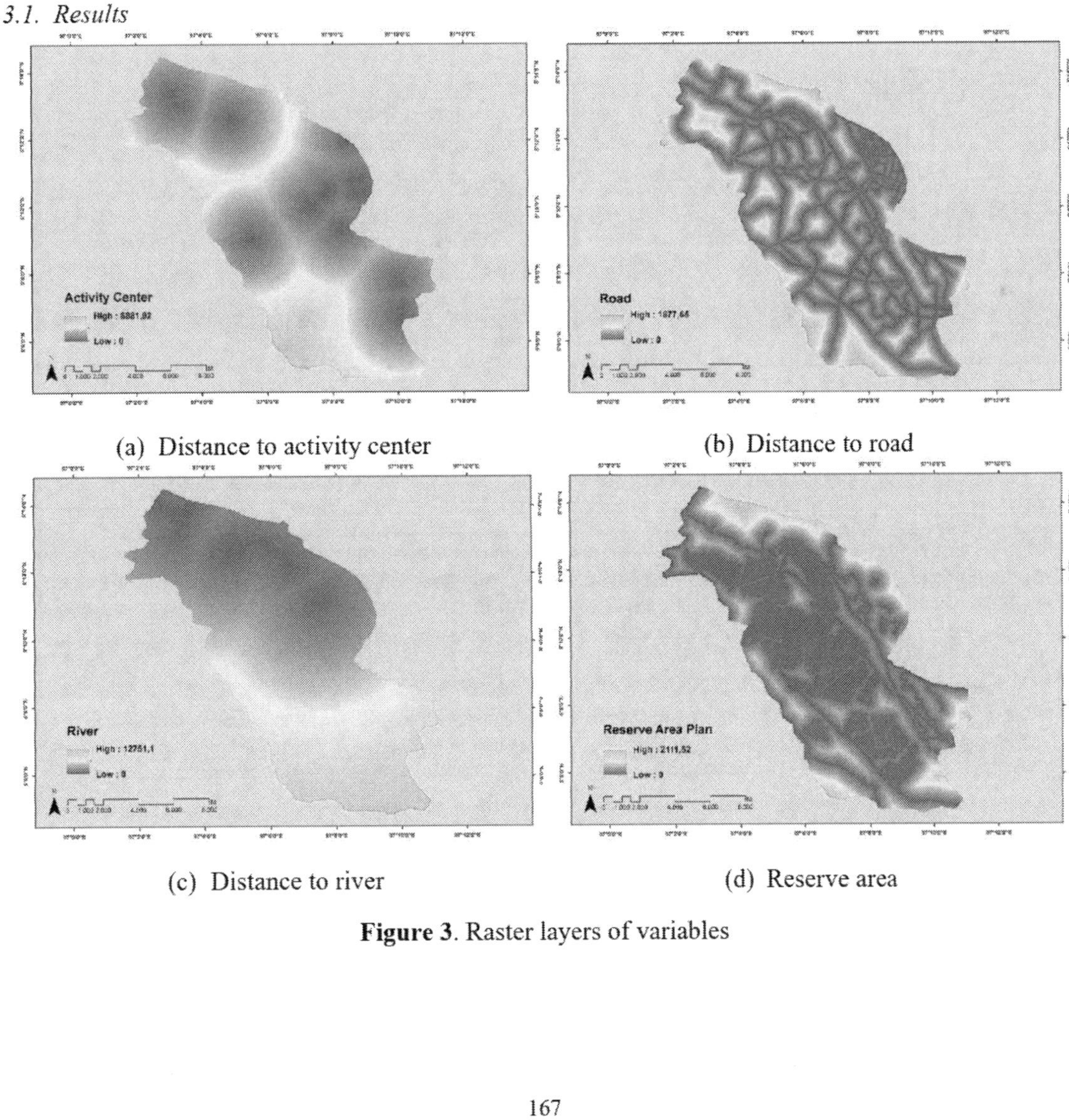

(a) Distance to activity center (b) Distance to road

(c) Distance to river (d) Reserve area

Figure 3. Raster layers of variables

IOP Conf. Series: Earth and Environmental Science **847** (2021) 012022 doi:10.1088/1755-1315/847/1/012022

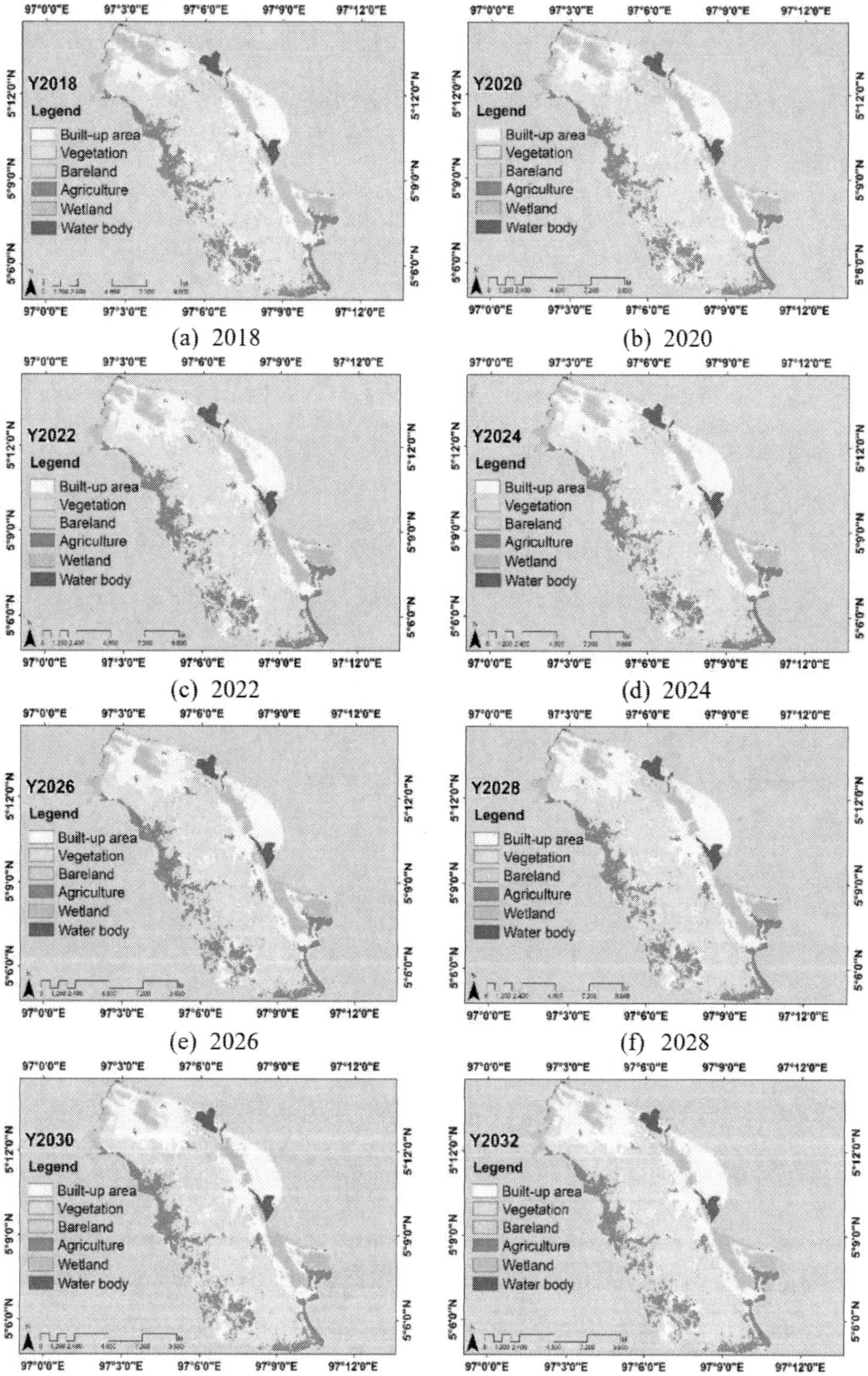

(a) 2018 (b) 2020

(c) 2022 (d) 2024

(e) 2026 (f) 2028

(g) 2030 (h) 2032

Figure 4. LUC maps using the BAU scenario

IOP Conf. Series: Earth and Environmental Science **847** (2021) 012022 doi:10.1088/1755-1315/847/1/012022

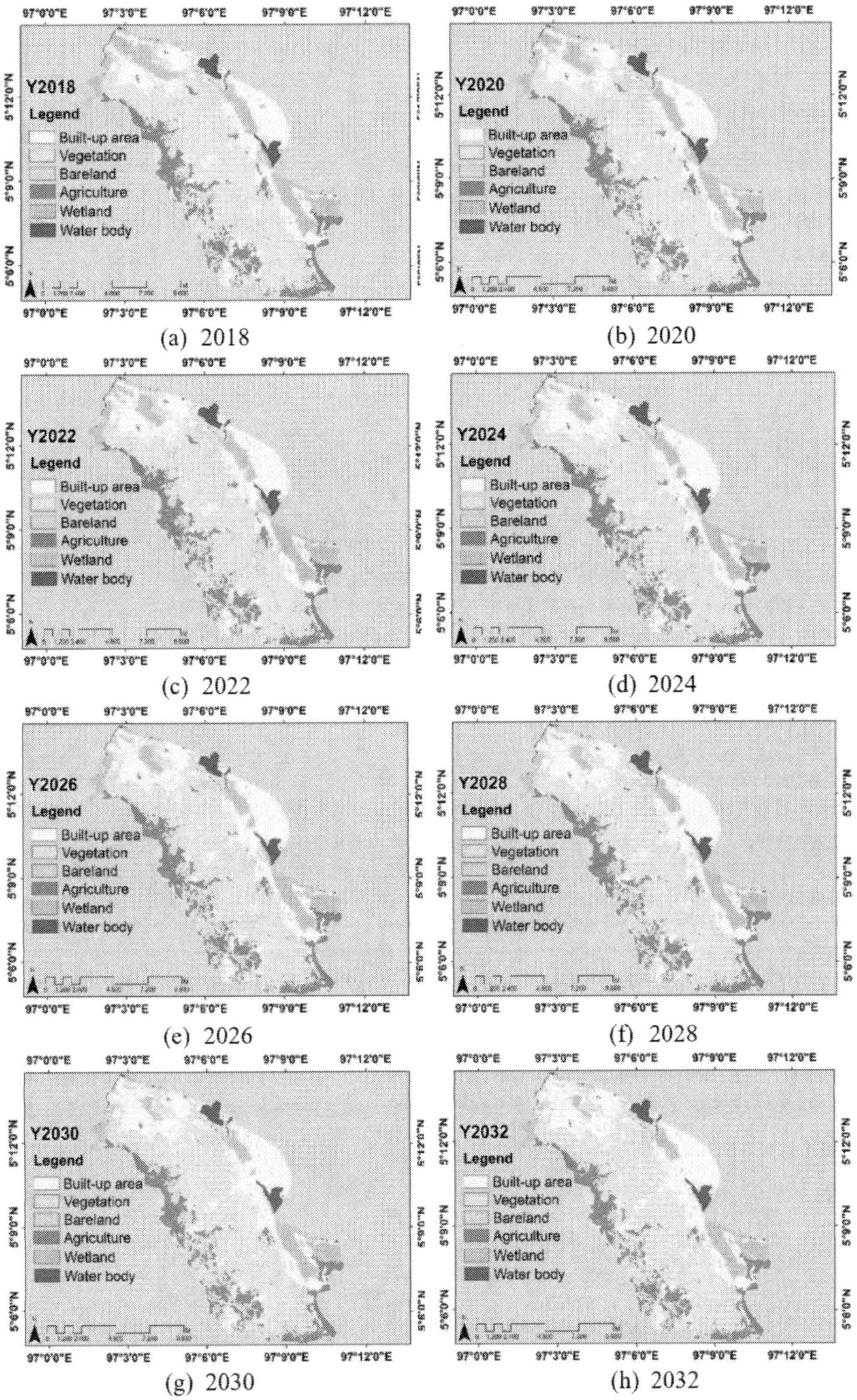

(a) 2018 (b) 2020 (c) 2022 (d) 2024 (e) 2026 (f) 2028 (g) 2030 (h) 2032

Figure 5. LUC maps using the RAI scenario

There are four raster maps produced, which are the variables in this study, namely DAC, DRD, DRV, and RAR. RAR is an inhibiting factor, while other variables are driving factors. The results of making this raster can be seen in Fig. 3.1. (a) is a raster visualization of the distance to activity center variable calculated in meters. The different colors in the image indicate the distance of each pixel to the city center, in case the locations of the activity centers in the Lhokseumawe City area. This variable is a continuous variable.

Figure 3 (a) is a raster visualization of the distance variable to locations that are centers of economic activity, which are calculated in meters. The different colors in the image show the distance of each pixel to the closest centers of economic activity. Determination of the center points of economic activity is based on Qanun no. 4 of 2009. This variable is a variable continuous.

Figure 3 (b) is a raster visualization of the DRD variable, both arterial and collector roads, which are calculated in meters. The different colors in the figure show the distance of each pixel to the nearest road line. This variable is a variable continuous. Figure 3.1 (c) is a raster visualization of the variable distance to the river, which is calculated in meters. The rivers chosen are large rivers that divide the city of Lhokseumawe. The different colors in the figure show the distance of each pixel to the nearest river line. This variable is a variable continuous.

Whereas in Figure 3.1 (d), you can see a raster visualization of the tightly controlled area variable, namely RAR. This distance is also calculated in meters. The different colors in the image indicate the distance of each pixel to the shoreline and this variable is a variable continuous.

The simulation results using BAU and RAI scenarios in 2032, produce a LUC map as can be seen in Figure 4 and Figure 5. A comparison of the two scenarios can also be seen in Figure 6 and Figure 7, which describes the quantitative. The built-up area has the same increase, in both scenarios in the period 2018 to 2032. The percentage of increase that occurs is 19.55%.

In the vegetation category, there are different predictions of widespread change. In the BAU scenario, vegetation experiences a wide decline each year, which at the end of the predicted year, 2032, is only 46.25%. In the RAI scenario, a vegetation is used as a reserve area variable, so that there is no change or decrease in the area again, still 48.85% of the total area of the city. Likewise with the water body, which in the RAI scenario is included in the reserve area, so that it does not experience a change in area, where the area remains as in 2018, which is 3.023 ha.

The bare land category also experienced broad declines in both scenarios. In the 2018-2032 period, the broad decline in the BAU scenario reached 11.67%, while in the RAI scenario a drastic reduction in area reached 25.95%. The area of bare land in each scenario is 1.81% and 1.51% of the total area of Lhokseumawe City.

In the agriculture category, the simulation results also show the differences between the two assessed scenarios. In the BAU scenario, the agriculture area decreased by 1.27%, while in the RAI scenario, the area of this category decreased by 5.92%. The decrease in area also occurred in the wetland category, both in the BAU and RAI scenarios. The BAU scenario simulation shows that the wetland category decreased by 18.33% and 35.06% in the 2018–2032-time frame.

3.2. Discussion

Based on the two prediction scenarios that have been analyzed, it shows that the city has experienced a change in the LUC pattern in the 2018-2032 period. The growth of the built-up area continues. This happened because the population was still increasing, which also triggered an increase in buildings to accommodate additional community activities. The BAU scenario shows the future urban growth will run without any control [9,22], which is only triggered by driving factors, namely DAC, DRD, and DRV. The RAI scenario shows the growth of the future city, apart from the driving factor, also includes an element of control, where there is an area that is used as a reserve area.

Under the BAU scenario, many vegetation will disappear in the future and urban areas will occupy 29.19% of the area in 2032 [9,23]. Using the RAI scenario, by maintaining a reserve area, vegetation can be maintained up to 3,580 ha. In the BAU scenario, the built-up area develops using vegetation land, while in the RAI scenario, the built-up area uses bare land and wet land in its development. In the water body category, each scenario gave similar results. In fact, the use of this BAU scenario can maintain a water body area of up to 72 ha. Under the BAU scenario, the LUC agriculture category can be saved as much as 803 ha compared to the simulation results in the RAI scenario. The same condition is also experienced by LUC in the wet land category. The BAU scenario can save the wetland category by up to 26.19% compared to the RAI scenario.

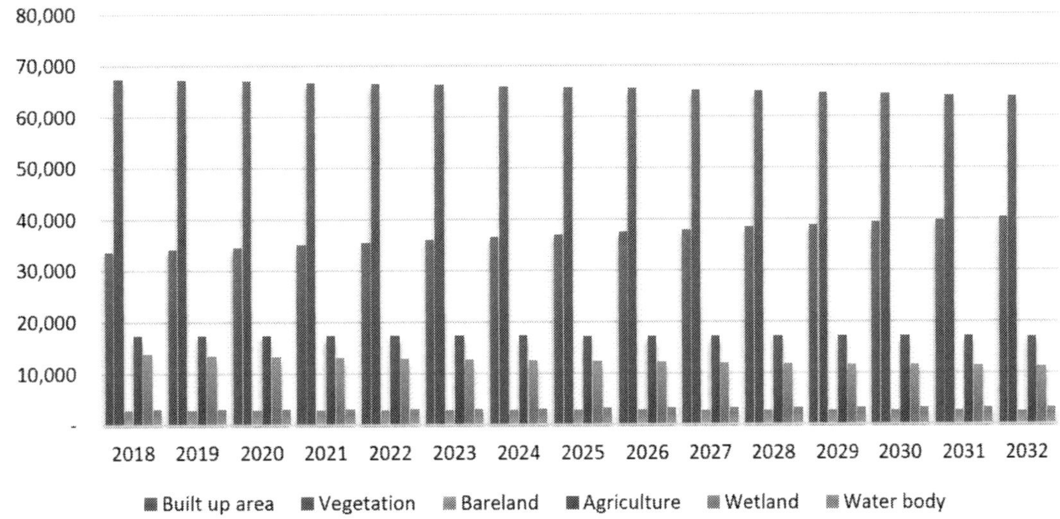

Figure 6. Area of each LUC category in the BAU scenario

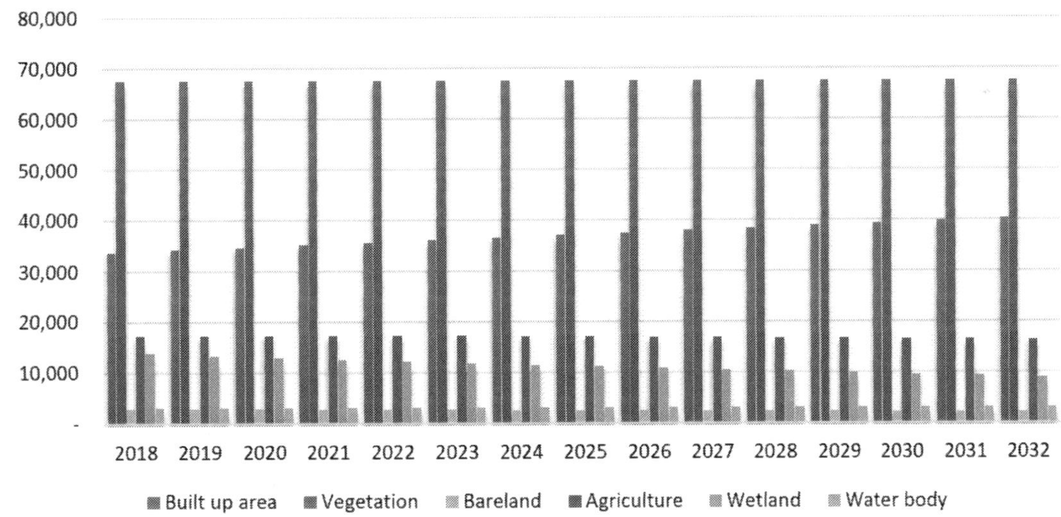

Figure 7. The area of each LUC category in the BAU

In each scenario, there are land categories that decrease and increase in size. The built-up area category has significantly increased, while the vegetation category has decreased. This is caused, among other things, by the increase in population and activities. If it is based on the value of ecosystem services [24,25], the greatest value is on forests, which in this study are categorized as vegetation. Even though wetland can be saved in the BAU scenario, using the principle of ecosystem service value (ESV), the

RAI scenario is considered the best-fit scenario, because forests in this study are categorized as vegetation are designated as the largest ESV [26].

The results of this scenario prediction will be able to guide and support urban planners to make sound policies to achieve sustainable development. The RAI scenario also directly protects limited land resources and promotes sustainable development, by increasing ESV [27]. This study provides a design and methodology in implementing predictive scenarios and presents an approach to analyzing future urban growth patterns with the aim of producing a sustainable spatial plan, and determining the best-fit scenario as a direction and regulation [28,29]. Further quantitative analysis, considering both ESV and other methods, as well as comparisons with other scenarios, needs to be continuously studied.

4. Conclusion

The use of CA with its transition rules allows for future urban growth scenarios to be generated, which in this study predicts to be carried out until 2032. Assessment of land use change in the coming year is carried out in two scenarios, namely BAU and RAI. The BAU scenario eliminates a lot of vegetation land categories, while under the RAI scenario, it removes a lot of wet land and bare land. Because the largest ESV is in the vegetation category, in order to achieve a sustainable spatial plan, the RAI scenarios are selected best-fit. The use of Geographic Information System (GIS) and Remote Sensing (RS) in this study is effective in producing predictive models for LUC growth scenarios, which assesses the LUC dynamics and associated LUC changes driving factors.

Acknowledgement

The authors would like to thank Universitas Syiah Kuala for supporting this research.

References

[1] Shafizadeh Moghadam H and Helbich M 2013 Spatiotemporal urbanization processes in the megacity of Mumbai, India: A Markov chains-cellular automata urban growth model **40** 140–9
[2] Estoque R C 2013 Spatial Analysis of Ecosystem Service Value **13**
[3] Mendoza-Ponce A, Corona-Núñez R, Kraxner F, Leduc S and Patrizio P 2018 Identifying effects of land use cover changes and climate change on terrestrial ecosystems and carbon stocks in Mexico. **53** 12–23
[4] Achmad A, Irwansyah M, Nizamuddin N and Ramli I 2019 Land Use and Cover Changes and Their Implications **145** 1–7
[5] Achmad A, Zainuddin and Muftiadi M 2019 The relationship between land surface temperature and water index in the urban area of a tropical city **365**
[6] Achmad A, Ramli I and Irwansyah M 2020 The impacts of land use and cover changes on ecosystem services value in urban highland areas **447**
[7] Muhyi A 2016 Mekanisme perkembangan fisik kota dan arah perkembangan fisik kota untuk mengetahui proses perkembangan **8** 71–8
[8] Vaz E de N, Nijkamp P, Painho M and Caetano M 2012 A multi-scenario forecast of urban change: A study on urban growth in the Algarve **104** 201–11
[9] Wu X, Hu Y, He H, Xi F and Bu R 2010 Study on forecast scenarios for simulation of future urban growth in Shenyang City based on SLEUTH. **13** 32–9
[10] Feng H H, Liu H P and Lü Y 2012 Scenario Prediction and Analysis of Urban Growth Using SLEUTH Model **22** 206–16
[11] Liu Y, He Q, Tan R, Liu Y and Yin C 2016 Modeling different urban growth patterns based on the evolution of urban form: A case study from Huangpi, Central China **66** 109–18
[12] Schulp C J E, Levers C, Kuemmerle T, Tieskens K F and Verburg P H 2019 Mapping and modelling past and future land use change in Europe's cultural landscapes **80** 332–44
[13] Geospasial untuk Negeri | (ina-sdi.or.id)
[14] DEMNAS and Batimetri indonesia.go.id

[15] Achmad A, Fadhly N and Deli A 2020 Land use/cover changes and socio-economic development on ecosystem services value of the Aceh Special Economic Zone (LPPM USK)
[16] Earth Explorer, *USGS*
[17] Achmad A, Fadhly N and Deli A 2020 Impact of land use/cover changes and socio-economic development on ecosystem services value of Aceh Special Economic Zone
[18] Barreira-González P, Gómez-Delgado M and Aguilera-Benavente F 2015 From raster to vector cellular automata models: A new approach to simulate urban growth with the help of graph theory **54** 119–31
[19] Arsanjani J J, Helbich M, Kainz W and Boloorani A D 2012 Integration of logistic regression, Markov chain and cellular automata models to simulate urban expansion **21** 265–75
[20] Burhan I M, Achmad A and Rizkiya P 2020 Forecasting The Land-Use Change of Urban Coastal Area in Banda Aceh and Its Impact on Urban Sustainability Using Landusesim Cellular Automata Simulation Model **9** 120–31
[21] Zhang Q, Ban Y, Liu J and Hu Y 2011 Simulation and analysis of urban growth scenarios for the Greater Shanghai Area, China. **35** 126–39
[22] Wang Y, Li X, Zhang Q, Li J and Zhou X 2018 Projections of future land use changes: Multiple scenarios-based impacts analysis on ecosystem services for Wuhan city, China. **94** 430–45
[23] Islam M S and Ahmed R 1970 Land Use Change Prediction In Dhaka City Using Gis Aided Markov Chain Modeling. **6** 81–9
[24] Costanza R and Kubiszewski I 2012 The authorship structure of "ecosystem services" as a transdisciplinary field of scholarship **1** 16–25
[25] Estoque R C and Murayama Y 2012 Examining the potential impact of land use/cover changes on the ecosystem services of Baguio city, the Philippines: A scenario-based analysis **35** 316–26
[26] Costanza R, D'Arge R, De Groot R, Farber S, Grasso M, Hannon B, Limburg K, Naeem S, O'Neill R V., Paruelo J, Raskin R G, Sutton P and Van Den Belt M 1997 The value of the world's ecosystem services and natural capital **387** 253–60
[27] Wang C, Wang Y, Wang R and Zheng P 2018 Modeling and evaluating land-use/land-cover change for urban planning and sustainability: A case study of Dongying city, China **172** 1529–34
[28] Feng H and Liu H 2012 Scenario prediction of urban growth based on the SLEUTH model **117 AISC** 945–52
[29] Vermeiren K, Van Rompaey A, Loopmans M, Serwajja E and Mukwaya P 2012 Urban growth of Kampala, Uganda: Pattern analysis and scenario development **106** 199–206

Application of overlay method in interpreting of traffic noise distribution in land use

I Lakawa[1], S Sufrianto[1] and J Jusrin[2]

[1] Civil Engineering Dept, Faculty of Engineering, University of Sulawesi Tenggara, Indonesia
[2] Student of Civil Engineering Dept, Faculty of Engineering, University of Sulawesi Tenggara

ironelakawa@gmail.com

Abstract. Overlay studies with geographic information systems (GIS) are usually only to describe the condition of land elevation points, and various situations in land use. The purpose of this study was to analyze the intensity of noise in various types of land use and visualization of noise distribution maps. The research method uses a mathematical and overlay approach. The results showed that the average traffic flow was 1931 veh/h with a composition of motorcycle 61%, light vehicle 35%, and heavy vehicle 4%. Traffic noise level on Tuesday is 75.2 dB, while on Sunday it is 75.5 dB. There is no significant difference in noise intensity on Tuesday and Sunday at Ahmad Yani, MT. Haryono, Sao-Sao, and Abunawas Street. On Tuesday the highest noise level occurs in front of the education area while on Sunday in front of the shopping area. The intensity of noise distribution that reaches the highest land use occurs in the shopping area, followed by the office area, and the lowest in the school area. This is due to the increase in distance from the roadside.

1. Introduction

Traffic noise is one of the strategic issues that has become a transportation problem in urban areas due to the high growth of motorized vehicles. At a certain intensity, traffic noise can have a negative impact on human health and comfort in settlements and road users themselves. The impact is in the form of psychological disorders that tend to lead to stress and hearing loss. Therefore, prevention and control of environmental pollution should be carried out to meet the provisions of environmental quality standards. Where, the main source of noise in big cities comes from the noise of motorized vehicles whose number continues to increase every year [1]. In addition, noise pollution is an environmental problem in industrialized and developed countries. Industrial noise is a source of noise pollution that interferes with the daily activities of workers [2].

Previous studies conducted by Wedagama (2012) at bypass I Gusti Ngurah Rai and Sunset Road have identified factors that affect traffic noise levels using the variables of vehicle volume, speed, traffic density, and observation distance. The results showed that the motorcycles volume had the highest effect on the noise level, which was around 26.7% [3]. But theoretically, the intensity of noise on the highway is not determined by traffic factors but also other factors, namely physical characteristics and environmental conditions road, and building density.

The density of buildings, whether in form of shops or housing, can influence the level of traffic noise on the highway. This is caused by the building uses quite hard material so that the sound that hits the walls bounces back. Likewise, air has mass, fills the empty space above the earth and used by sound to propagate. But that doesn't always happen. The highest traffic flow occurs when the density of truck vehicles is optimal. After this value, the flow starts to decrease, even with a lot of passing traffic queues form and speed drops [4].

Kendari City in the last five years, the vehicle growth is quite high an average of 19% per year. From the results of initial observations at several points on arterial and collector roads, it was identified that the noise level had reached a value of 72 dB, where this value had exceeded the noise threshold applicable in Indonesia, namely 55 dB for school areas, 65 dB for office areas, and 70 dB. for trading area [5].

All this time, information about traffic noise distribution on the highway is only based on numerical data by displaying trend data, Where the information is still poorly understood by most people. In addition, exposure to noise in various land uses using the overlay method, especially in Kendari City, has never been done. So this study will elaborate on the traffic noise variable as vector data that interprets the digitized noise elevation point, namely fluctuations in traffic noise exposure.

2. Materials and methods

Highways are a source of noise in urban areas. This is due to the large use of motorized vehicles compared to other types of vehicles, both two-wheeled, four-wheeled, and those with more than four wheels. Heavy vehicles (truck, bus) and passenger cars are the main sources of noise on the highway, although motorcycles are more dominant in composition. The variability of the noise level is determined by vehicles volume of 75.2% for the roads of arterial and 77% for collector roads [6]. Heavy vehicles with a composition of 4% greatly affect noise on arterial roads. While on the collector road with a composition of 3% it does not really affect noise fluctuations. Noise is more dominantly influenced by motorcycles and light vehicles [7].

Noise has a bad effect on living things. The effects of noise include hearing loss, sleep disturbances, communication disorders, decreased efficiency and work disturbances. The effort that has been made is to develop a fuzzy-based model to see the effect of traffic noise in the city of Agartala, India. From the study shows that there is a correlation between traffic noise and everyday disturbances for residents on the roadside [8]. High-income households react more to noise exposure than low-income households [9]. Based on research, urban residents in the United States are at high risk of exposed to traffic noise. However, other research is needed to obtain information about the level of noise exposure. Therefore, it is very difficult to avoid traffic noise, especially in the big cities of the world [10].

Noise mapping that examines noise pollution due to heterogeneous traffic conditions and the effect of honking vehicles parked on the roadside in the Surat city (India). The mapping takes multiple points and the propagation distance to the receiver. The variables used are vehicle volume, speed, noise level, road geometry, illegal parking, and horn sound. The results showed that vehicle volume, speed, road geometry, and noise did not have a significant relationship with noise due to heterogeneous traffic conditions. Therefore, an integrated noise prediction model with multiple linear regression approach was developed for heterogeneous traffic conditions in India [11].

Development of a road traffic noise model with a graph theory approach, where the variables used include vehicle speed, traffic volume, road width, number of heavy vehicles, and number of horn sounds. The interaction of the selected variables is entered in the form of a permanent noise function in the form of a matrix. The model obtained gives satisfactory results, but the results are rather large. This method proves useful for predicting the noise level in the future [12].

The TNM 2.5 model is used to predict traffic volume and its distribution on Vakilabad highway, because it gives the same rating for areas with low or high noise, and produces a Leq value that is almost the same as the recorded noise value [13].

2.1 Noise level

Noise is unwanted sound from an activity at a certain level and time which can cause disturbance to human health and environmental comfort. So that no matter how small the sound is heard, if it is not desired, it is called noise. The noise level standard is the maximum level of noise that is allowed to be discharged into the environment from a business or activity so as not to cause harm to human health and environmental comfort. The standard noise level in Indonesia which is intended for the area/environment of activities can be seen in Table 1.

Table 1. Noise level quality standard

Description	Noise Level (dB)
a. Appropriation of region	
1. Housing and Settlements	55
2. Trade and Services	70
3. Office and Commerce	65
4. Green open space	50
5. Industry	70
6. Government and Public Facilities	60
7. Recreation	70
8. Specifically:	
- airport *	
- Railway station *	
- harbor	70
- Cultural heritage	60
b. Surrounding Activity	
1. Hospital or the like	55
2. School or the like	55
3. Worship place or the like	55

Equivalent continuous noise level (L_{eq}) is the value of the noise level that varies (fluctuating) over a certain time which is equivalent to the steady noise level measured at the same time interval. The equivalent noise level is a model used to express the average sound pressure level in a certain time interval, which is obtained from the results of field measurements using an SLM (Sound Level Meter) measuring instrument. Equation for calculating noise from motor vehicles (highways):

$$L_{eq} = L_{50} + 0{,}43\ (L_1 - L_{50}) \quad (1)$$

Where, L_{eq}: equivalent noise level (dB), L_{50}: noise indicator number 50% (dB), and L_1: noise indicator number 1% (dB).

2.2 Overlay

Overlays are an important step in GIS analysis. The overlay overlays a digital map on another digital map and its attributes and produces a combined map of the two that has the attribute information of both maps. Overlay is the process of combining data from different layers. In simple terms overlay is referred to as a visual display that requires more than one layer to be physically combined.

There are two techniques used to overlay maps in GIS, namely union and intersect. If it is analogous to the language of Mathematics, then union is a union, intersect is a slice Geographic Information System can be broken down into several subsystems, namely input data, output data, management data, data manipulation and analysis [14].

Noise is a factor that must be considered in urban planning. Noise maps can be generated when data has been recorded, including map numbering, road facilities database, and traffic volume data [15]. Road traffic noise model using a graph theory approach with road traffic subsystem parameters. raffic subsystem variables include vehicle speed, vehicle volume, road width, horn sound. The model is quite satisfactory although the results are somewhat higher. This method is very useful for predicting noise levels [16].

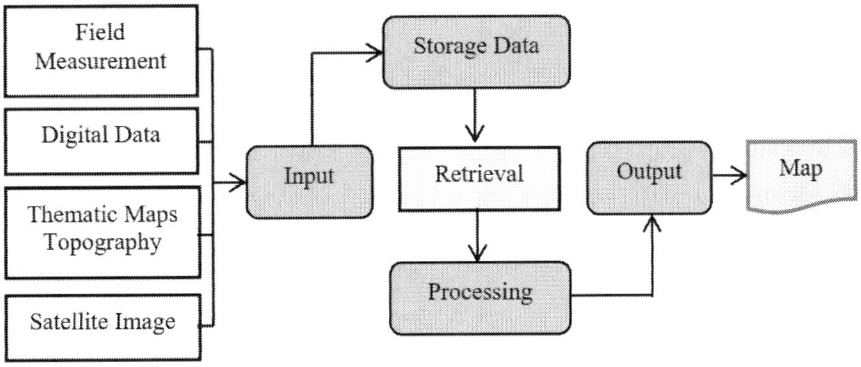

Figure 1. GIS sub system

2.3 Traffic survey and noise measurement

The traffic survey was conducted for 2 (two) days, namely routine days represented by Tuesdays and holidays represented by Sundays. Data sampling was carried out for 10 minutes at peak hours in the morning, afternoon, and evening along with the noise survey time at each predetermined sample point. Noise measurements were carried out simultaneously with a traffic survey for 2 (two) days. Data sampling was carried out for 10 minutes, with a reading time of 5 seconds. The SLM microphone is placed at a distance of 1 m from the edge of the pavement. The noise survey was carried out in conjunction with the traffic survey. The scheme for the placement of measuring instruments and equipment in noise measurement can be seen in Figure 2.

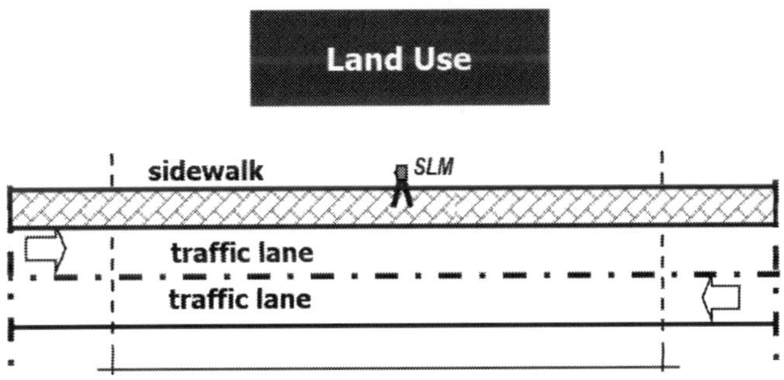

Figure 2. Noise measurement scheme

3. Results and discussion

Ring road A. Yani, MT. Haryono, Sao-Sao, and Abunawas generally have a road type of 2/2 UD and 2/2 D with an average pavement width of 6 - 11 meters, a shoulder width of 2 meters and the type of land use in front of the road is a shopping area, offices, and schools. The distance from the building to the edge of the road pavement is between 9.3m - 91.2m. The traffic referred to in this study is the volume of motorized vehicles. Calculation based on the number of vehicles passing the research location during the observation time which has been converted in units of passenger cars per hour. Based on the results of traffic volume analysis, it can be visualized in the form of a bar chart as shown in Figure 3.

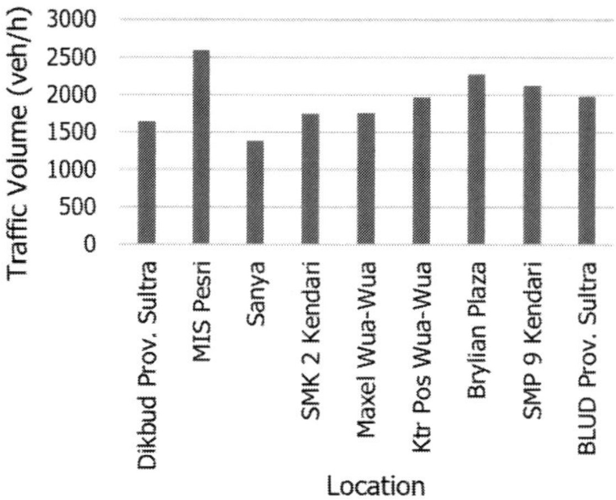

Figure 3. Tuesday Traffic Volume

In Figure 3 the traffic volume on the Ahmad Yani, MT. Haryono, Sao-Sao, and Abunawas street on Tuesday with an average of 1941 veh/h. Where the highest volume occurs in front of the MIS Pesri school, which is 2592 veh/h. This happens because the school is currently one of the main educational goals that are in demand by the community. Meanwhile, the lowest traffic volume occurred in front of the Southeast Sulawesi Province Education and Culture office, which was 1644 veh/h.

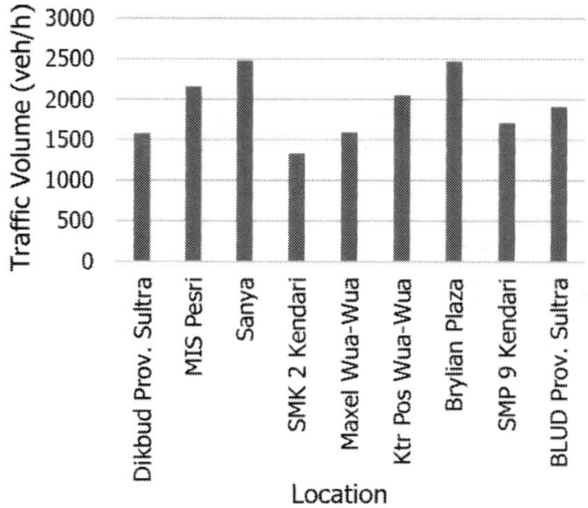

Figure 4. Sunday Traffic Volume

In Figure 4 the traffic volume on the Ahmad Yani, MT. Haryono, Sao-Sao, and Abunawas street on Sunday with an average of 1921 veh/h. Where the highest volume occurs in front of the Sanya shopping area, which is 2478 veh/h. This is because the shopping area is currently one of the main travel destinations that people are interested in. Meanwhile, the lowest traffic volume occurred in front of SMK 2 Kendari, which was 1332 veh/h. In general, it can be seen that the highest volume of motorized vehicle traffic on Tuesdays generally occurs in front of the education area, while on Sundays it occurs in the shopping area.

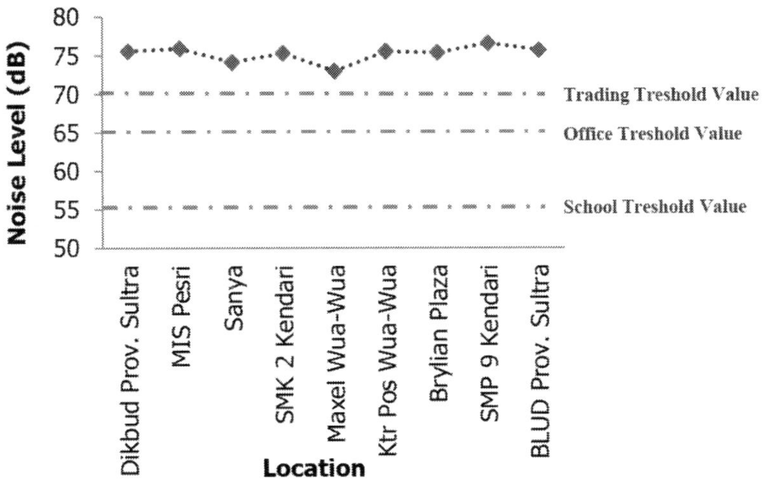

Figure 5. Tuesday Traffic Noise

In Figure 5 it can be seen that the level of traffic noise on the Ahmad Yani ring road, MT. Haryono, Sao-Sao, and Abunawas on Tuesday exceeded the environmental quality standards according to the Decree of the state minister for the Environment of the Republic of Indonesia Number: 48/MENLH/11/1996, both for the designation of school areas, office areas, and trade areas. Where the average noise level on the roadside is 75.2 dB.

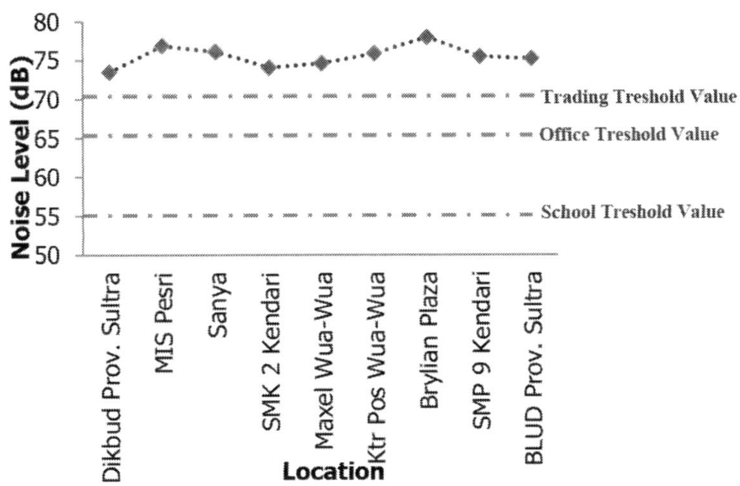

Figure 6. Sunday Traffic Noise

In Figure 6 it can be seen that the level of traffic noise on the Ahmad Yani ring road, MT. Haryono, Sao-Sao, and Abunawas on Sunday had environmental quality standards in accordance with the Decree of the state minister for the Environment of the Republic of Indonesia, both for the designation of school areas, office areas, and trade areas. Where the average noise level on the roadside is 75.5 dB.

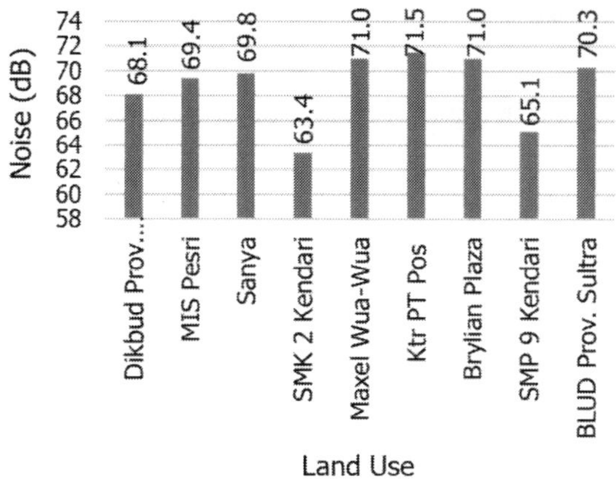

Figure 7. Noise Intensity in Land Use

Figure 7 shows that the noise intensity that reaches the highest land use occurs in the PT. POS Wua-Wua is 71.5 dB, while the lowest is in the SMKN 2 Kendari area, which is 63.4 dB. Overall, the shopping area ranks first in the highest noise area, followed by the office area, and the lowest in the school area.

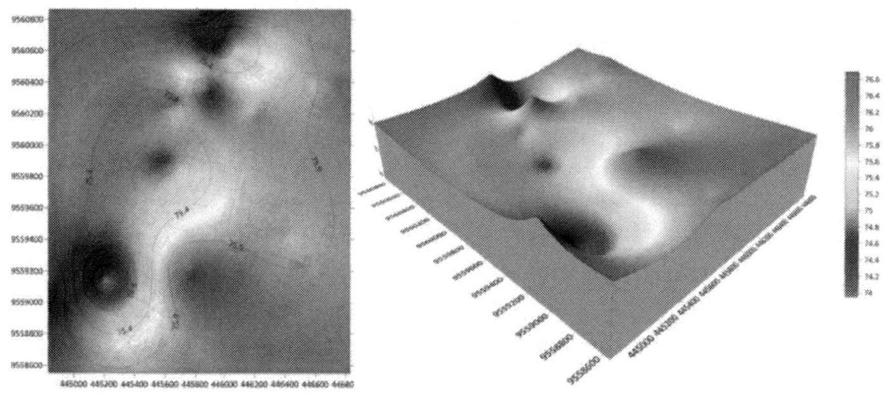

Figure 8. Roadside Noise

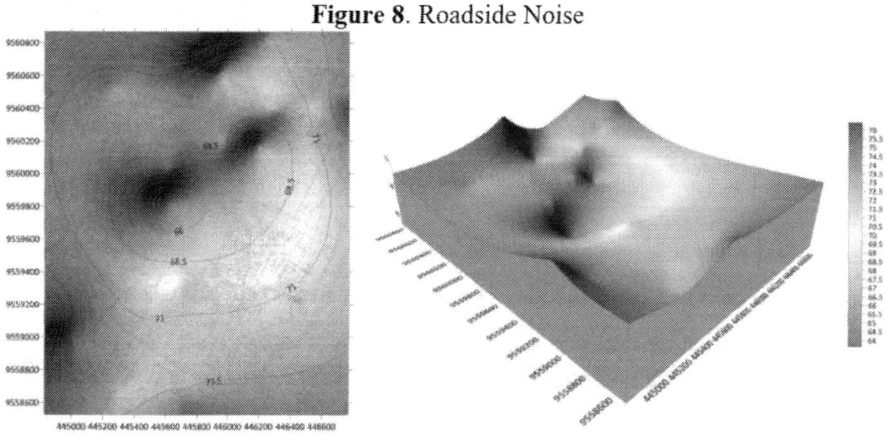

Figure 9. Land Use Noise

Through the overlay method, it can be seen that in Figures 8 and 9 there is a difference in the distribution of noise. Where the red color indicates a high noise value, while the green color indicates a low noise value. So that the arrangement of color gradations is adjusted to the noise level. Figure 8 shows a map of the noise distribution based on the value of noise level on the roadside, while Figure 9 shows a map of the noise distribution based on the value of noise level on land use. By using the overlay method in mapping noise, people will easily understand the distribution of noise in the surrounding environment.

The results of the analysis show that the farther the receiver is from the noise source, the smaller the noise intensity will be. Therefore, the determination of exposure to noise levels is seen based on the intensity that occurs in the area and the length of exposure time. The difference in noise reduction in the nine areas is not only due to the increase in distance and the amount of roadside noise level, it is also influenced by the type of soil surface from the roadside to land use, as well as the density and density of green arrangements/tree plants on the roadside.

4. Conclusion

The average traffic flow is 1931 vehicles/hour with a composition of motorcycles 61%, light vehicles 35%, and heavy vehicles 4%. The traffic noise level on Tuesday is 75.2 dB, while on Sunday it is 75.5 dB. There is no significant difference in noise intensity on Tuesday and Sunday. On Tuesday the highest noise level occurs in front of the education area while on Sunday in front of the shopping area. The intensity of noise distribution that reaches the highest land use occurs in the shopping area, followed by the office area, and the lowest in the school area. This is due to the increase in distance from the roadside.

Acknowledgements

The author expresses his gratitude to Prof. Dr. Ir. Andi Bahrun, M.Sc.Agric as the Chancellor of the University of Sulawesi Tenggara, who has provided support so that this research can be carried out. We also convey the same words to the academic community of the Engineering Faculty, University of Sulawesi Tenggara, as well as all parties who helped in this research.

References

[1] Shvetsov A 2021 Aspects of Traffic Noise Reduction **39** 27-29.
[2] Al-Dosky B H M 2014 Noise level and annoyance of Industrial factories in Duhok city **8(5)** 01-08.
[3] Wedagama D M P 2012 The Influence of Local Traffic on Noise Level (Case Study: By Pass Ngurah Rai and Sunset Road, Bali) **12(1)** 24-31.
[4] Rodrigues F, Resende C, Nassi C D and Kahn S 2010 Traffic Engineering Indicators Analysis as Explanatory Variables of Traffic Noise.
[5] Decree of the State Minister of Environment of the Republic of Indonesia Number: 48/MENLH/11/1996 Concerning Noise Level Standards.
[6] Mediastika C E 2005 *Akustika Bangunan:Prinsip-Prinsip dan Penerapannya di Indonesia.*
[7] Lakawa I, Samang L, Selintung M and Hustim M 2015 Relationship Models of Traffic Volume Vs Noise Level in Arterial and Collector Roads **5(9)** 5463-5466.
[8] Bhattacharya D and Pal D 2012 A Study of Road Traffic Noise Annoyance on Daily Life in Agartala City Using Fuzzy Expert System and Multiple Regression Analysis **2(5)** 1-7.
[9] Kuehnel N, Ziemke D and Moeckel R 2021 Traffic noise feedback in agent-based Integrated Land-Use/Transport Models **14(1)** 325-344.
[10] Kim M, Chang S I, Seong J C, Holt J B, Park T H, Ko J H and Croft J B 2012 Road Traffic Noise Annoyance, Sleep Disturbance, and Public Health Implications **43(4)** 353–360.
[11] Sonaviya D R and Tandel B N 2020 Integrated road trafic noise mapping in urban Indian context **7** 99-113.
[12] Gilani T A and Mir M S 2021 Modelling Road Traffic Noise Under Heterogeneous Traffic

IOP Conf. Series: Earth and Environmental Science **847** (2021) 012023 doi:10.1088/1755-1315/847/1/012023

Conditions Using The Graph-Theoretic Approach.
[13] Khayami E, Mohammadi M, Bahadori M S, Hasani F and Ghorbani A 2019 Prediction of Highway Noise Pollution Level by Model FHWA -TNM (Case Study: Vakilabad Highway in Mashhad-Iran) **7(1)** 1395-1402.
[14] Prahasta E 2009 *Konsep-Konsep Dasar Sistem Informasi Geografis* Informatika, Bandung.
[15] Jeong J H, Che Din N, Otsuru T and Kim H C 2010 An Application of a Noise Maps for Construction and Road Traffic Noise in Korea **5(7)** 1063-1073.
[16] Golmohammadi R, Abbaspour M, Nassiri P and Mahjub H 2009 A compact model for predicting road traffic noise **6(3)** 181-186.

Angle of inclination and radiation intensity variation effects on the flat plate solar collector's performance using graphene oxide (GO)-water nanofluid

A A Permanasari[1,2], S A Putra[1], P Puspitasari[1,2], S Sukarni[1,2], S N A Zaine[3], and W Wahyunengsih[4]

[1]Department of Mechanical Engineering, Faculty of Engineering, Universitas Negeri Malang, Jl. Semarang No 5, Malang 65145, Indonesia
[2]Centre of Advanced Materials for Renewable Energy (CAMRY), Universitas Negeri Malang, Jl. Semarang No 5, Malang 65145, Indonesia
[3]Fundamental and Applied Sciences Department, Universiti Teknologi PETRONAS, 32610, Bandar Seri Iskandar, Perak, Malaysia
[4]Development of Islamic Society Department, State Islamic University Syarif Hidayatullah Jakarta, Indonesia

avita.ayu.ft@um.ac.id

Abstract. Flat Plate Solar Collector is a special heat exchanger instrument that transforms solar radiation energy into heat energy. Various attempts have been completed to improve the flat plate solar collector's performance. One of them is by developing advanced heat transfer fluid with better thermophysical properties than base fluid. Further, the flat plate solar collector's inclination angle and radiation intensity were investigated to identify the most optimal solar radiation angle of incidence. This study discusses the use of a 0.006% graphene oxide-water nanoparticle toward the nanofluid's thermophysical properties, sufficient radiation intensity of flat plate solar collector performance, as well as the inclination angle of flat plate solar collector toward the collector's performance. The used radiation intensity variations were 415 W/m^2, 515 W/m^2 and 615 W/m^2. Meanwhile, the radiation used in this study was from a sun simulator 750 W/m2. In each radiation intensity variation, the inclination angle test of 30°, 45°, and 60° for the flat plate solar collector were carried out. This study was expected to discover the best performance of graphene oxide-water nanofluid usage, collector's inclination angle variation, and radiation intensity variation.

1. The first section in your paper

Solar energy is one of the renewable energy alternatives with a massive capacity that remains minimally utilized. Solar energy usage in the industrial and daily fields is classified into photovoltaic, photothermal, and photochemistry [1]. As a tropical country passed by the sun the whole year, Indonesia has the opportunity to apply photothermal daily; thereby, solar-powered water heater or solar collector is one of the alternative energy usage massively investigated by experts, recently [2].

Solar collector operates using solar radiation heat energy to infuriate the liquid fluid within the collector pipe and is divided into three categories of flat plate collector, pipe collector, and HPSC [3]. Flat plate solar collector is the most commonly used water heater system classified as a low-temperature device (<100° C). The flat plate solar collector's performance depends on the solar radiation absorbance, which later transferred into the working fluid within the solar collector pipe in the form of energy [4]. The low heat conductivity on conventional heat transfer fluid, such as water, oil, and ethylene glycol, severely limits solar collector performance; thus, an advanced heat transfer fluid with a higher conductivity is developed in this research [5]. The small solid particle suspension within the fluid is an innovative means that escalates the fluid's thermal conductivity. Various types of powder, such as metallic, non-metal, and polyamic particles, can be added to the solution to expand the solar collector's heat transfer efficiency [6].

Nanofluid and nanoparticle were first introduced by Choi in 1995 to enhance the thermal conductivity by dispersing the 1-100nm metal particle. They were also used to improve the base fluid's heat transfer. Researchers investigating the effect of nanoparticle concentration report the intensified thermal conductivity and the increased nanoparticle concentration [8]. Several nanoparticles, such as CuO, MgO, MWCNT, TiO_2, Al_2O_3, and ZnO, have been investigated using basic water fluid. ZnO nanoparticle encounters 22% widen thermal conductivity from an additional 0.005% ZnO concentration to the base fluid [9]. The significant increase of fluid's thermal conductivity and heat transfer coefficient are known as exceptional nanofluid's physical effect. The solid phase metal possesses a higher thermal conductivity than conventional fluid [10]. Therefore, metal nanoparticles suspended within the fluid is a solution that better improve the thermal conductivity than the pure fluid [4].
According to the ideas mentioned above, this research is required to escalate the solar collector's performance. This study uses 0.006% concentration Graphene Oxide (GO) nanofluid base fluid combined with distilled water. This study was carried out by various radiation intensity and inclination angles on a 30cm2 area dimension solar collector instrument. The effects of radiation intensity and inclination angle variation on the flat plate solar collector's heat transfer using Graphene Oxide (GO)-water nanofluid was analyzed.

2. Method

2.1. Flat plate solar collector

This experimental study used a laboratory-scale flat plate solar collector. The flat plate solar collector's scheme is illustrated in Figure. 1. This study was carried out using base fluid and 0.006% graphene oxide-water nanofluid, as well as radiation intensity and inclination angle variation of flat plate solar collector.

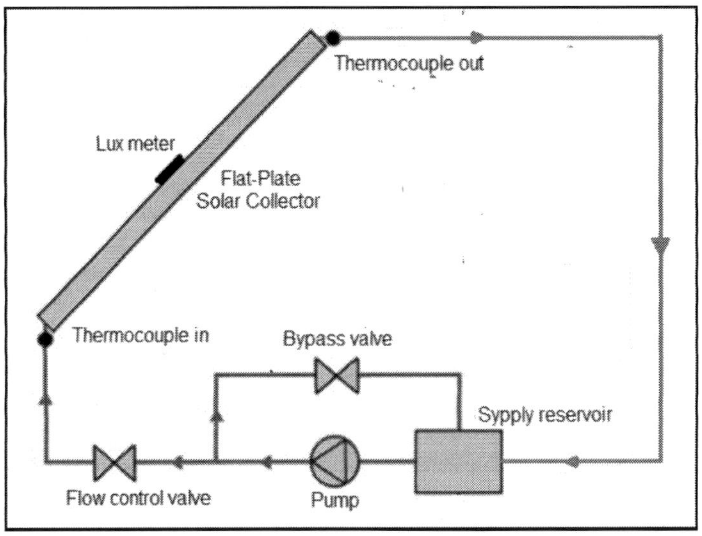

Figure 1. Flat plate solar collector's scheme

Table 1. Specification of Flat Plate Solar Collector

Specification	Dimension	Unit
Length	330	mm
Wide	320	mm
High	35	mm
Collector's Total Area	105600	mm^2
Absorber's Area	93310	mm^2
Protective Glass' Thickness	5	mm
Tube Diameter	9.94	mm
Protective Glass' Emissivity (τ)	0.95	unitless
Plate Absorber's Emissivity (α)	0.88	unitless
Inclination Angle	30; 45; 60	degree (°)

2.2. *Graphene oxide*
The nanoparticle used in this study was graphene oxide. It is a thick layer of the single atom from a highly stable carbon [11]. Graphene has a hydrophobic property, but with the O function group's presence, graphene oxide can be dispersed into the water [12]. Its hydrophilic feature made graphene oxide becomes a suitable material for nanofluid. The specification of graphene oxide is presented in Table 2.

Table 2. Specification of graphene oxide

Specification	Description
Purity	99.9wt%
Thickness	0.6-1.0 nm
Sheet diameter	0.5-5 μm
Number of layers	1-2
Specific surface area	1000-1217m^2/g
Appearance	Black or brown flake

2.3. *Base fluid*
The base fluid used as nanofluid in this research was distilled water (H_2O). The thermophysical properties of distilled water are shown in Table 3.

Table 3. Thermophysical properties of water

Characteristic	Value	Unit
Density	976,512	kg/m^3
Specific Heat	4178,10	J/kg. °C
Thermal Conductivity	0,557	W/m.°C
Viscosity	0,00081	kg/m.s

2.4. *Preparation of Nanofluid*
The nanofluid was constructed by blending nanoparticle into the base fluid. This research used 500ml of liquid and 0.006% of graphene oxide as the nanoparticle. The nanofluid production was divided into two phases of magnetic stirrer and ultrasonic sonication. During the magnetic stirrer process, the base fluid and nanoparticle were mixed and stirred using a magnetic stirrer for 30 minutes. After that, the nanofluid was sonicated on ultrasonic sonication for 2 hours with a 100% amplitude to minimize the nanoparticle aggregation in the second stage. Consequently, the 0.006% concentration nanofluid was obtained.

Figure 2. The 0.006% graphene oxide-water nanofluid

2.5. Experimental Preparation

This study involved two stages of the test. The first stage was the thermophysical properties test of base fluid and 0.006% graphene-oxide-water nanofluid; meanwhile, the second test was the flat plate solar collector test. On the first test, the density, viscosity, specific heat, and thermal conductivity were tested. The density test was carried out using a digital scale and measuring cup to reveal the used fluid's mass per unit volume. Simultaneously, the viscosity test was completed using viscometer NDJ-8S DRV to identify the viscosity of both base fluid and 0.006% graphene oxide-water nanofluid. Meanwhile, the thermal conductivity and specific heat tests were conducted using a KD2 pro thermal analyzer. After the thermophysical properties of the base fluid and graphene oxide nanofluid were obtained, the flat plate solar collector test was performed with those fluids acting as its working fluid. The test was carried out indoor using a sun simulator 750 watt, for 60 minutes, with 1 l/m discharge, inclination angles of 30°, 45°, and 60°, and radiation intensity of 415 W/m^2, 515W/m^2, and 615 W/m^2.

2.6. Flat Plate Solar Collector's Performance

The thermal performance of flat plate solar collectors can be tested by concluding the instantaneous efficiency at a different value from the occurring radiation, the surrounding temperature, and the temperature of heat transfer fluid inlet. The instantaneous efficiency is defined as the ratio of attained energy used for solar energy passed on the plate absorber, as shown by Formula (1).

$$\eta_c = \frac{Q_u}{A_c . I} \qquad (1)$$

in which η_c represents the collector efficiency, Q_u is the energy obtained by the collector, Ac is the collector's surface area, and I is the radiation intensity attained by the solar collector.

The heat energy absorbed by the collector is defined as the heat energy extraction rate from the working fluid flowing to the flat plate solar collector (Q_u) through Formula (2) [12] [13].

$$Q_u = m \, . C_p \, (T_o - T_i) = \rho \, . v . A \, . C_p \, (T_o - T_i) \qquad (2)$$

where m represents the occurring mass flow rate, C_p is the specific heat, and $T_o - T_i$ is the temperature difference between input and output.

The highest heat loss coefficient is the function of the various parameter, including the absorber instrument's temperature, temperature ambient, wind velocity, the emissivity of the absorber and cover glass plate, the inclination angle, and so forth [14]. All of the loss can be estimated using Formula (3-5).

$$U_L = U_b + U_t + U_e \tag{3}$$

$$U_t = \left(\frac{N}{\frac{C}{Tpm}\left[\frac{Tpm-Ta}{(N+f)}\right]^e} + \frac{1}{h_\omega} \right)^{-1} + \frac{\sigma(T_{pm}+T_a)}{\frac{1}{\varepsilon_p+0{,}0059Nh_\omega} + \frac{2N+f-1+0{,}133\varepsilon_p}{\varepsilon_g} - N} \tag{4}$$

$$U_b = \frac{k}{L} \tag{5}$$

$$U_e = \frac{k}{L} x \frac{A_{edge}}{A_{collector}} \tag{6}$$

in which U_b is the heat loss coefficient below the collector, U_t is the heat loss coefficient above the collector, and U_e is the heat loss coefficient at the side of the collector.

Heat removal factor portrays the ratio of actual beneficial energy increase and the attainment of useful energy if all collector is located at fluid inlet temperature [15]. The F_R value as the heat removal factor can be calculated using Formula (3).

$$F_R = \frac{m \cdot C_p\,(T_o - T_i)}{A_c[I(\tau.\alpha) - U_L(T_i - T_a)]} \tag{7}$$

where F_R is the heat removal factor, τ is the plate's emissivity, α is the glass emissivity, and U_L is the overall heat loss coefficient.

Efficiency is the ratio of profitable energy addressed to the occurring energy at the collector aperture. The value of efficiency is dominated by parameters, such as transmittance product and absorption glaze of plate's absorption rate, global radiation passed on the collector intensity, water line temperature, and the surrounding air temperature [13] [14]. The collector efficiency can be estimated using Formula (8).

$$\eta_c = \frac{A_c F_R[I(\tau.\alpha) - U_L(T_i - T_a)]}{A_c . I(\tau.\alpha)} \tag{8}$$

with A_c representing the collector's surface area, F_R is the heat removal factor, I is the radiation intensity, τ is the plate's emissivity, α is glass emissivity, U_L is the total heat loss, T_i is the inlet temperature, and T_a is the ambient temperature.

3. Results and discussion

3.1. Thermophysical Properties

The flat plate solar collector working parameter requires the utilized working fluid's thermophysical properties. Table 4 presents the results of the thermophysical properties test for the working fluid.

Table 4. Results of Thermophysical Properties Test

Working Fluid	Density (kg/m3)	Specific heat (J/kg. K)	Dynamic Viscosity ($x10^{-3}$ kg/m.s)	Thermal Conductivity (W/m.K)
Water	976.512	4178.10	0.81	0.557
GO-Water 0,006%	987.620	4131.36	0.89	0.593

Table 4 demonstrates the expansion of nanofluid's density prior to the addition of graphene oxide. Nanofluid density increases following the rise of nanoparticle volume fraction [16] [17]. In most cases, nanofluid presents a more substantial density increase than base fluid. The more considerable density of the nanoparticle used improves the nanofluid's density. It happens due to the graphene oxide nanoparticle density is higher than the base fluid; therefore, it escalates the nanofluid's density. A decrease of specific heat after the 0.006% graphene oxide nanoparticle was added due to graphene oxide has a smaller specific heat than the basic water fluid. Thus, the addition of nanoparticle into the base fluid provokes a linear reduction toward the nanofluid volume fraction with the declined used nanoparticle's specific heat. The reduced nanofluid's specific heat decreases the energy required to increase the substance's temperature in that nanofluid [18] [19].

The data also shows that the escalated viscosity due to the addition of graphene oxide nanoparticles. It is caused by graphene oxide nanoparticle inclusion that leads to the collision among nanofluid molecules due to the rise of nanoparticles in the fluid. Thus, the addition of 0.006% graphene oxide carries a more considerable effect of the increasing GO-water nanofluid's viscosity value. Researchers also verify that the increase of nanoparticle volume enhances the nanofluid's viscosity level [20] [21]. The fluid's thermal conductivity escalates as the graphene oxide was added. It happens due to the graphene oxide's thermal conductivity is higher than the pure water base fluid. Therefore, the addition of 0.006% graphene fluid brings a linear effect toward the GO-water nanofluid's thermal conductivity. Many researchers have proven that adding a small number of solid particles increases the thermal conductivity significantly, linear to the volume concentration [21] [23].

3.2. Flat Plate Solar Collector's Performance

The highest heat loss coefficient is the function of various parameters, involving the absorber instrument's temperature, temperature ambient, wind velocity, the absorber's emissivity, cover glass plate, the inclination angle, and so forth [14] [13]. In which UL represents the total loss of the flat plate solar collector (FPSC) that is equal to the above, below, and side loss. Table 3 presents the calculation of the overall heat loss coefficient.

Table 5. The Estimation Results of Overall Heat Loss Coefficient

Intensity (W/m^2)	β (°)	U_t (W/m^2.K)	U_b (W/m^2.K)	U_e (W/m^2.K)	U_L (W/m^2.K)
	30	4.871025	0.9	0.179	5.95
415	45	4.796483	0.9	0.179	5.87
	60	4.687639	0.9	0.179	5.76
	30	5.015327	0.9	0.179	6.09
515	45	4.940032	0.9	0.179	6.01
	60	4.829892	0.9	0.179	5.90
	30	5.173208	0.9	0.179	6.25
615	45	5.097281	0.9	0.179	6.17
	60	4.986031	0.9	0.179	6.06

Table 5 shows that there is a decrease of overall heat loss coefficient following the increase of flat plate solar collector's inclination angle. This is caused by the Ut value affected by the amount of inclination angle. A bigger inclination angle reduces the top loss coefficient so that the heat obtained by FPSC is more significant, and the UL is also declined following the reduced heat loss [23]. During the FPSC test, the highest UL value of 6.25 W/m^2.K was obtained at the inclination angle of 30° and radiation intensity of 615 W/m^2. In contrast, the lowest UL value of 5.76 W/m^2.K was found at the inclination angle of 60° and 415 W/m^2 intensity.

Table 6 illustrates that the upsurge of Qu that is in line with the rise of occurring radiation intensity. The Qu value is affected by three primary aspects, namely mass flow rate, specific heat, and ΔT between the input and output. The increase of Qu value toward radiation intensity occurs due to the rising temperature that escalates FPSC's surface temperature; that high-temperature difference results in bigger energy being passed [13]. Besides, the addition of graphene oxide-water nanofluid also influences the Qu value. The graphene oxide-water enhances the thermal conductivity so that the heat grows bigger, causing a considerable temperature difference and affecting the Qu value. At the same time, the heat removal factor's rise following the more substantial light intensity. Heat removal value increase is affected by the expanded Qu value following the escalated FPSC radiation intensity. The heat removal factor represents the ratio of acceptable actual heat value and the ideal useful heat value. Additionally, the input temperature also influences the heat removal factor; thus, a higher input fluid temperature means a more considerable heat removal factor value (FR) [23].

Table 6. The Estimation Results of Q_u and FR

Fluid	Inclination Angle (°)	Intensity (W/m²)	Q_u (Watt)	FR
Graphene oxide-water nanofluid	30	415	22.223	0.668
		515	28.118	0.675
		615	34.467	0.687
	45	415	22.902	0.676
		515	28.798	0.692
		615	35.601	0.701
	60	415	23.583	0.716
		515	29.932	0.721
		615	36.508	0.739
Base Fluid (water)	30	415	16.099	0.459
		515	21.314	0.493
		615	26.302	0.506
	45	415	17.232	0.501
		515	22.674	0.533
		615	28.116	0.551
	60	415	17.91306	0.518
		515	22.90151	0.537
		615	28.34345	0.552

Table 7. Results of Flat Plate Solar Collector's Efficiency Estimation

Fluid	Inclination Angle (°)	Intensity (W/m²)	Efficiency (%)
Graphene oxide-water nanofluid 0.006%	30	415	60.65
		515	61.84
		615	63.48
	45	415	62.51
		515	64.11
		615	65.15
	60	415	64.36
		515	65.83
		615	67.24
Base fluid	30	415	43.94
		515	46.88
		615	48.44
	45	415	47.03
		515	49.87
		615	51,78
	60	415	48.89
		515	50.37
		615	52.21

Data in Table 7 indicates that the improved efficiency following the heightened flat plate solar collector's inclination angle. That happens because the heat removal factor and overall heat loss coefficient (UL) values positively affect efficiency. Consequently, a higher FR value means a bigger FPSC's efficiency value due to the increased ratio between the FPSC's acceptable heat and the ideal heat [24] [25]. The broadening FSPC's efficiency value is also linear to the extended light intensity and FPSC's inclination angle due to the reduced overall heat loss coefficient. It is caused by the decreased wasted heat that improves the FPSC's efficiency value [23]. The highest intensity value has been identified at 615 W/m^2 intensity with a 60° inclination angle and 67.24 % of nanofluid. In contrast, the lowest intensity is at 415 W/m^2 intensity with a 30° inclination angle and a base fluid of 43.94 %. The substantial FPSC's value represents its improved performance, faster water heating process, and higher temperature. Therefore, graphene oxide-water nanofluid is beneficial to increase the flat plate solar collector's efficiency.

4. Conclusion

This present research concludes that the addition of 0.006% graphene oxide-water nanofluid increases the nanofluid's thermophysical properties that improve the flat plate solar collector's performance. The most substantial efficiency of using the amount of 0.006% graphene oxide-water (67.24%) is found at a 60° inclination angle and 615 W/m^2 radiation intensity.

References

[1] Ma F and Zhang P 2019 "Performance investigation of the direct absorption solar collector based on phase change slurry," Appl. Therm. Eng. **162** 114244.
[2] Irawan R 2018 "Pembuatan Dan Pengujian Kolektor Surya Pelat Datar (Flat-Plate Collectors) Kondisi Steady Berdasarkan Standar En 12975," **12975** 61.
[3] Dehaj M S and Mohiabadi M Z 2019 "Experimental investigation of heat pipe solar collector using MgO nanofluids," Sol. Energy Mater. Sol. Cells **191** 91–99.
[4] Barone G, Buonomano A, Forzano C and Palombo A 2019 Solar thermal collectors. Elsevier Inc.
[5] Firlianda D A, Permanasari A A, Puspitasari P and Sukarni S 2019 "Heat transfer enhancement using nanofluids (MnFe2O4-ethylene glycol) in mini heat exchanger shell and tube," AIP Conf. Proc. **2120.**
[6] Tong Y, Lee H, Kang W and Cho H 2019 "Energy and exergy comparison of a flat-plate solar collector using water, Al2O3 nanofluid, and CuO nanofluid," Appl. Therm. Eng. **159** 113959.
[7] Choi S U S 1995 "Enhancing thermal conductivity of fluids with nanoparticles," Am. Soc. Mech. Eng. Fluids Eng. Div. FED **231** 99–105.
[8] Islam R and Shabani B 2019 "Prediction of electrical conductivity of TiO2 water and ethylene glycol-based nanofluids for cooling application in low temperature PEM fuel cells," Energy Procedia **160** 550–557.
[9] Shoghl S N, Jamali J and Moraveji M K 2016 "Electrical conductivity, viscosity, and density of different nanofluids: An experimental study," Exp. Therm. Fluid Sci. **74** 339–346.
[10] Puspitasari P, Permanasari A A, Shaharun M S and Tsamroh D I "Heat Transfer Characteristics of NiO Nanofluid in Heat Exchanger."
[11] Schäffel F (2013) The Atomic Structure of Graphene and Its Few-layer Counterparts.
[12] Klevinskis A and Bučinskas V 2012 "Analysis of a Flat-Plate Solar Collector / Plokščiojo Saulės Kolektoriaus Tyrimas," Moksl. - Liet. ateitis **3** 39–43.
[13] Akram N by et al. 2019 "An experimental investigation on the performance of a flat-plate solar collector using eco-friendly treated graphene nanoplatelets – water nanofluids," J. Therm. Anal. Calorim **4**.
[14] William A B and John A D 2013 Wiley: Solar Engineering of Thermal Processes, 4th Edition - John A. Duffie, William A. Beckman.
[15] Montoya-Márquez O and Flores-Prieto J J 2018 "Heat Removal Factor in Flat Plate Solar Collectors :," 1–12.
[16] Pastoriza-Gallego M J, Casanova C, Legido J L and Piñeiro M M (2011) "CuO in water nanofluid: Influence of particle size and polydispersity on volumetric behaviour and viscosity," Fluid Phase Equilib. **300** 188–196.
[17] Mahbubul I M 2019 Thermophysical properties of nanofluids. preparation, characterization, properties and apllication of nanofluid **c**.
[18] Strandberg R and Das D K 2010 "Influence of temperature and properties variation on nanofluids

in building heating," Energy Convers. Manag. **51** 1381–1390.

[19] Choi J and Zhang Y 2012 "Numerical simulation of laminar forced convection heat transfer of Al 2O 3-water nanofluid in a pipe with return bend," Int. J. Therm. Sci. **55** 90–102.

[20] Das S K, Putra N and Roetzel W 2003 "Pool boiling characteristics of nano-fluids," Int. J. Heat Mass Transf. **46** 851–862.

[21] Prasher R, Song D, Wang J and Phelan P 2006 "Measurements of nanofluid viscosity and its implications for thermal applications," Appl. Phys. Lett. **89** 67–70.

[22] Murshed S M S, Leong K C and Yang C 2008 "Investigations of thermal conductivity and viscosity of nanofluids," Int. J. Therm. Sci. **47** 560–568.

[23] Montoya-Marquez O and Flores-Prieto J J 2017 "The effect of the angle of inclination on the efficiency in a medium-temperature flat plate solar collector," Energies **10**.

[24] Verma S K and Arun K T D S C 2017 "Experimental evaluation of flat plate solar collector using nanofluids," Energy Convers. Manag. **103–115** 12.

[25] Permanasari A A, Kuncara B S and Puspitasari P 2019 "Convective heat transfer characteristics of TiO2-EG nanofluid as coolant fluid in heat exchanger Convective Heat Transfer Characteristics of TiO 2 -EG Nanofluid as Coolant Fluid in Heat Exchanger," AIP Conf. Proc. **2120** 05001.

IOP Conf. Series: Earth and Environmental Science **847** (2021) 012025 doi:10.1088/1755-1315/847/1/012025

Optimum operational conditions for mixotrophic microalgae growth and nutrient recovery

C Oz[1], L Fletcher[2] and M A Camargo-Valero[2,3]

[1] Çanakkale Onsekiz Mart University, Faculty of Engineering, Department of Environmental Engineering, 17020, Merkez/Çanakkale, Turkey
[2] BioResource Systems Research Group, School of Civil Engineering, University of Leeds, Leeds, LS2 9JT, UK
[3] Departamento de Ingeniería Química, Universidad Nacional de Colombia, Campus La Nubia, Manizales, Colombia

cigdem.oz@comu.edu.tr

Abstract. Microalgae have been presented as microorganisms with great potential to recover nutrient from wastewater. Mixotrophic cultivation of microalgae in nutrient rich wastewater can help eliminating the deficiencies of both phototrophic and heterotrophic growth by allowing the independent optimisation of respiration and photosynthesis processes. Nutrient control and uptake by mixotrophic microalgae can be achieved either in a single or two-stage process using sequential reactors in a continuous flow system. Therefore, this work aims at studying mixotrophic microalgae growth in a two-stage biological process under continuous flow conditions with biomass recycle to recover nutrients from wastewater, considering the effects of different operational conditions (hydraulic retention time (HRT), cell retention time (CRT) and different nitrogen sources). The optimum operational conditions for algal nutrient uptake were identified to be 48 h HRT and 14 d CRT, using a mix of nitrogen sources (Ammonium-N to Nitrate-N ratio of 1:1) with 40.0% and 93.2% of phosphorus of nitrogen recovery in algal biomass, respectively.

1. Introduction

Microalgae are considered as a promising solution for nutrient control and recovery at wastewater treatment works due to their capacity to uptake nutrients and high algal biomass production with the additional benefit of enhanced biofuel production [1]. Microalgae can be separately cultivated under phototrophic, heterotrophic and mixotrophic conditions, as well as under a combination of them [2-3]. The most common microalgae cultivation mode is phototrophic. Microalgae utilise CO_2 and/or HCO_3^- and light as carbon and energy sources; molecular oxygen (O_2) and new algal cells are produced as a result of the photosynthesis process [4]. Organic carbon is consumed as an energy and carbon source by heterotrophic microalgae under fully dark conditions, using oxygen as a final electron acceptor. Despite the fact that heterotrophic microalgae growth overcomes the requirements for light to support phototrophic cultivation, it also has inherent disadvantages: (i) higher CO_2 emissions; (ii) the risk of bacterial and fungal contamination; and (iii) the need for an organic carbon source and a terminal electron acceptor (oxygen) [5]. Mixotrophic cultivation means that microalgae are able to grow either phototrophically or heterotrophically or both, in which mixotrophic metabolism can occur either simultaneously or separately. Both organic and inorganic carbon are assimilated as carbon sources.

Microalgae release CO_2 by heterotrophic assimilation which is subsequently used during photosynthesis [6]. Light and organic materials are the energy sources for the fixation of inorganic carbon and aerobic respiration, respectively, and therefore microalgae growth does not solely depend on photosynthesis because organic carbon can support cell growth as well [7-8]. Different growth regimes were investigated to identify which of them resulted in higher algal biomass and biodiesel production rates. The results revealed that Chlamydomonas reinhardtii had the greatest growth rate and lipid accumulation under mixotrophic cultivation [9].

Mixotrophic cultivation of microalgae can occur either in a single reactor or in separated, sequential reactors. With separated reactors, microalgae take up nutrients under phototrophic conditions and further nutrient uptake and lipid accumulation can take place in subsequent heterotrophic reactors [2]. Batch cultures have been used for microalgae growth on the industrial scale; however, low biomass productivity, high harvesting costs and uncertain product quality can limit its applicability. In the last few decades, continuous flow systems have caught the attention of the industry due to their ability to maintain growth rates at close to maximum values, reduce harvesting costs and stabilise the characteristics of any final products [10].

In the current literature, there is a limited number of research works investigating the use of continuous flow systems for microalgae cultivation with the simultaneous benefit of nutrient control and recovery via microalgae uptake from wastewaters. In order to fill this gap, the work reported herein is aimed at identifying the optimum operational conditions for a two-stage biological process combining phototrophic and heterotrophic microalgae cultivation, and biomass recirculation under continuous flow conditions. In order to assess the performance of the proposed system, a series of experiments were performed under different operational conditions to test the effects of hydraulic retention time (HRT), cell retention time (CRT) and different nitrogen sources on microalgae growth, nutrient control and recovery.

2. Materials and methods

2.1. Microalgae species and Culture Conditions

A pure culture of *Chlamydomonas reinhardtii* 11/32C was ordered from the Culture Collection of Algae and Protozoa, Scotland (CCAP), and propagated in 500 mL conical flasks containing 300 mL of Bold's Basal Media (BBM) which consists of 250 mg $NaNO_3$ L^{-1} , 75 mg K_2HPO_4 L^{-1}, 175 mg KH_2PO_4 L^{-1}, 25 mg $CaCl_2.2H_2O$ L^{-1}, 75 mg $MgSO_4.7H_2O$ L^{-1}, 25 mg NaCl L^{-1}), 50 mg EDTA L^{-1}, 31 mg KOH L^{-1}, 0.001 mL H_2SO_4, 11.42 mg H_3BO_3 L^{-1}, 8.82 mg $ZnSO_4.7H_2O$ L^{-1}, 1.44 mg $MnCl.4H_2O$ L^{-1}, 0.71 mg MoO_3 L^{-1}, 1.57 mg $CuSO_4.5H_2O$ L^{-1}, 4.98 mg $FeSO_4.7H_2O$ L^{-1} and 0.49 mg $Co(NO_3)_2.6H_2O$ L^{-1}, using a shaking incubator (Infors Multitron). Culture media was autoclaved at 121°C and 1 bar for 15 minutes. Controlled environmental conditions for temperature, photoperiod and light intensity during cultivation (axenic microalgae culture) were set at 25°C, 24 h of light and 40 μE/m2S, respectively.

2.2. Experimental setup for a two-stage biological process

The two-stage biological process setup includes seven main components as presented in the process flow diagram and the picture depicted in Figure 1.a. and b, respectively. The two-stage microalgal process was fed by synthetic wastewater (SWW) including 25 mg NH_4^+-N L^{-1}, 25 mg NO_3^--N L^{-1}, 15 mg PO_4^{3-}-P L^{-1}, 2800 mg $NaHCO_3$ L^{-1}, 567 mg $CH_3COONa.3H_2O$ L^{-1} which are typically found in the effluent of a conventional activated sludge process and the same concentrations of inorganic salts, trace elements and pH conditioners were used as those in the BBM media with the addition of 20 mg L^{-1} of cationic polymer (Zetag 50). The experimental setup for mixotrophic microalgae growth comprises a photobioreator (PBR) and a heterotrophic reactor (HTR) to which is connected in series, having an effective volume: 2.3 L (diameter: 7 cm; height: 61 cm). The effluent from the HTR was transferred by gravity to an Imhoff cone that allows algal biomass sedimentation and recycling to the PBR by pumping, while the supernatant is collected for characterisation. Settled microalgal biomass was recycled into the PBR at a specific rate according to the predefined VSS target concentration (mixed liquor volatile suspended solids – MLVSS). A set amount of settled microalgal biomass (algal sludge) was discarded daily according to the predefined cell retention time.

Water samples were collected every other day from three sampling points including: PBR effluent (2); HTR effluent (3); and, Imhoff cone supernatant (final effluent) (6) (see Figure 1.a). Collected data was processed to assess process performance and monitor biomass growth and nutrient uptake. Microalgae growth was determined using gravimetric analysis for volatile suspended solids (VSS) (SM 2540E). Ammonium and phosphorus were analysed following the methods (SM 4500-NH_3.B and SM 4500-P.B, respectively). All analytical tests were conducted following standard methods reported by [11]. Nitrate was determined by Ion Chromatography. Ion Chromatographer (Metrohm 850 Professional IC) using a with Metrosep A supp 5 column (length: 150 mm, diameter: 4 mm) and an eluent comprising 1 mM NaHCO3 and 3.2 mM Na2CO3. Samples were passed through 0.45 μm Minisart syringe filters and diluted to adjust nitrate concentration to < 30 ppm.

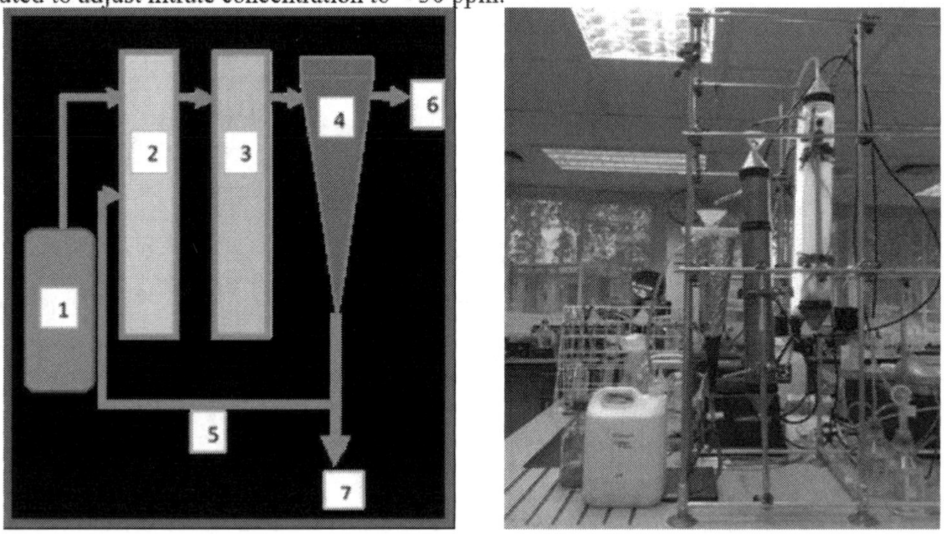

(a) (b)

Figure 1. (a) Process diagram, (b) and picture of the experimental setup of the two-stage biological process.

2.3. Operational Conditions for a Two-Stage Biological Process

The two-stage biological process was operated under different operational conditions to optimise process performance based on operational parameters (hydraulic retention time (HRT) and cell retention time (CRT)) and nitrogen source availability in the SWW. The tested conditions are listed in Table 1.

Table 1. Experimental conditions.

Number of experiments	Variables			
	HRT (hours)	CRT (days)	NH4+-N concentration (mg L-1)	NO3--N concentration (mg L-1)
1	36	14	25	25
2	48	14	25	25
3	72	14	25	25
4	48	7	25	25
5	48	21	25	25
6	48	14	50	-
7	48	14	-	50

Each experiment was operated for approximately one month, which gave the possibility of producing plenty of data to test the stability and performance of the process under the set condition, i.e., an experiment with an HRT of 72 h (3 d) run for 30 days is the equivalent of running 10 replicates in a

IOP Conf. Series: Earth and Environmental Science **847** (2021) 012025 doi:10.1088/1755-1315/847/1/012025

batch reactor with a three-day residence time. Biomass concentrations in each reactor with different operational conditions at steady-state conditions are reported as average ± one standard deviation.

2.4. Calculations

2.4.1. Hydraulic Retention Time (HRT)

HRT is calculated dividing the volume of the reactor by the flowrate feeding the system (equation (1)). Flowrates for operating the two-stage biological process were calculated via equation (1). HRTs tested in the two-stage biological process were 36, 48 and 72 h, which corresponded to feeding flowrates of 2.5, 1.7 and 0.9 L d^{-1}, respectively.

$$\mathrm{HRT} = \frac{V}{Q} \tag{1}$$

Where, HRT is hydraulic retention time (d), V is working volume of each reactor (L), and Q is flowrate of each reactor (L d^{-1}).

2.4.2. Cell retention time (crt)

The system was tested at 7, 14 and 21 d of CRT by carefully controlling the amount of algal biomass daily purged from the system (algal sludge); the corresponding volume of algal sludge purged daily was 660, 330 and 220 mL, respectively. CRTs and the volume of algal sludge purged were calculated via equation (2) and (3), respectively.

$$\mathrm{CRT} = \frac{V\,(Xpbr+Xhtr)}{(Q-Qw)Xe+QwXr} \tag{2}$$

$$Q_w = \frac{Vt}{\mathrm{CRT}} \tag{3}$$

Where; CRT is cell retention time (d), V is working volume of each reactor (L), V_T is total volume of system (L), X_{PBR} is average biomass concentration in PBR (mg L^{-1}), X_{HTR} is average biomass concentration in HTR (mg L^{-1}), Q is flowrate of each reactor (L d^{-1}), Q_w is flowrate of wasted algae (L d^{-1}), Xe is biomass concentration in final effluent (mg L^{-1}), and X_r is biomass concentration in the recycle line (mg L^{-1}). Operational parameters for the two-stage biological process are listed in Table 2.

Table 2. Operational parameters for the two-stage biological process.

Parameter	Value		
VT	4.6 L (2*2.3 L)		
HRT	36 h, 48 h, and 72 h (18 h, 24 h, and 36 h for each reactor)		Equation (1)
Q	2.5 L d-1 1.7 L d-1 0.9 L d-1	36 h 48 h 72h	
CRT	7 d, 14 d and 21 d		
Qw	660 mL d-1 330 mL d-1 220 mL d-1	7 d 14 d 21 d	Equation (3)
Xe	75 mg VSS L-1		MLVSS concentration in PBR and HTR were assumed 1 and 1.5 g L-1, respectively (Equation (2))
Xr	750 mg VSS L-1		

2.5. Statistical analysis

Data processing included statistical analyses for descriptive statistics and variable comparisons; all data were processed using IBM's SPSS Statistics 22 software suite. One-way analysis of variance (ANOVA) was used to evaluate the differences among the treatments. If ANOVA effects were significant, comparisons between the different means were made using *post hoc* least significant differences (LSD).

3. Results and discussion

3.1. The Influence of HRT on Microalgae Growth and Nutrient Recovery

Algal biomass concentrations in the effluent of the PBR, HTR and sedimentation unit (final effluent - FE) as a response to different initial feeding flowrates in the two-stage biological process are presented in Figure 2. It can be clearly seen that a slightly higher biomass concentration was achieved in the PBR than in the HTR for all HRTs tested. Microalgae concentration in the PBR fluctuated between 370 and 680 mg VSS L^{-1}, 330 and 600 mg VSS L^{-1} and 380 and 570 mg VSS L^{-1} at 36 h, 48 h and 72 h HRT, respectively. Biomass concentrations in the HTR varied from 320 to 560 mg VSS L^{-1}, 340 to 560 mg VSS L^{-1} and 370 to 600 mg VSS L^{-1} at 36 h, 48 h and 72 h HRT, respectively. However, it was expected that a higher biomass concentration would be achieved in the heterotrophic reactor due to the fact that heterotrophic microalgae a have higher growth rate in comparison with the growth rate of phototrophic microalgae [12]. This can be attributed to the synthetic wastewater containing low organic carbon concentrations and bacterial contamination under heterotrophic conditions. There was a similar trend in the change of biomass concentration with HRTs under both phototrophic and heterotrophic conditions. The average dry weight concentration of microalgae declined with increasing HRT from 36 h to 48 h. This could be attributed to the reduced nutrient supply at higher HRT due to the decrease in flowrate [13]. A further increment in HRT to 72 h did not result in any significant difference in algal biomass concentration for both the PBR and HTR ($p > 0.05$).

With regard to effluent quality, algal biomass in the final effluent showed different patterns with regard to the HRTs tested. At 48 h HRT, biomass concentration (dry weight) fluctuated between 90 and 160 mg VSS L^{-1}. While there was a constant decrease from 100 to 40 mg VSS L^{-1} at 36 h HRT, a continuous increase was observed from 80 to 150 mg VSS L^{-1} at 72 h HRT. The biomass concentration increased almost double with increasing HRT for 36 h to 48 h HRT; after that, dry weight concentrations of biomass in the FE remained stable at 72 h of HRT. Solid concentrations were quite high, even though a cationic polymer was added to the feeding media to increase the settling ability of the microalgae. Therefore, the additional process was required to improve the settlement quality of the algae and reduce solids concentration in the effluent for the implementation of a two-stage biological process in the wastewater treatment plant. Harvesting is considered the main limitation to the large-scale cultivation of microalgae due to its limited settlement capacity. Sedimentation, centrifugation and filtration are the most common methods to separate solid and liquid phases. All harvesting techniques have some deficiencies such as the land area requirement for sedimentation, and high energy consumption for the others [14].

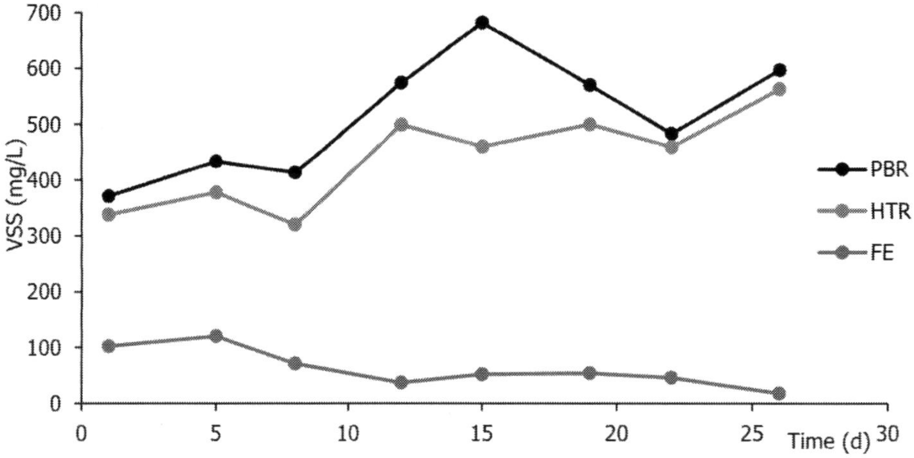

(a)

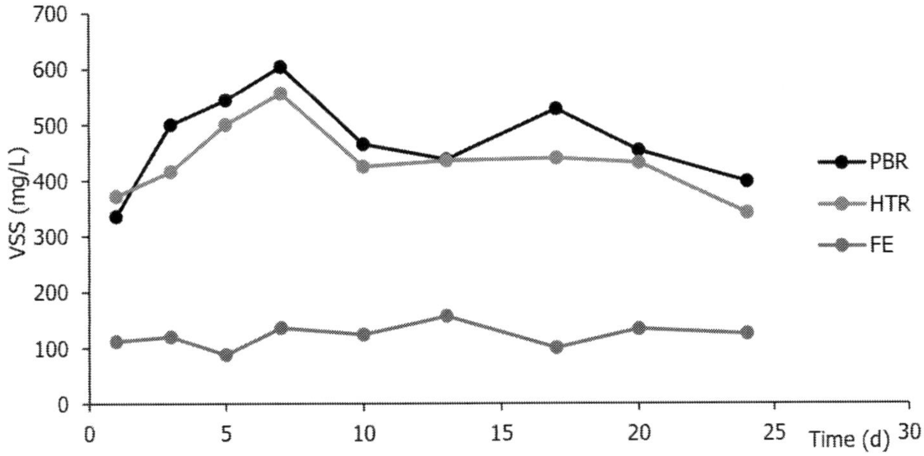

(b)

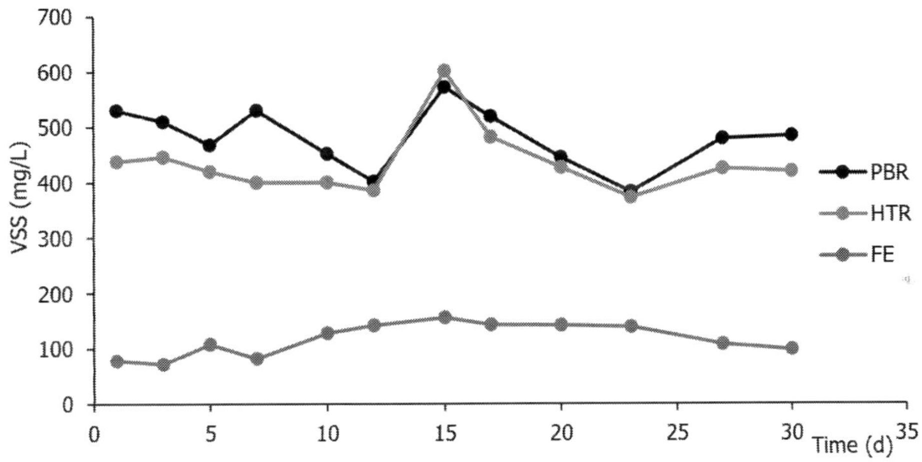

(c)

Figure 2. Microalgae concentration under different hydraulic retention times: (a) 36 h HRT, (b) 48 h HRT, and (c) 72 h HRT.

The uptake of nutrients from the two-stage biological process was determined by analysing phosphate, ammonium and nitrate in the PBR, HTR and FE samples. Phosphorus concentrations in each reactor at 36 h, 48 h and 72 h HRT are presented in Figure 3 and similar trends were observed. At 36 h HRT, phosphorus concentrations in FE increased from 9.8 to 11.3 mg P L^{-1} at Day 12, thereafter it remained stable while there was a constant decline in phosphorus concentration in FE from 15.0 to 9.0 mg P L^{-1} at 48 h HRT. At 72 h HRT, the phosphorus concentration in FE fluctuated around 9.0 mg P L^{-1}. P in the effluent at 36 h was significantly different than the ones at 48 h and 72 h ($p < 0.05$), whereas P in the effluent did not change significantly between 48 h and 72 h ($p = 0.348$). Phosphorus uptake efficiency incremented from 24.6% by 40% with the increase of HRT from 36 h to 48 h due to the increased time, providing more time to assimilate phosphorus [13][15]; it then remained invariant (40%) at 72 h HRT. Therefore, 48 h HRT was chosen as the optimum because prolonged HRT requires a larger reactor volume.

Ammonium recovery rates in the PBR were found to be 90.1, 88.4 and 89.4% at 36 h, 48 h and 72 h HRT, respectively, while nitrate was fully consumed under phototrophic conditions. There were not any changes in ammonium concentration under heterotrophic conditions.

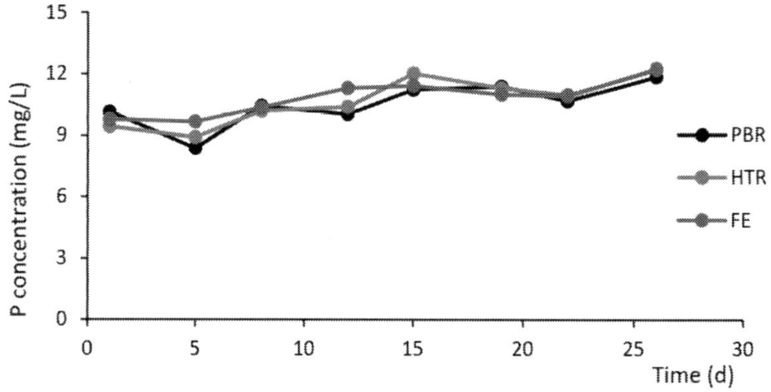

(a)

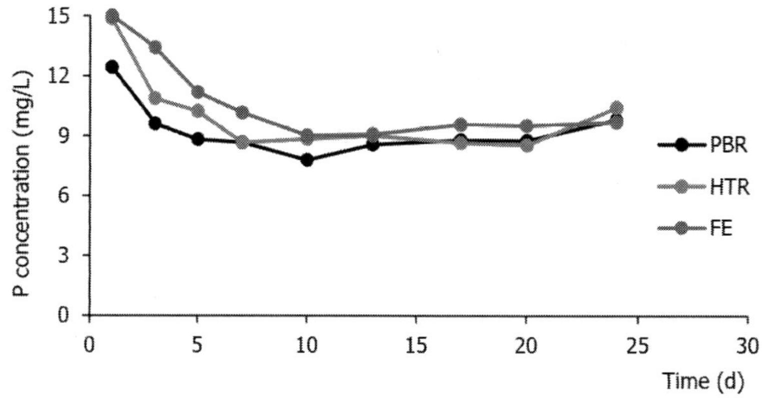

(b)

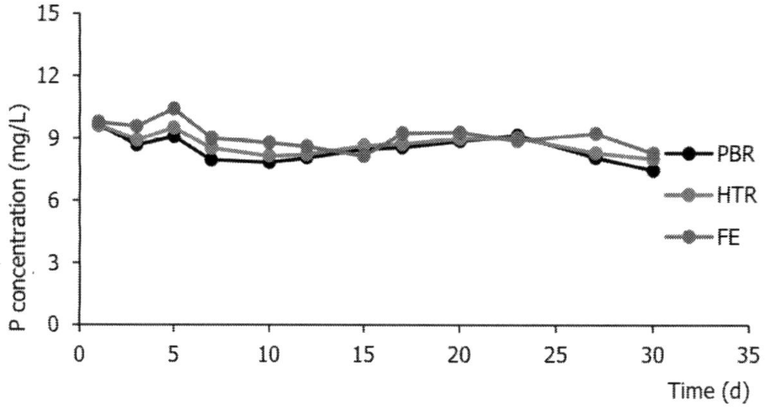

(c)

Figure 3. Changes in P concentration during the cultivation period for different HRTs: (a) 36h HRT, (b) 48h HRT and (c) 72h HRT.

3.2. The Impact of CRT on Microalgae Growth and Nutrient Recovery

The two-stage biological process was operated at different CRTs of 7 d, 14 d and 21 d at 48 h HRT using a mix of different nitrogen sources (Figure 4). The same growth trend was observed in the PBR and HTR throughout the experiments for cell retention times ($p = 0.103$). A slightly higher microalgae concentration was obtained in the PBR than in the HTR. Dry weight concentrations fluctuated from 330 to 600 mg VSS L^{-1}, from 330 to 600 mg VSS L^{-1} and from 260 to 810 mg VSS L^{-1} for the PBR and from 250 to 500 mg VSS L^{-1}, from 340 to 560 mg VSS L^{-1} and from 230 to 860 mg VSS L^{-1} for the HTR at 7 d, 14 d and 21 d CRT. Between 7 d and 14 d CRT, average algal biomass concentrations in both the PBR and the HTR increased. Algal biomass concentrations in the PBR significantly increased at 21 d CRT ($p < 0.05$) whereas a slight increase was observed in HTR ($p = 0.397$). It can be clearly anticipated that biomass concentration increased with incrementing cell retention times at fixed hydraulic retention times under phototrophic and heterotrophic conditions There is a good agreement with [16], reporting that lower biomass concentrations were obtained at lower CRTs with *Chlamydomonas reinhardtii*. Biomass concentration in the final effluent differed from 50 to 120 mg VSS L^{-1}, from 90 to 160 mg VSS L^{-1} and from 40 to 80 mg VSS L^{-1} at 7 d, 14 d and 21d CRT along with the cultivation period. FE included an approximate two-fold dry weight concentration at 14 d HRT (128 ± 20 mg VSS L^{-1}) which was significantly different from figures obtained at 7 d and 21 d CRT (76 ± 22 and 59 ± 13 mg VSS L^{-1}, respectively) ($p < 0.05$).

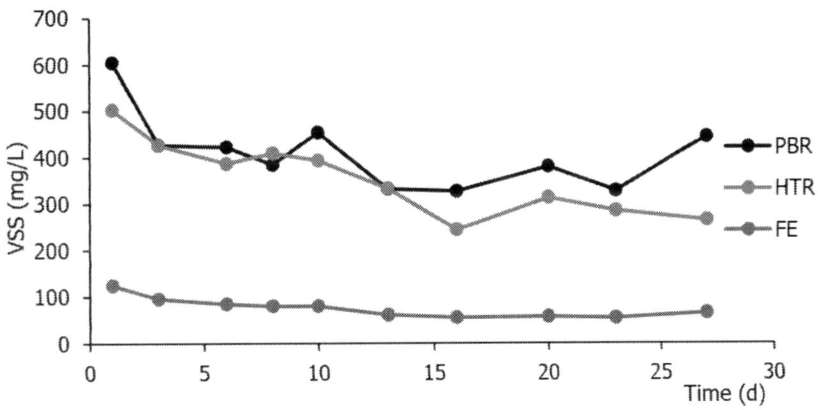

(a)

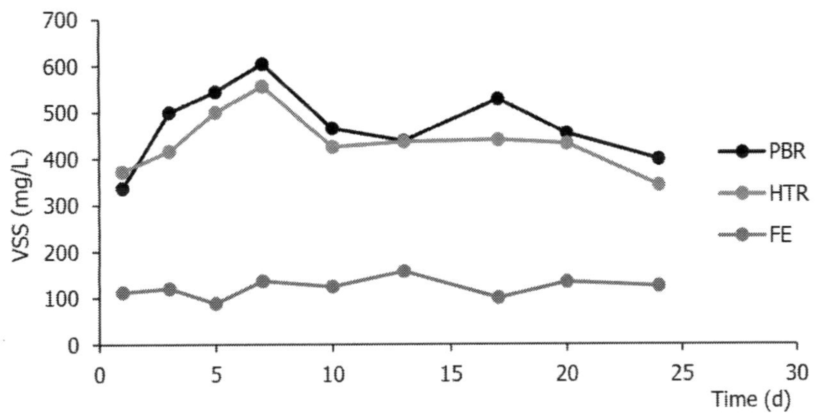

(b)

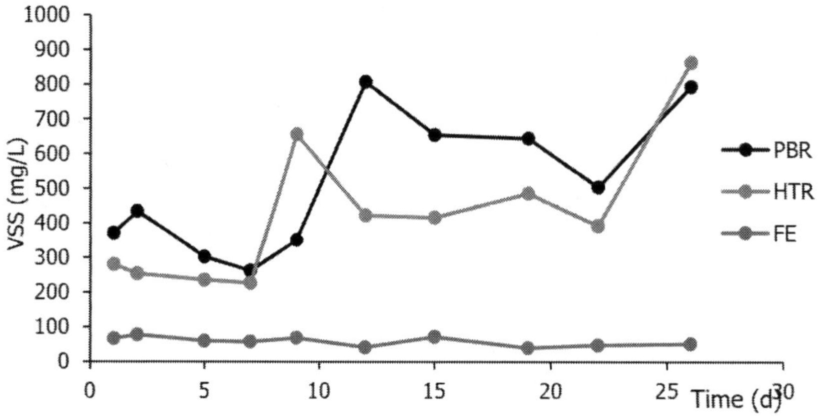

(c)

Figure 4. Algal biomass concentration variation with different CRTs: (a) 7 d CRT, (b) 14 d CRT, and (c) 21 d CRT.

Phosphorus concentrations over the two-stage biological process at the tested CRTs at 7 d, 14 d and 21 d are demonstrated in Figure 5. Overall, any significant differences in P concentration throughout the continuous flow system was not reported with varying CRT, meaning that the P in each reactor showed a similar trend ($p > 0.05$). Average P concentrations under steady-state conditions at 7 d and 21d CRT were 11.3 and 11.9 mg P L^{-1}, consistent with lower P recovery efficiencies via *Chlamydomonas reinhardtii* which were obtained as 24.7% and 20.7%, respectively ($p = 0.408$). There was a significantly difference in P concentration in the final effluent and recovery efficiency of P uptake at 14 d CRT (9.0 mg L^{-1} P and 40%, sequentially) ($p < 0.05$). Greater phosphorus uptake at 14 d CRT could be attributed to higher biomass concentration at this time compared to the one reported at 7 d CRT. Although biomass concentration continued to increase when further increasing CRT to 21 d, there was a decrease in P uptake which could be the result of biomass decay and excretion of phosphorus from algae cells into the media [16]. Hence, 14 d CRT was considered the optimum producing the lowest content in the final effluent.

The ammonium recovery rates achieved in the PBR were 100.0%, 88.4% and 87.7% at 7 d, 14 d and 21 d CRT, respectively, and nitrate was fully consumed by microalgae under phototrophic conditions. Heterotrophic conditions offered only limited contribution to nitrogen recovery.

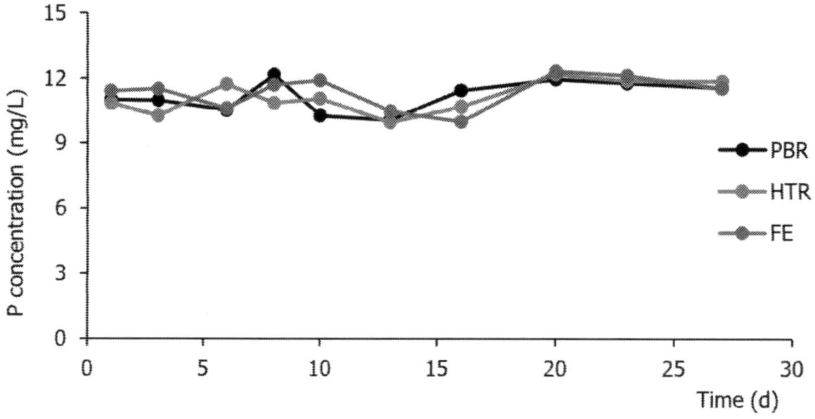

(a)

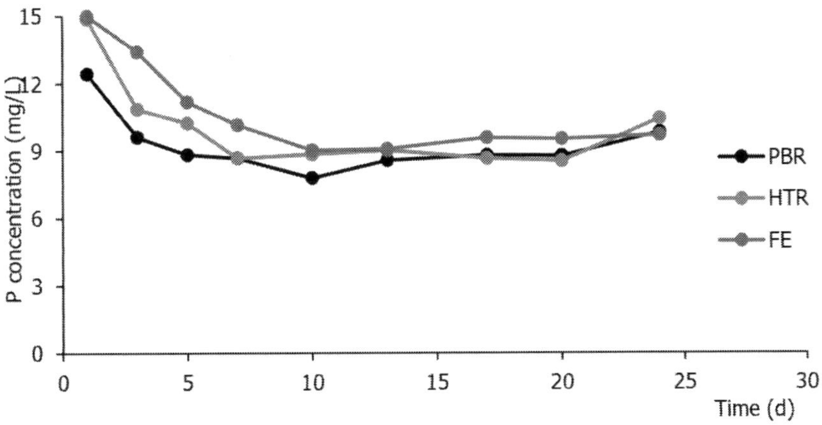

(b)

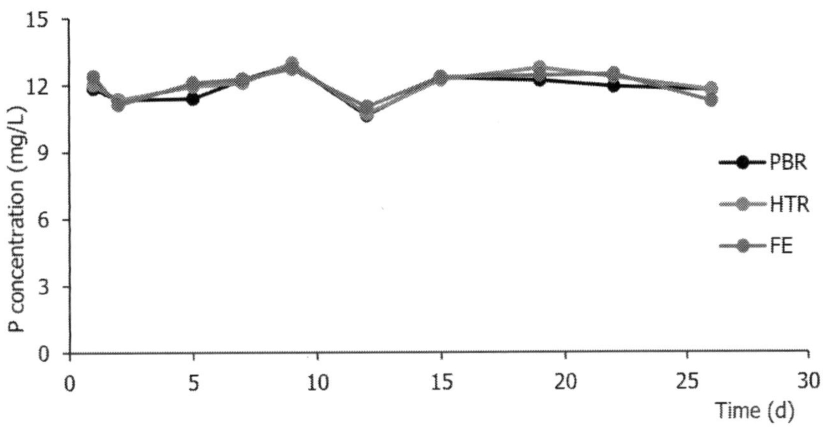

(c)

Figure 5. The effect of CRT on phosphate uptake and recovery: (a) 7 d CRT, (b) 14 d CRT, and (c) 21 d CRT.

3.3. The Effect of Different Nitrogen Sources on Microalgae Growth and Nutrient Recovery

Chlamydomonas reinhardtii was cultivated in a feeding medium including different nitrogen sources as a mix of ammonium and nitrate, ammonium alone, and nitrate alone (Figure 6). Similar microalgae concentrations were achieved in the PBR and HTR. Nitrogen sources did not show any significant effect on algal biomass concentration ($p = 0.605$ for PBR and $p = 0.243$ for HTR). Biomass growth fluctuated between 330 and 600 mg VSS L^{-1}, between 230 and 540 mg VSS L^{-1}, and between 300 and 490 mg VSS L^{-1} for the PBR, and from 340 to 560 mg VSS L^{-1}, from 270 to 550 mg VSS L^{-1}, and from 290 to 470 mg VSS L^{-1} for the HTR when a mix of ammonium and nitrate, solely ammonium and nitrate were used as the nitrogen sources, respectively. The dry weight of biomass in the final effluent differed from 90 to 160 mg VSS L^{-1}, from 50 to 200 mg VSS L^{-1} and from 30 to 80 mg VSS L^{-1} with different nitrogen sources as a mix of ammonium and nitrate, solely ammonium and nitrate, respectively. There was a similar effluent quality, as based on average solid concentration, between a mix of NH_4^+-N and NO_3^--N, (128 ± 20 mg VSS L^{-1}) and solely NH_4^+-N (107 ± 19 mg VSS L^{-1}) ($p < 0.05$). With the use of NO_3^--N in the synthetic wastewater, average microalgae concentration was decreased by half to 55 ± 16 mg VSS L^{-1} VSS, respectively ($p = 0.502$).

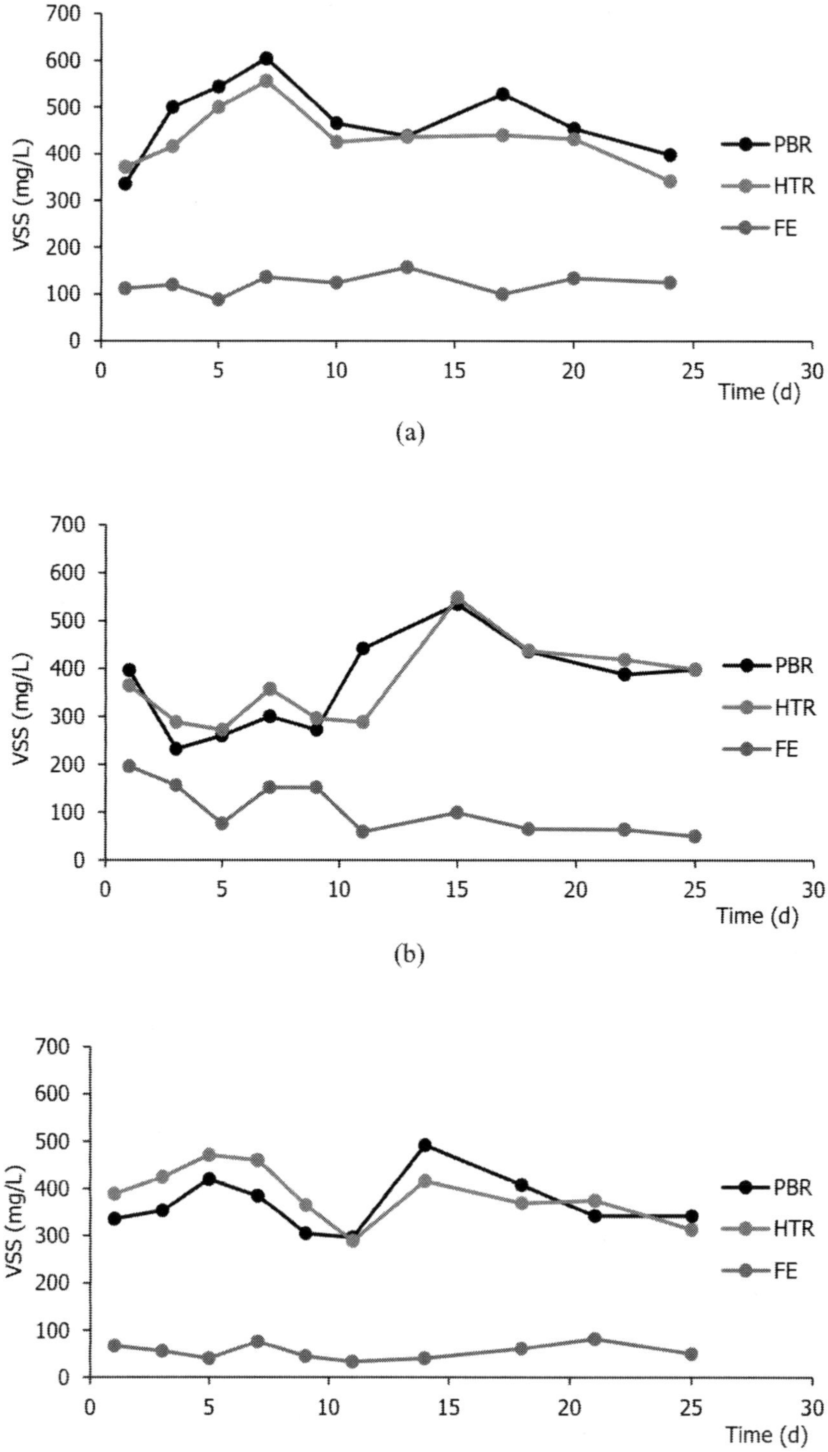

(a)

(b)

(c)

Figure 6. Biomass growth during cultivation with different nitrogen sources: (a) a mix of NH_4^+-N and NO_3^--N (25:25), (b) NH_4^+-N, and (c) NO_3^--N.

Figure 7 exhibits the influences of different nitrogen sources on P concentration in the continuous flow system for mixotrophic microalgae growth. It can be clearly seen that there was not any significant differences in phosphorus uptake by *Chlamydomonas reinhardtii* in each reactor when cultivating with different nitrogen sources ($p = 0.764$). The lowest P concentration in the effluent achieved was 9.0 mg P L^{-1} using a mixture of ammonium and nitrate, correspond to 40% P uptake efficiency. This was significantly different from when NH_4^+-N and NO_3^--N were used as the nitrogen source ($p = 0.001$). Similar P concentrations in the FE were obtained as 11.3 mg P L^{-1} when *Chlamydomonas reinhardtii* was cultivated in ammonia culture and nitrate culture, consistent with the P uptake efficiencies attained being 24.7% ($p = 0.924$). Thus, a mix of ammonium and nitrate was the favoured nitrogen source for the two-stage biological process. The mixture of nitrogen sources was preferred for *Chlamydomonas reinhardtii* rather than ammonia and nitrate in the view of biomass growth and phosphorus uptake [17].

An ammonium recovery rate of 88.4% was achieved and nitrate was taken up via *Chlamydomonas reinhardtii* at phototrophic cultivation when a mixture of ammonium and nitrate were used as the nitrogen source. Ammonium uptake efficiency was reduced to 66.4% when ammonium alone was used. Nitrate was fully consumed via microalgae uptake under phototrophic conditions in nitrate alone culture. There was a small contribution to nitrogen recovery in the HTR.

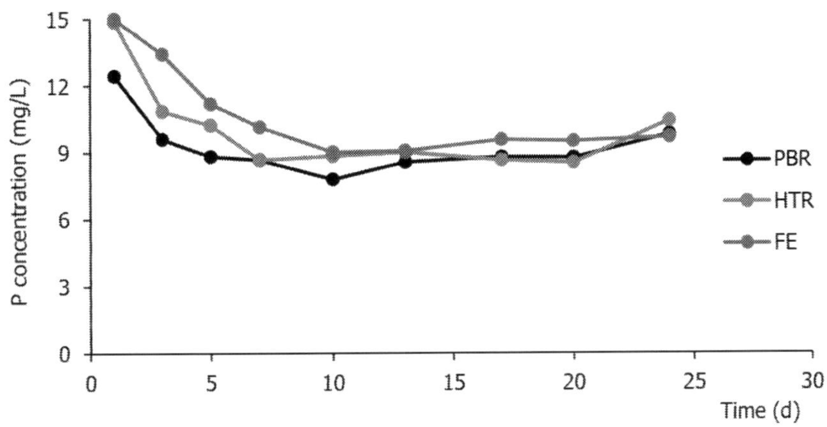

(a)

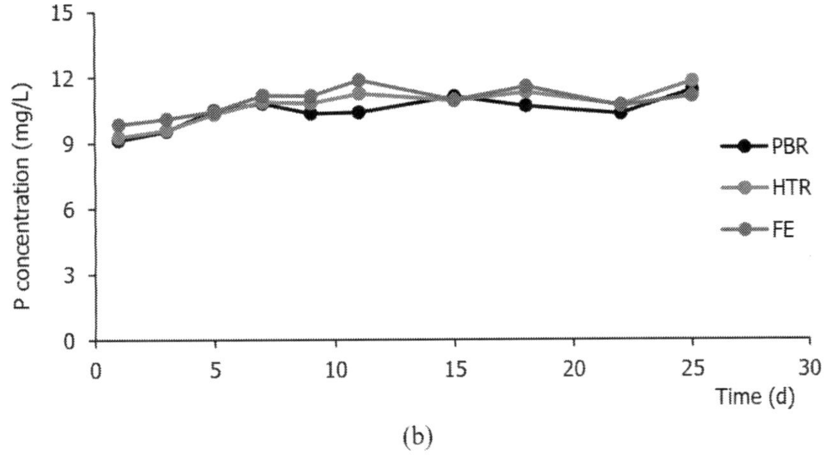

(b)

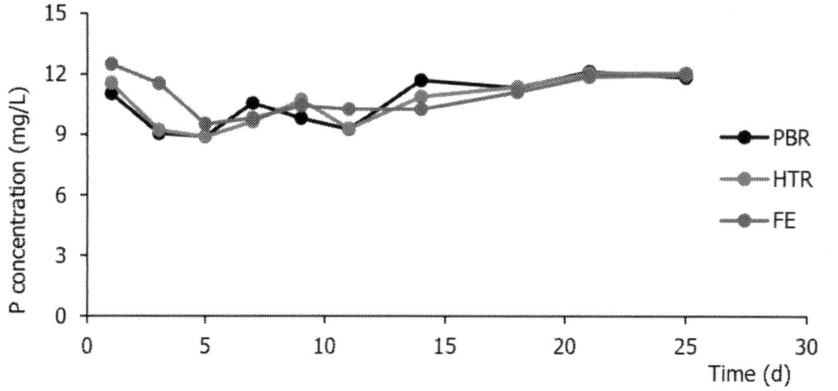

(c)

Figure 7. P concentration throughout the cultivation period with different nitrogen sources: (a) a mix of NH_4^+-N and NO_3^--N (25:25), (b) NH_4^+-N, and (c) NO_3^-- N.

4. Conclusion

This research aimed to identify the optimum operational conditions for the two-stage biological process combining phototrophic and heterotrophic microalgae cultivation with biomass recirculation under continuous flow conditions with regard to nutrient control and recovery via biological uptake. Although the contribution from the HTR to nutrient recovery was limited under all experimental conditions, *Chlamydomonas reinhardtii* recovered phosphorus and nitrogen at 40% and 93.2%, respectively, under optimum operational conditions of 48 h HRT, 14 d CRT and a mix of nitrogen sources.

Acknowledgment

The authors would like to acknowledge the Republic of Turkey Ministry of National Education for financial support for Ms. Cigdem Oz as a part of her PhD Scholarship at University of Leeds.

References

[1] Hu J, Nagarajan D, Zhang Q, Chang J S and Lee D J 2018 *Biotechnol Adv* **36** 54-67
[2] Zhou W, Min M, Li Y, Hu B, Ma X and Cheng Y 2012 *Bioresour Technol* **110** 448-55
[3] Sakarika M and Kornaros M 2016 *Bioresour Technol* **219** 694-701
[4] Perez-Garcia O, Escalante F M E, de-Bashan L E and Bashan Y 2011 *Water Res* **45** 11-36
[5] Lowrey J, Brooks M Sand McGinn P J 2015 *J Appl Phycol* **27** 1485 -98
[6] Chen C Y, Yeh K L, Aisyah R, Lee D J and Chang J S 2011 *Bioresour Technol* **102** 71–81
[7] Andrade M R and Costa J A V 2007 *Aquaculture* **264** 130-4
[8] Brennan L and Owende P 2010 *Renew Sustain Energy Rev* **14** 557-77
[9] Moon M, Kim C W, Park W K, Yoo G, Choi Y E and Yang J W 2013 *Algal Res* **2** 352-7
[10] Fernandes B D, Mota A, Teixeira J A and Vicente A A 2015 *Biotechnol Adv* **33** 1228-45
[11] Baird R B, Eaton A D, Rice E W and Bridgewater L 2017 *Standard Methods for the Examination of Water and Wastewater* 23rd ed. American Public Health Association Washington DC
[12] Morales-Sánchez D, Martinez-Rodriguez O A, Kyndt J and Martinez A 2015 *World J Microbiol Biotechnol* **31** 1-9
[13] Gao F, Peng Y Y, Li C, Cui W, Yang Z H and Zeng G M 2018 *Chem Eng J* **335** 169-175
[14] Gerardo M L, Van Den Hende S, Vervaeren H and Coward T 2015 Skill SC *Algal Res* **11** 248-62
[15] Iman Shayan S, Agblevor F A, Bertin L and Sims R C 2016 *Bioresour Technol* **211** 527-33
[16] Gardner-Dale D A, Bradley I M and Guest J S 2017 *Water Res* **121** 231-9
[17] Yulistyorini A and Camargo-Valero M A 2020 *Int J Integr Eng* **12** 253-60

IOP Conf. Series: Earth and Environmental Science **847** (2021) 012026 doi:10.1088/1755-1315/847/1/012026

Performance analysis of a H-Darrieus wind turbine for a Naca 0015 Airfoil: CFD and RSM-based design optimization

A A Permanasari[1,2], A B Budiearso[1], S Sukarni[1,2], P Puspitasari[1,2], S N A Zaine[3], and W Wahyunengsih[4]

[1]Department of Mechanical Engineering, Faculty of Engineering, State Universitas Negeri Malang, Jl. Semarang 5 Malang 65145, East Java, Indonesia
[2]Centre of Advanced Materials for Renewable Energy (CAMRY), Universitas Negeri Malang, Jl. Semarang No 5, Malang 65145, East Java, Indonesia
[3]Fundamental and Applied Sciences Department, Universiti Teknologi PETRONAS, 32610, Bandar Seri Iskandar, Perak, Malaysia
[4]Development of Islamic Society Department, State Islamic University Syarif Hidayatullah Jakarta, Indonesia

avita.ayu.ft@um.ac.id

Abstract. With the increasing demand for electrical energy and the depletion of fossil fuel reserves, this situation forces humans to look for alternative renewable energy. The Indonesian region has a very large wind energy potential because it is located in the tropics. The wind turbine is alternative energy based on kinetic energy into electricity. The purpose of this study was to analyze the effect of the number of blades, wind speed, and angle of attack on turbine performance in the form of the efficiency of the H-Darrieus wind turbine using Airfoil NACA 0015. Variations used in this study were the number of blades (3; 4; 5 pieces) and angle of attack (5°; 15°; 25°). The method used in this research is the Computational Fluid Dynamics (CFD) method and the response surface methodology (RSM). The CFD method is used to find the moment pressure and tangential velocity values, while the RSM method is used to find the most optimum variation. The results of this study indicate that a wind turbine with a number of blades 3 pieces, and a turbine angle of attack 11° is the most optimum H-Darrieus wind turbine design with an efficiency of 5.38%.

1. Introduction

Energy is one of the primary commodities needed by all people in the world. Due to the rapid growth of population and economic growth in the world, this is also in line with the increasing demand for energy. One of them is the need for electrical energy. With the increasing human needs, the sources of electrical energy will also increase. Especially for developing countries or regions, where most of what humans do requires electrical energy starting from small needs to large industrial worlds [1]. Wind as an energy source in abundance is a renewable energy source and does not cause air pollution because it does not produce exhaust gases that can cause the greenhouse effect. Indonesia is an archipelagic country which has about 17,500 islands with a coastline of more than 81.29 km and is located in a tropical area where monsoon winds pass each season. Indonesia has a very large wind energy potential, which is around 9.3 GW [2]. One way to utilize wind energy is to use a wind turbine.

Wind turbines are able to convert the kinetic energy of the wind into electrical energy with the help of a generator. One type of wind turbine is the H-Darrieus type wind turbine. H-Darrieus wind turbine is a vertical axis wind turbine that utilizes lift force when converting wind kinetic energy. The Darrieus-H turbine is also capable of receiving wind from all directions and rotating at low wind speeds [3]. Response Surface Methodology (RSM) is a mathematical and statistical method that is useful for analyzing problems, where several independent variables affect the response variable and the goal is to optimize the response. The basic idea of this method is to use a statistical-based experimental design to find the optimum value of a response [4]. While, CFD (Computational Fluid Dynamic) is a science that studies how to predict fluid flow, heat transfer, chemical reactions and other phenomena by solving mathematical equations or mathematical models. CFD can minimize time and cost in designing a product if the design process is carried out by experimental tests with high accuracy [5].

Hybrid turbine is based on a combination of two turbine technologies, namely Savonius and Darrieus [6] [7]. The main point of the study is that the asynchronous clutch (identical direction of rotation for the two rotors) provides the best performance compared to other hybrid configurations, as the built-in Savonius makes starting Darrieus easy, by providing additional power. The numerical results show that the hybrid design has important strengths over the range of tip speed ratios under consideration. The effect of variations in angle of attack on vertical axis wind turbines. It was found that the most optimum angle of attack on the turbine was between 0°-16°, the largest value was obtained at the angle of attack 6° [8] [9]. The Vertical Axis Wind Turbine (VAWT) has a good ability to extract more power from low winds than the Horizontal Axis Wind Turbine (HAWT). It is proven that 3 blades VAWT is more efficient than 2 or 4 blades. Solidity 0.2-0.4 is more efficient. Using the three-blade turbine power coefficient as a reference, it shows a 5% performance reduction for the four-blade design, while the five-blade design shows a 15% performance reduction [10].

2. Methodology

2.1. Optimization Method using RSM

The first step in this research is to enter the independent variables, namely the number of blades (3; 4; 5 pieces) and angle of attack (5°; 15°; 25°) mentioned above into the Minitab software with the RSM method. The wind velocity applied to the turbine is 5 m/s. It aims to test some variations that will be generated randomly variations. The dependent variables used in this study are moment pressure and tangential velocity, then the calculation is carried out until the torque, turbine power and efficiency of the wind turbine are obtained to obtain the optimum design. Design of experiment (DOE) obtained from the optimization process can be seen in Table 1. This study uses the RSM method with a second order model on the central composite design (CCD). Equation 1 shows the second order model of CCD method. In this study also used the CCD method with 3 factors and 3 levels which resulted in the regression equation in equation 2.

$$\hat{y} = \beta_0 + \sum_{i=0}^{k} \widehat{\beta_i} x_i + \sum_{i=0}^{k} \widehat{\beta_i} {x_i}^2 + \sum_i \sum_j \widehat{\beta_{ij}} x_i x_j \, , i < j \quad (1)$$

$$\hat{y} = b_0 + b_1 x_{i1} + b_2 x_{i2} + b_3 x_{i1}^2 + b_4 x_{i2}^2 + b_5 x_{i1} x_{i2} \quad (2)$$

Where x represents the independence variable, y is the response variable, and is a constant. In equation 2, the code x_1 indicates the number of blades, x_2 indicates the angle of attack, i indicates the repeated number, and b is a coefficient from the test results. The independent variable codes were -1,0,1, and ±1.42. The value of ±1.42 was the result of the rotatability value $(2^k)^{1/4} = (2^3)^{1/4} = 1.42$, where k is the total independent variable. Table 2 shows the independent variable code of x_1, and x_2.

Table 1. Design of Experiment (DOE) using CCD Method

Std Order	X_1 Code	X_2 Code	Number of Blade (pcs)	Angle of Attack (^{0})
1	-1	-1	3	5
2	1	-1	5	5
3	-1	1	3	25
4	1	1	5	25
5	-1.42	0	2	15
6	1.42	0	6	15
7	0	-1.42	4	1
8	0	1.42	4	29
9	0	0	4	15
10	0	0	4	15
11	0	0	4	15
12	0	0	4	15
13	0	0	4	15

2.2. H-Darrieus Wind Turbine Design and CFD Method

The dimensions of the blades of the H-Darrieus wind turbine are showed in Table 2. Figure 1 shows the design of the H-Darrieus Wind turbine using the Autodesk Inventor and then imported into the ANSYS application to obtain the moment pressure and tangential velocity values using the CFD method. Figure 2 shows the design of the NACA 0015 airfoil used in a wind turbine. The steps in the CFD method are preprocessing, processing and post processing. The preprocessing step is making a model according to the design parameters that have been provided, meshing, and applying boundary conditions according to the fluid properties. The processing step is the input of material and wind speed, which is then applied to mathematical calculations. The postprocessing step is to organize and interpret the simulation data in the form of numbers, contour displays, vector displays, pathline displays, curves, and animations.

Table 2. Design Spesification of H-Darrieus wind Turbine

Design	Spesification
Diameter of Blade	1 m
Height of Blade	1 m
Length of Cord	100 m
Type of Blade	*Airfoil* NACA 0015
Material	Aluminium

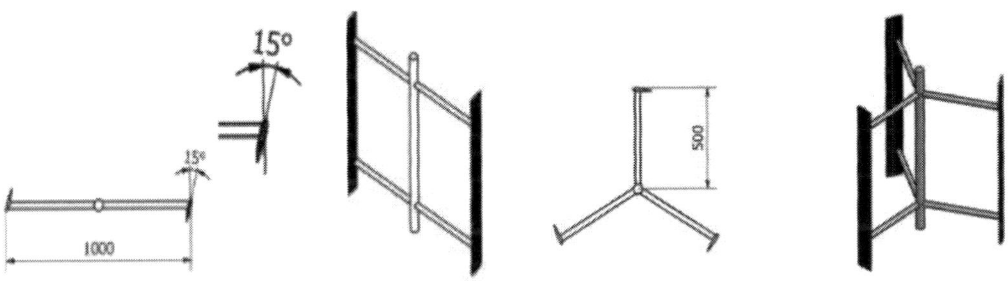

(a) (b)

IOP Conf. Series: Earth and Environmental Science **847** (2021) 012026 doi:10.1088/1755-1315/847/1/012026

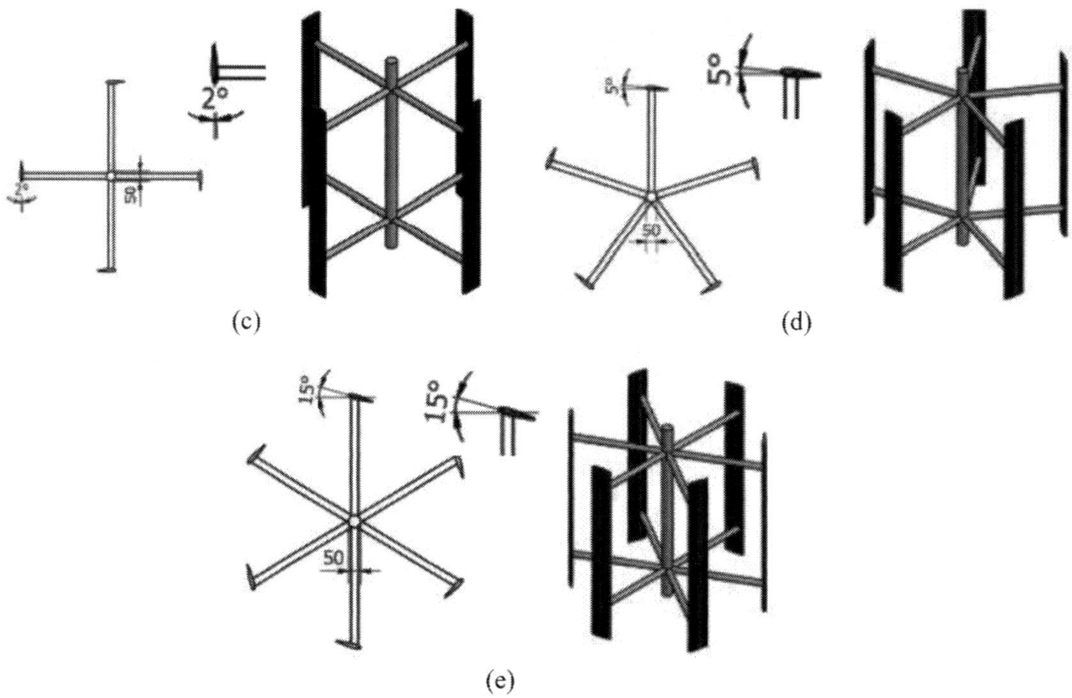

Figure 1. H-Darrieus Wind Turbine Design with a number of blade variation (a) 2 pcs; (b) 3 pcs; (c) 4 pcs; (d) 5 pcs; and (e) 6 pcs

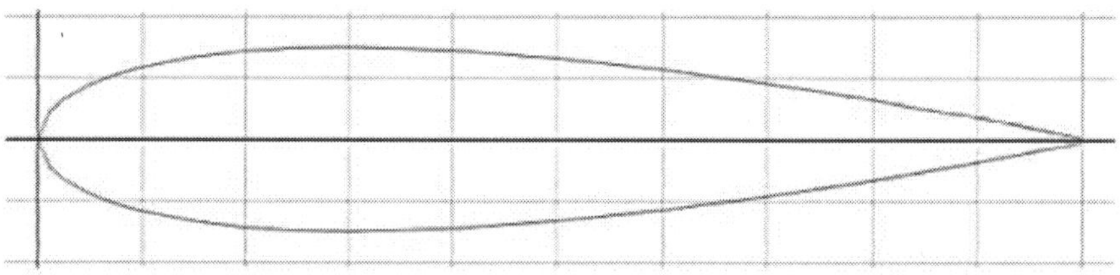

Figure 2. Airfoil NACA 0015

2.3. Performance Analysis

The value of moment pressure and tangential velocity obtained at the post-processing stage in the CFD method, then calculate the performance of the H-darrieus wind turbine according to equations 3-6 to calculate the torque, power and efficiency of the turbine. Then it was analyzed again using Minitab software using the RSM method to see which was the optimum variation of the 20 trials. Equation 3 shows the wind power generated from the rotational speed of the wind turbine [11] [12].

$$P_{wind} = \frac{1}{2}\rho v A^3 \quad (3)$$

where p is the density of air (kg/m^3), v is the wind speed (m/s) and A is the sweep area of the turbine blades (m2). Equation 4 shows the value of the mechanical power of the turbine [11] [12].

$$P_{mechanic} = \frac{2\pi n \tau}{60} \quad (4)$$

where $P_{mechanic}$ is the mechanical power of the turbine, n is the turbine rotational speed (rpm), and τ is the torque. Equation 5 shows the wind turbine power coefficient [11] [12].

$$C_p = \frac{P_{mechanic}}{P_{wind}} \quad (5)$$

where C_p is the power coefficient, $P_{mechanic}$ is the turbine mechanical power (Watts) and P_{wind} is the wind power (Watts). Equation 6 shows the efficiency of the wind turbine. Wind turbine efficiency can be calculated using the ratio of the power coefficient to the Betz limit, Betz limit = 16/27 = 0.593. The value of 0.593 is the maximum value of the efficiency of a wind turbine that converts kinetic energy into mechanical energy [13].

$$\eta = \frac{C_p}{16/27} \tag{6}$$

3. Results and discussion

The experimental design using Central Composite Design (CCD) method with 2 factors and 3 levels is shown in Table 1 which is then carried out using ANSYS simulation process to obtain moment pressure and tangential velocity to determine the performance of the H-Darrieus wind turbine, namely efficiency. Table 3 shows the test results using the design of experiment (DOE). Experiments were carried out for 13 variations and used a wind speed of 5 m/s. Efficiency contour from program simulation is shown in Figure 3.

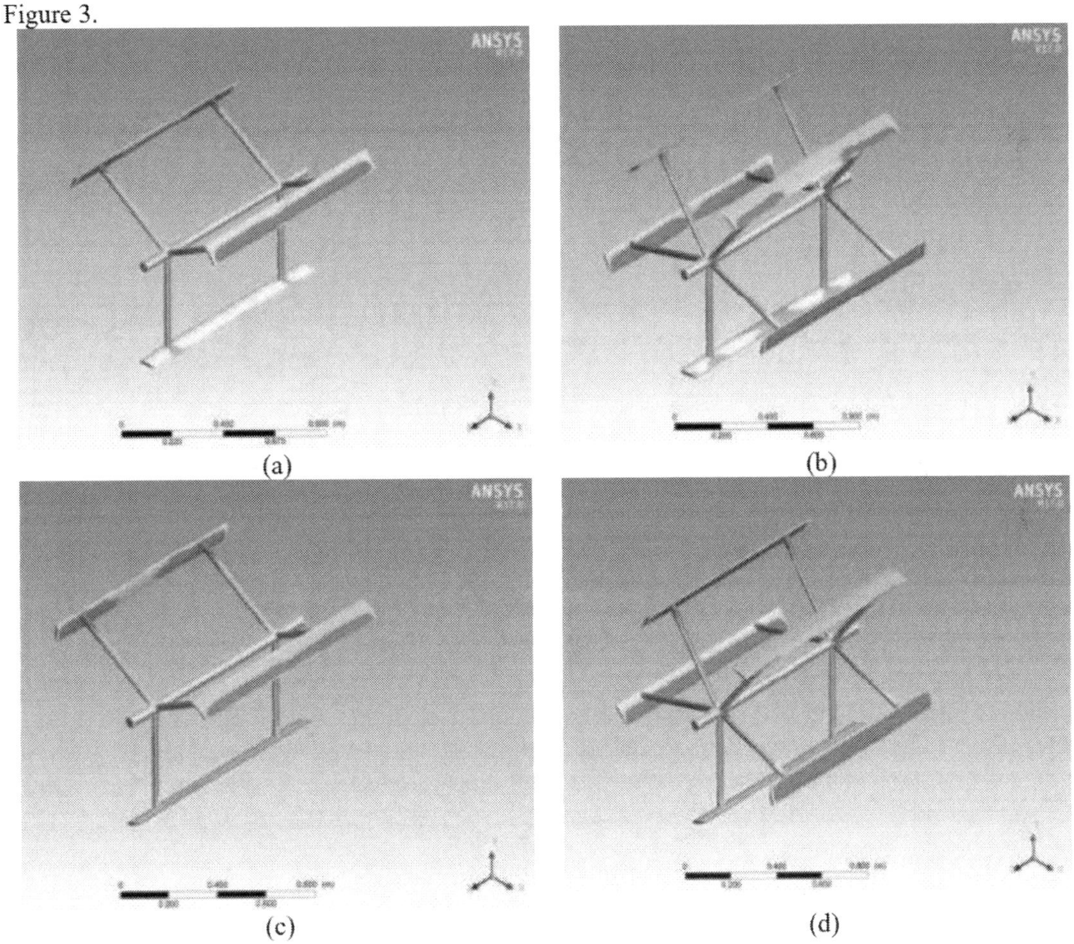

Figure 3. Pressure Contour of H-Darrieus Wind Turbine with (a) Number of blade 3 pcs and angle of attack 5^0; (b) Number of blade 5 pcs and angle of attack 5^0; (c) Number of blade 3 pcs and angle of attack 25^0; (d) Number of blade 5 pcs and angle of attack 25^0

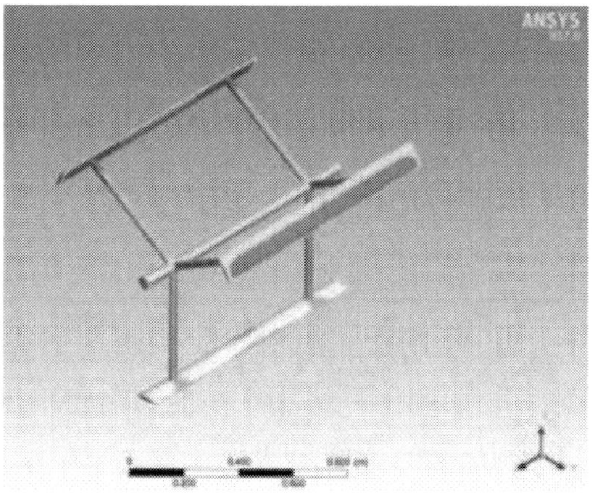

(e)

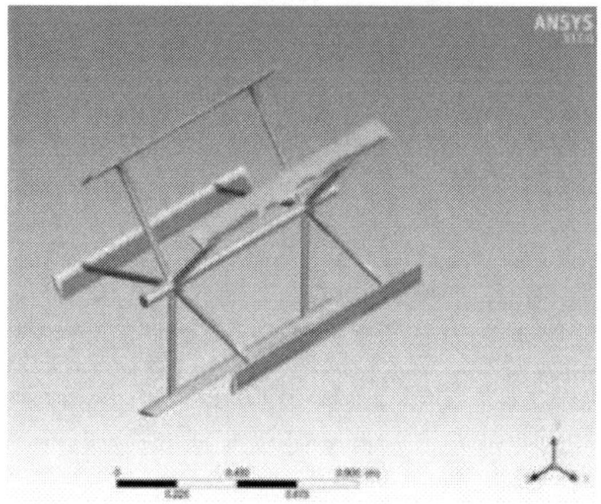

(f)

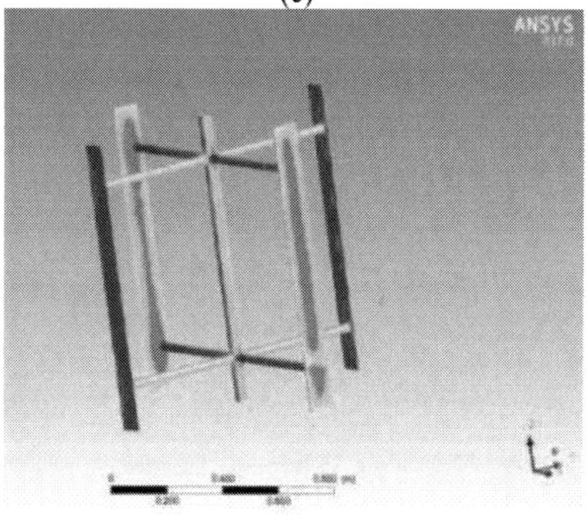

(g)

(h)

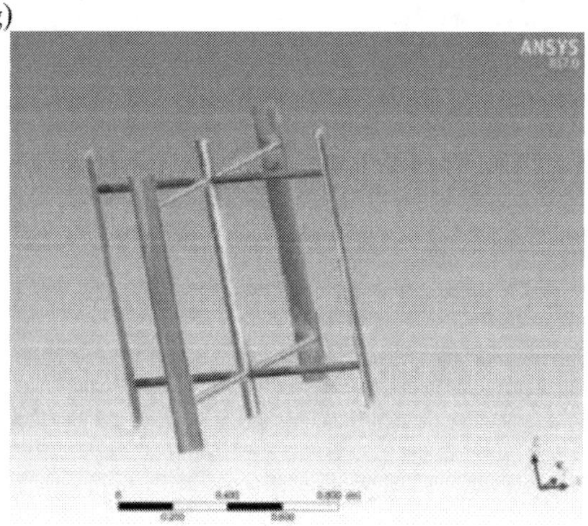

(i)

Figure 3. Pressure Contour of H-Darrieus Wind Turbine (e) Number of blade 3 pcs and angle of attack 15^0; (f) Number of blade 5 pcs and angle of attack 15^0; (g) Number of blade 4 pcs and angle of attack 1^0; (h) Number of blade 4 pcs and angle of attack 29^0; and (i) Number of blade 4 pcs and angle of attack 15^0

Moment pressure simulation results in Figure 3, based on the standard order in Table 3, are used to find the efficiency value of the wind turbine. The optimum factors generated from the statistical software are shown in Figures 4 (a) and (b). The contour plot shows the maximum efficiency value indicated by a red contour. These results are in line with the contour plots and optimization plots where the results of the number of blades are 3 pieces and angle of attack 11.46° which are shown in Table 4. Equation 7 is a regression equation that can be used to determine the predicted turbine efficiency without performing simulation process to determine the value of efficiency. The comparison of actual and predicted efficiency is shown in Table 3 and Figure 5, where the efficiency value is not more than 5%.

$$\hat{y} = 30.88 - 14.40x_1 + 0.302x_2 + 1.793x_1^2 - 0.01204x_2^2 - 0.0087x_1x_2 \quad (7)$$

Table 3. Performance test results of H-Darrieus wind turbine

Std Order	Factors		Response			
	Number of Blade (pcs)	Angle of Attack (⁰)	Moment Pressure (Nm)	Tangential Velocity (m/s)	Efficiency Actual (%)	Efficiency Predictive (%)
1	3	5	0.42	2.4	5.65	4.89
2	5	5	0.48	2	5.35	4.69
3	3	25	0.19	2.2	2.39	3.18
4	5	25	0.09	3.4	1.75	2.64
5	2	15	0.42	2.2	5.25	5.25
6	5	15	0.29	3	5.07	4,87
7	4	1	0.05	3.95	1.17	3.03
8	4	29	0.04	2.7	0.73	1.12
9	4	15	0.18	3.18	3.23	3.27
10	4	15	0.18	3.18	3.23	3.27
11	4	15	0.18	3.18	3.23	3.27
12	4	15	0.18	3.18	3.23	3.27
13	4	15	0.18	3.18	3.23	3.27

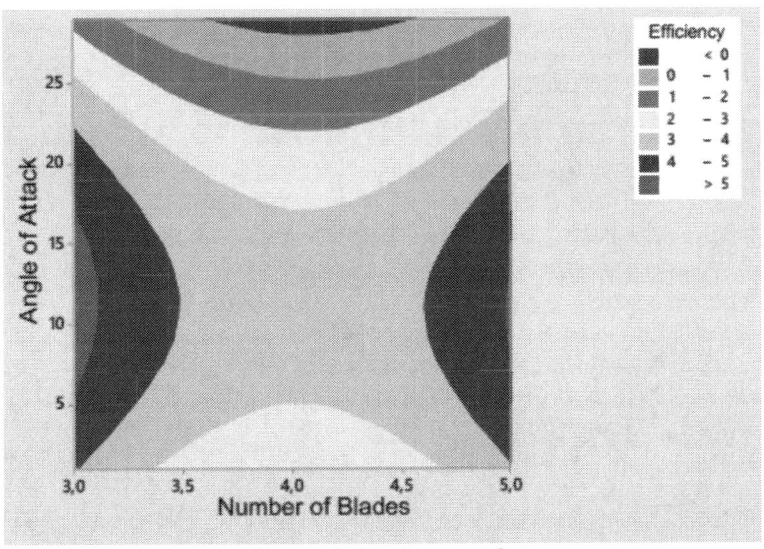

Figure 4. (a) Contour plot

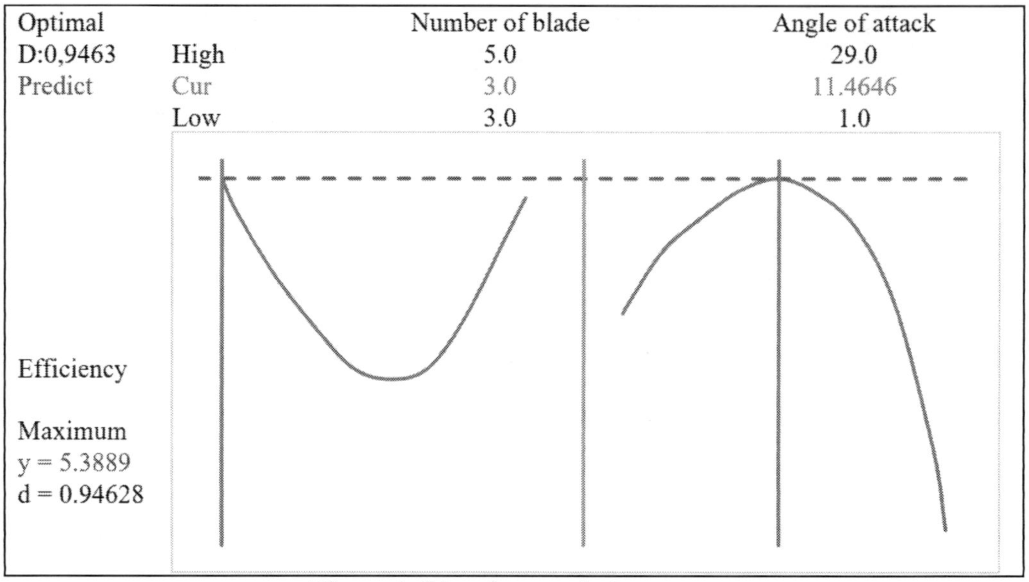

Figure 4. (b) optimum solution

Table 4. Optimum Solution using CCD Method

Number of Blades	Angle of Attack	Efficiency
3	11.46	5.39%

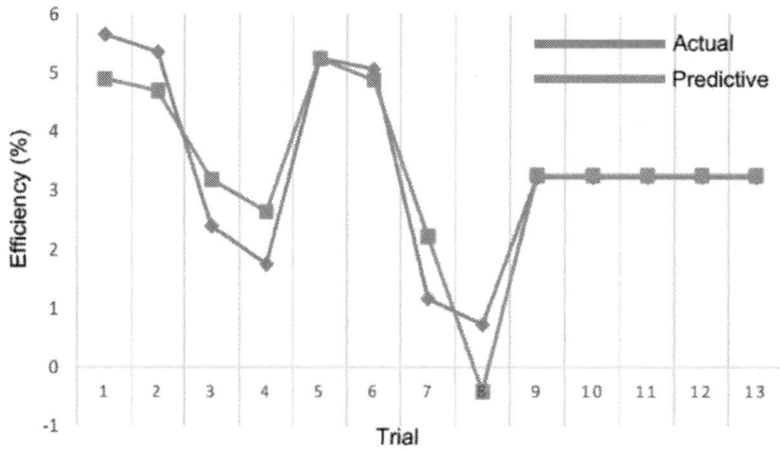

Figure 5. Comparison of actual and predicted H-Darrieus wind turbine efficiency values

Based on data in Table 3, In Efficiency with 3 blades, the lowest value is 2.39% in an angle of attack of 25° and the highest value is 5.65% in an angle of attack of 5°. Meanwhile, with the number of blades 4, the lowest value is 0.73% in an angle of attack of 25° and the highest value is 3.23% in an angle of attack of 15°. The efficiency value on the number of blades 5 pieces, obtained the lowest value of 1.74% in an angle of attack of 25° and the highest value of 5.07 % in an angle of attack 15°. The number of turbine blades has a relationship with the turbine power coefficient (Cp). Where the fewer the number of blades, the Cp will increase, which indirectly also increases the turbine's effective power. Based on data from Figure 4(b), it was found that with the number of blades 3 pieces, it became the optimal condition in the Darrieus-H type turbine design on the Airfoil NACA 0015. In addition, the number of turbine blades also affects the rotation of the turbine rotor, where the more the number of blades it causes movement the turbine rotor will slow down and become heavier, whereas if the number of blades is small, it causes the turbine rotor to rotate faster, so that the effective power generated from the turbine increases [14] [15] [16].

The angle of attack, lift (lift) and drag (drag) that work as the turbine rotates also changes. Lift and drag will form a resultant that can help the rotation or inhibit the rotation. In this study, as shown in Figure 6. Based on the data from the ANSYS simulation results of the CFD method and the optimization results of the RSM method, the highest efficiency performance was obtained on a turbine with an angle of attack of 15^0. However, in angle of attack of 25^0, efficiency decreased because the angle of attack has passed the critical angle or also called the critical angle of attack. The critical angle is the angle of attack where the lift generated reaches the maximum, above that angle the lift will decrease while the drag will increase rapidly, because the turbulent flow increases [17] [18].

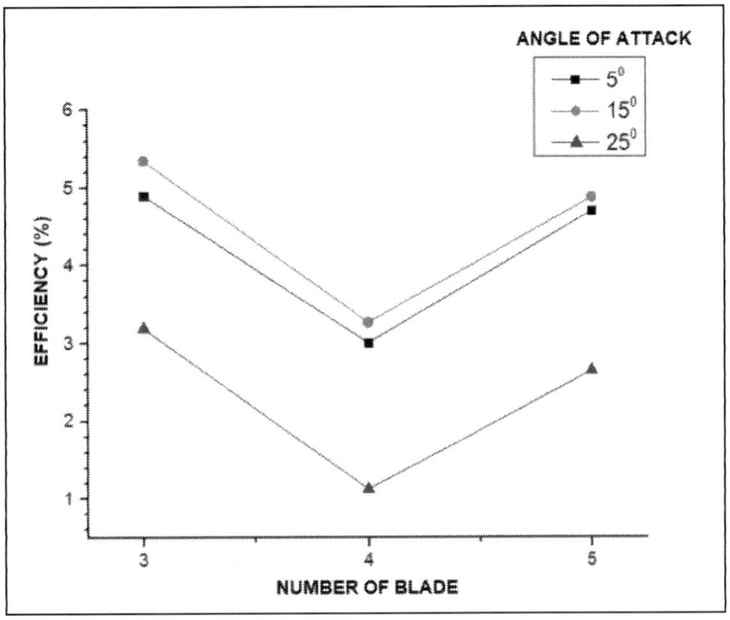

Figure 6. The performance of H-Darrieus wind turbine for efficiency

4. Conclusion

In the test results of variations in the number of blades, it was found that an increase in the number of blades tends to reduce the efficiency of the Darrieus-H wind turbine. In the tests carried out using the CFD and RSM approaches, the best efficiency values were found in a turbine with 3 blades and an angle of attack of 11^0. The best optimization design results according to the Darrieus-H wind turbine RSM method on Airfoil NACA 0015 is a number of blade 3 pieces and an angle of attack 11^0 which has maximum efficiency of 5.3889%.

References

[1] Ali S, Lee S and Jang C 2018 Effects of instantaneous tangential velocity on the aerodynamic performance of an H-Darrieus wind turbine Energy Conversion and Management **171** 1322–1338

[2] Beyhaghi S and Amano R S 2018 Aerodynamics A parametric study on leading-edge slots used on wind turbine airfoils at various angles of attack Journal of Wind Engineering & Industrial Aerodynamics **175** 43–52

[3] Castelli M R, Betta S D and Benini E 2012 Effect of Blade Number on a Straight-Bladed Vertical-Axis Darreius Wind Turbine **6(1)** 68–74

[4] Davane V H and Thorat M L 2017 Performance Analysis of Giromill Vertical Axis Wind Turbine

with NACA 63618 Airfoil 566–572
[5] Hashem I and Mohamed M H 2017 Aerodynamic Performance Enhancements of H-rotor Darrieus Wind Turbine Energy Elsevier B.V https://doi.org/10.1016/j.energy.2017.10.036
[6] Hau E and von Renouard H 2006 Wind Turbines https://doi.org/10.1007/3-540- 29284-5
[7] Manwell J F, Mcgowan J G, A L R 2009 Wind Energy Explained. Theory, Design and Application Second Edition.
[8] Permanasari A A, Puspitasari P and Utama S B 2019 Experimental Investigation and Optimization of Floating Blade Water Wheel Turbine Performance Using Taguchi Method and Analysis of Variance (ANOVA) IOP Conference Series Material Science and Engineering
[9] Permanasari A A, Kuncara B S and Puspitasari P 2019 Convective heat transfer characteristics of TiO2-EG nanofluid as coolant fluid in heat exchanger AIP Conf. Proc. **2120** 05001
[10] Khuri A I, Carter W H and Myers R H 1986 Respone Surface Metodology 1966–1986
[11] Mathew S 2006 Sathyajith Mathew Wind Energy Fundamentals Resource Analysis and Economics
[12] Douglas M 2005 Design and Analysis of Experiments Response Surface Method and Designs I. John Wiley and Sons Ed. New Jersey
[13] Mustak R 2017 Design and Construction of NACA-4415 Airfoil with Various Shaped Surface Modifications 1–8
[14] Firlianda D A, Permanasari A A, and Puspitasari P 2019 Heat transfer enhancement using nanofluids (MnFe2O4-ethylene glycol) in mini heat exchanger shell and tube AIP Conf. Proc. **2120** 05001
[15] Myers R H and Montgomery D C 1995 Response Surface Methodology: Process and Product Optimization Using Designed Experiments New York.
[16] Xu H, Qiao C, Yan H and Ye Z 2016 Delayed Detached Eddy Simulation of The Wind Turbine Airfoil S809 For Angles Of Attack Up To 90 Degrees Journal of Energy

Elemental compositions analysis of *Arthrospira platensis*-activated carbon blend using energy-dispersive X-Ray spectroscopy

A Y Aminullah[1], S Sukarni[1,2], A Prasetiyo[1], Y Zakaria[1], A A Permanasari[1,2] and P Puspitasari[1,2]

[1]Center for Renewable and Sustainable Energy Engineering (CRSEE), Department of Mechanical Engineering, Universitas Negeri Malang, Jl.Semarang No 5, Malang, 65145, Indonesia
[2]Centre of Advanced Materials for Renewable Energy, Universitas Negeri Malang, Jl.Semarang No 5, Malang, 65145, Indonesia

sukarni.ft@um.ac.id

Abstract. The biomass combustion in the boiler furnace faces a common problem related to slagging and fouling. The slag deposits and fouling of boiler components have been significantly influenced by inorganic matter in biomass origin. Therefore, chemical composition greatly determines the potential of biomass as a fuel. This study was conducted to determine the chemical content of *Arthrospira platensis* (AP) microalgae, activated carbon (AC), and a mixture of AP and AC with a mass ratio of 10:1 (AP/AC:10/1) by using an Energy Dispersive X-Ray (EDX) spectroscopy. The results showed that the chemical content of AP microalgae was dominated by C and O elements by 63.25 and 27.99 (wt.%), respectively. It was also shown that activated carbon was dominated by C and O elements at 89.91 and 10.08 (wt.%), respectively. Moreover, the addition of AC to the AP led to a decrease in the percentage of O content and increased the percentage of C from 63.25 to 91.20 (wt.%). It is expected that adding AC can improve the performance of AP biomass during the thermal conversion process.

1. Introduction

Energy demand will be linear to an increase in population and industrial activity. Energy sources in Indonesia are dominated by coal for electricity supply and petroleum for transportation [1]. Fossil fuels are non-renewable energy sources, so they will run out due to intense use, concurrently with the pollution problem emitted by burning fossil fuels increasing SOx, CO, CO_2, and NOx compounds in the earth's atmosphere [2]. Thus, natural phenomena such as climate change and global warming are increasing. When no restrictions are made, they can cause the extinction of many species and complicate human life [3]. Renewable energy is predicted to be able to answer all problems of energy accessibility. It can be obtained from several sources, including biomass. Based on the Ministry of Energy and Mineral Resources data in 2019, biomass contributed 20% of the new and renewable energy sources for electricity generation. Its sources come from the organism and micro-organisms. Microalgae is a biomass source that is mass-produced potentially and classified as a third-generation biofuel with a much higher energy density than the previous generation of fuel [4] and relatively large bio-oil content [5].

One of the microalgae species is *Arthrospira platensis* (AP) from the phylum cyanobacteria. AP microalgae are widely cultivated since they have high nutritional value, rich in chemical content, and can be processed into biomass for various purposes [5].

The physicochemical properties are the primary references for proper furnace design to achieve efficient thermal conversion processes [6]. However, the chemical content of biomass is also an important parameter since it determines the amount of contained energy in the fuel. In addition, inorganic elements in the fuel affect the formation of slag deposits, ash, aerosols, and corrosion on boiler components during the thermal conversion process [7]. Therefore, researching the chemical constituents of microalgae is essential to know its potential as a fuel and as a basis for designing suitable reactors and proper handling of byproducts.

This work has analyzed the chemical elements of the AP, AC, and mixtures of AP/AC:10/1 (wt,%). All analyzes were performed using Energy Dispersive X-Ray (EDX) spectroscopy. These research results being fundamental data to serve a further consideration in the AP utilization as fuel feedstock and its conversion.

2. Materials and methods

2.1. Materials preparation

The study was initiated by preparing samples of AP microalgae from the Center of Brackish Water Aquaculture (BPBAP) Situbondo Regency and of activated carbon from Sigma-Aldrich (CAS Number 7440-44-0). First, a 60 mesh size sieve filtered AP microalgae powder to have a uniform size. Then, mixing AP and AC microalgae powder with a mass ratio of 10:1 (AP/AC:10/1) was carried out with a magnetic stirrer for 30 minutes at 1200 rpm to obtain a homogeneous mixture.

2.2. Methods

The test was carried out with a scanning electron microscope (SEM) instrument, FEI Inspect S50 equipment fitted out with the capability of X-ray microanalysis (AMETEK EDAX TSL). The sample was gold-coated to decrease the influence of surface-electron-charging, which might distort the image. EDX testing was performed three times for each sample in order to obtain valid results. The data from the three tests were processed so that the average value and standard error were known.

3. Results and discussion

The results of the EDX test for samples of AP, AC, and a mixture of AP/AC:10/1 were shown in Table 1, Table 2, and Table 3. The analyzed chemical elements were carbon (C), oxygen (O), sodium (Na), magnesium (Mg), phosphorus (P), sulfur (S), chlorine (Cl), and potassium (K). Table 1 showed the EDX test results on AP samples that are dominated by C and O. There were inorganic elements needed for microalgae growth. Besides, the presence of elements other than carbon and oxygen were assumed to originate from the AP microalgae habitat.

The results of the EDX test on AC shown in Table 2 indicated that there were only two elements, namely C and O. These results indicated that the activated carbon material had a high level of purity. Table 3 showed the results of the EDX test on the mixed sample AP/AC:10/1, showing that there were only two elements, namely C and O. The inorganic element content in the mixture was not detected, possibly because the very small amount was.

Table 1. EDX test results of *Arthrospira platensis*

Replications	Elemental compositions (wt,%)							
	C	O	Na	Mg	P	S	Cl	K
Test 1	64.20	26.29	1.83	0.53	1.09	1.15	3.54	1.38
Test 2	62.20	30.02	2.25	0.33	0.55	0.71	3.11	0.82
Test 3	63.35	27.67	1.84	0.88	0.00	1.04	3.61	1.61
Mean	63.25	27.99	1.97	0.58	0.54	0.96	3.42	1.27
Standard error	0.57	1.08	0.13	0.16	0.31	0.13	0.15	0.23

Table 2. EDX test results of activated carbon

Replications	Elemental compositions (wt.%)							
	C	O	Na	Mg	P	S	Cl	K
Test 1	87.66	12.34	0.00	0.00	0.00	0.00	0.00	0.00
Test 2	90.65	9.35	0.00	0.00	0.00	0.00	0.00	0.00
Test 3	91.44	8.56	0.00	0.00	0.00	0.00	0.00	0.00
Mean	89.91	10.08	0.00	0.00	0.00	0.00	0.00	0.00
Standard error	1.51	1.15	0.00	0.00	0.00	0.00	0.00	0.00

Table 3. EDX test results for mixture of *Arthrospira platensis* and activated carbon

Replications	Elemental compositions (wt.%)							
	C	O	Na	Mg	P	S	Cl	K
Test 1	90.96	9.04	0.00	0.00	0.00	0.00	0.00	0.00
Test 2	91.33	8.67	0.00	0.00	0.00	0.00	0.00	0.00
Test 3	91.33	8.67	0.00	0.00	0.00	0.00	0.00	0.00
Mean	91.20	8.79	0.00	0.00	0.00	0.00	0.00	0.00
Standard error	0.12	0.12	0.00	0.00	0.00	0.00	0.00	0.00

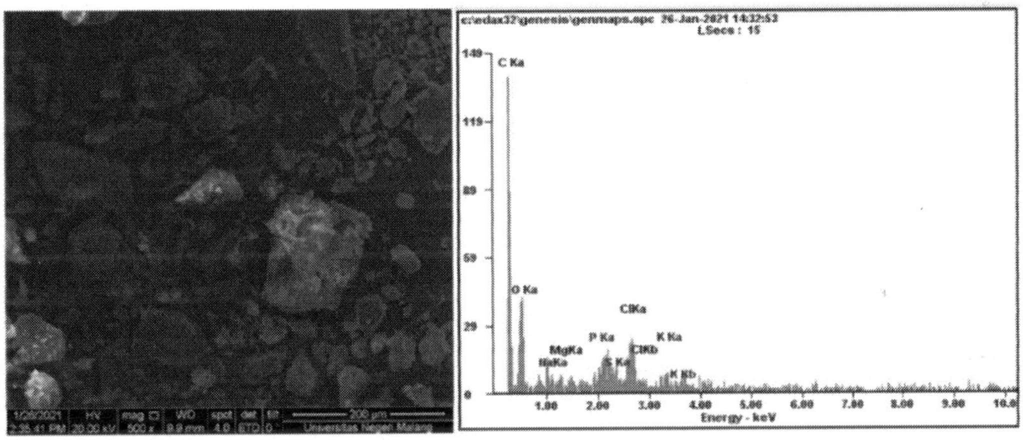

Figure 1. EDX spectrogram of AP sample

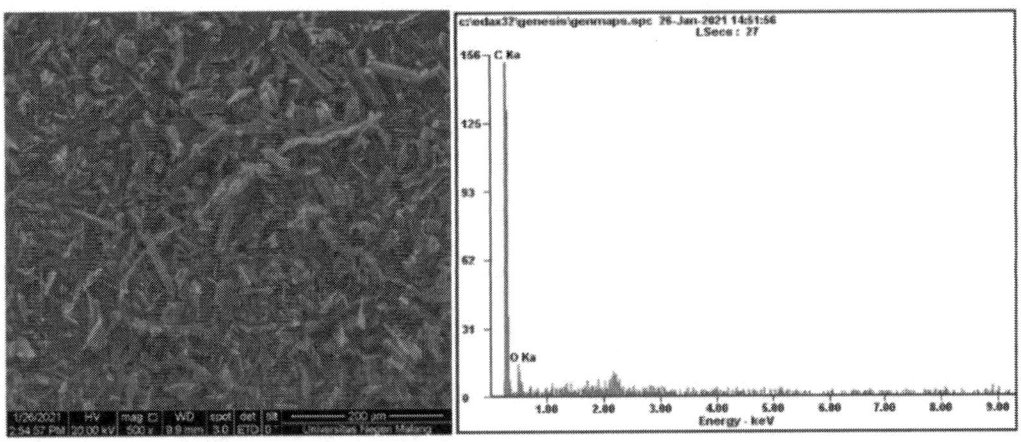

Figure 2. EDX spectrogram of AC sample

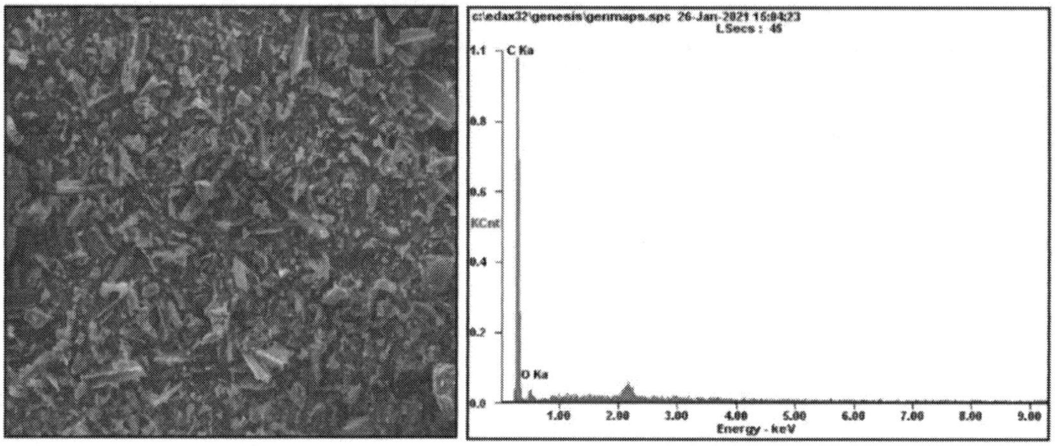

Figure 3. EDX spectrogram of mixture of AP/AC:10/1

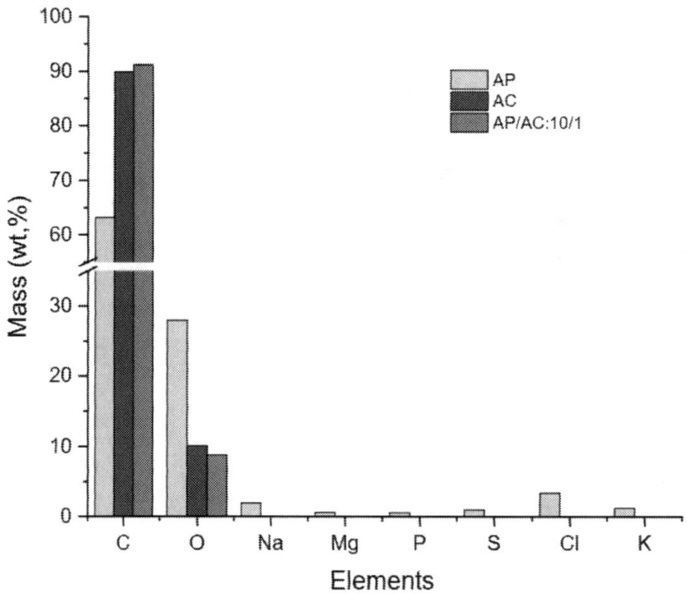

Figure 4. Elemental compositions comparison of AP, AC, and mixture of AP/AC:10/1

The carbon element (C) contributes to the production of calorific value, while for oxygen (O), especially in biomass, it functions as one of the supplies for the combustion reaction even though the oxygen content can reduce the higher heating value (HHV) [8–10]. The elements of chlorine (Cl) and sulfur (S) contained in AP can evaporate in the combustion process, which will bind to hydrogen (H) and form exhaust gases [11]. The content of elements other than C, O, and Cl in the sample contributed to the final product formation in combustion, namely ash.

Residues can be categorized based on their percentage, namely major ash-forming elements (>1%) and minor ones (1-0.1 %). If the combustion process is carried out in pure AP, the elements sodium (Na) and potassium (K) are classified as major ash-forming elements, while for minor elements, there are magnesium (Mg) and phosphorus (P). Chlorine content can lower the melting point of ash and cause corrosion. Potassium can cause corrosion, lower the melting temperature of ash, and can be used as fertilizer. Sodium causes corrosion and lowers the melting temperature of ash. Calcium causes an increase in the melting temperature of ash and can be used as fertilizer. Phosphorus produces ash which can be used as fertilizer [7,12,13].

The decrease in the percentage of O in the sample mixture AP/AC:10/1 was due to the nature of AC which was able to absorb elemental oxygen [14]. Likewise, AC can absorb elemental sulfur [15]. Thus, it was estimated that adding AC to AP could improve the performance of AP microalgae during the thermal conversion process.

4. Conclusion

The elemental composition in AP, AC, and AP/AC:10/1 mixture samples has been analyzed by EDX. The results showed that the AC could reduce the percentage of O in AP powder and increase the percentage of carbon elements by 91.2%, with a difference of 27.95%. Thus, based on the research results, it can be seen that activated carbon (AC) is possible to increase the performance of *Arthrospira platensis* (AP) microalgae during thermal conversion.

Acknowledgments

The author would like to thank DRPM, the Ministry of Research, Technology and Higher Education, Republic of Indonesia, for supporting this study in 2021.

References

[1] Sugiyono A 2018 Outlook Energi Indonesia 2015-2035 : Prospek Energi Baru Terbarukan OUTLOOK ENERGI INDONESIA 2015-2035 : PROSPEK
[2] Phillips M C K, Cinderich A B, Burrell J L and Will R G 2015 The Effect of Climate Change on Natural Disasters : A College Student Perspective The Effect of Climate Change on Natural Disasters : A College Student Perspective
[3] Kweku D W, Bismark O and Maxwell A 2018 Greenhouse Effect : Greenhouse Gases and Their Impact on Global Warming Greenhouse Effect : Greenhouse Gases and Their Impact on Global Warming
[4] Dahiya A 2015 Bioenergy: biomass to biofuels
[5] Kim S 2015 Bioremediation of Heavy Metals by Microalgae [IN:] Handbook of marine microalgae biotechnology advances. Kim S-K. [ed.]
[6] Sukarni S 2018 Physical and Chemical Properties of Water Hyacinth (Eichhornia crassipes) as a Sustainable Physical and Chemical Properties of Water Hyacinth (Eichhornia crassipes) as a Sustainable Biofuel Feedstock
[7] van Loo S and Koppejan J 2008 The Handbook of Biomass Combustion and Co-firing (London: Earthscan)
[8] Obernberger I, Brunner T and Ba G 2006 Chemical properties of solid biofuels — significance and impact Biomass and Bioenergy 30 973–82
[9] Sukarni, Sudjito, Hamidi N, Yanuhar U and Wardana I N G 2015 Thermogravimetric kinetic analysis of Nannochloropsis oculata combustion in air atmosphere Front. Energy 9 125–33
[10] McKendry P 2002 Energy production from biomass (part 1): overview of biomass Bioresour. Technol. 83 37–46
[11] Sukarni, Sudjito, Hamidi N, Yanuhar U and Wardana I N G 2014 Potential and properties of marine microalgae Nannochloropsis oculata as biomass fuel feedstock Int. J. Energy Environ. Eng. 5 279–90
[12] Sumarli S, Upendra C, Bandhana S and Himawan K 2019 Physicochemical Characteristics of Corn Silk as Biomass Fuel Feedstock 515 1–5
[13] McKendry P 2002 Energy production from biomass (part 2): Conversion technologies Bioresour. Technol. 83 47–54
[14] Zhou Y, Wei L, Yang J, Sun Y and Zhou L 2005 Adsorption of Oxygen on Superactivated Carbon 1068–72
[15] Liu J, Gao S, Jiang X, Shen J and Zhang H 2014 NO emission characteristics of superfine pulverized coal combustion in the O2/CO2 atmosphere Energy Convers. Manag. 77 349–55

Façade design modification in complying the Indonesia's national standard of energy conservation for tall building envelope – Case study: Green Office Park 9, Serpong, Indonesia

A M Hajji[1] and A R. Z. Hilmi[2]

[1]Civil Engineering Study Program, Universitas Negeri Malang, Indonesia
[2]Industrial Engineering Study Program, Bandung Institute of Technology, Indonesia

apif.miptahul.ft@um.ac.id

Abstract. Most tall buildings use glass as its façade and exterior materials, which makes the building's envelope is exposed to heat penetration. It is important to select and design the proportion between transparent and opaque materials for building façade to reduce the heat gain from outside, which is strategic to conserve the energy use for tall buildings. This study is aimed to observe the use of building façade materials in accordance with the strategic effort to reduce the heat gain in order to conserve the energy for AC system. The façade design modification is calculated by using Overall Thermal Transfer Value (OTTV) method in compliance with the Indonesia's National Standard of Energy Conservation for Tall Buildings. The results show that the modification on façade design can reduce the heat gain and improve thermal performance of the building. There is a dramatic change of OTTV from 124.40 w/m^2 on original design to 27.73 w/m^2 after several steps of façade modification. The biggest reductions of OTTV is obtained when modifications on openings are applied. Since the heat transmission is mainly coming into buildings through openings, modifications on glass and windows are the keys. The change of glass material, from clear-glass to those with lower shading coefficient (SC) gives reduction of 37.52 w/m^2.

1. Introduction

In building and construction industry, energy use and conservation is a big issue. The use of air conditioning (AC) system is the biggest portion of tall building energy use (50-70%), followed by lighting and other electrical system (10-25%), and building transportation system (2-10%) [1–3]. In very humid tropical climate like Indonesia, the AC system is commonly used in multi-tenant tall buildings to fulfill the needs for thermal comfort. The use of proper materials for building envelopes is the key for thermal comfort. Most tall buildings use glass as its façade and exterior materials, which makes the building's envelope is exposed to heat penetration [4]. The more penetrated heat coming into the buildings, the more energy needed to get the thermal comfort. It is important to select and design the proportion between transparent and opaque materials for building façade to reduce the heat gain from outside, which is strategic to conserve the energy use for tall buildings. This study is aimed to observe the use of building façade materials in accordance with the strategic effort to reduce the heat gain in order to conserve the energy for AC system. A case study has been selected to show how the modification of both materials configuration and external wall design could lead to lower energy use for

AC system. Green Office Park (GOP) 9 building is designed and constructed for rental office use (Figure 1). This five-story building, which is located in BSD, Serpong, Banten, is owned by PT Bumi Serpong Damai. It has one basement, one lobby-reception floor, and four typical floors. The façade design modification is calculated by using Overall Thermal Transfer Value (OTTV) method in compliance with the Indonesia's National Standard of Energy Conservation for Tall Buildings.

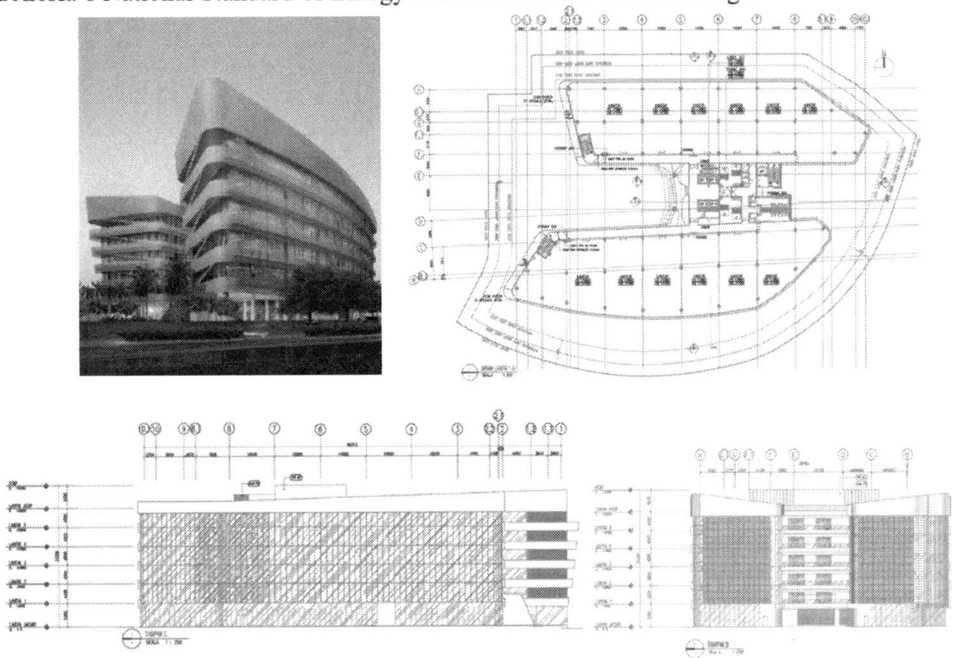

Figure 1. Design of Green Office Park (GOP) 9 Building, Serpong, Indonesia

2. Related studies

Building envelopes have significant impact on its thermal performance and total energy consumption, since it can affect the cooling load. The cooling load for building is determined by heat radiation through its openings, which is gained particularly from natural lighting design. OTTV method is widely used to measure the thermal performance of building envelopes, which is defined by the external heat gain transmitted through each area of building envelopes in watt/m^2. Generally, the external heat gain is transmitted mostly through openings rather than through solid walls. It requires careful design strategies for windows and openings to minimize heat gain by determining correct façade orientation, the area of openings, materials for windows (shading coefficients), and external shading construction [5]. Some studies have been carried out to calculate the external heat gain through building envelopes [6,7].

OTTV calculation can be conducted by using formula as detailed in Indonesia's National Standard SNI 03-6389 [8]. To refine and make this OTTV method more accurate and easier, some approaches have been applied particularly to support decision making process in building façade design [9]. The use of Building Information Modelling (BIM) can also be applied to help architects in using OTTV and determining the correct materials for the façade, especially for windows and openings [10]. The results of OTTV can also be used for measuring the thermal performance of building façade, observing the alternatives design for shading [11,12], windows and vents [13], exploring architectural design, and adapting local conditions in designing energy-efficient building façade [14]. The criteria in OTTV method can also determine national strategic policy on addressing the issues of energy regulation [15–17]. On policy level, the formulation of national energy conservation index and municipality regulation can also use the results of OTTV method [18].

3. Method

The concept of OTTV covers three basic elements of thermal transfer through building envelopes: heat conduction through solid walls, heat conduction through openings, and heat radiation through openings. Each material used on solid walls and openings has its specific heat transmission and absorption. OTTV is a number defined as a building envelope's design criteria to limit the value of heat gain, which is identified as external cooling load for AC system. To reduce this cooling load, Indonesia's National Standard through SNI 03-6389-2011[8] has defined that the energy-efficient value the OTTV is not more than 35 watt/m^2. OTTV in specific orientation is calculated by the following Equation 1 and 2.

$$OTTV = (Qw + Qg + Qsol) / Ai \quad (1)$$

$$OTTV = ((Aw*Uw*TDeq) + (Af*Uf*\Delta T) + (Af*SC*SF)) / Ai \quad (2)$$

Once each OTTV of specific orientation is obtained, the OTTV of whole building envelope is calculated by Equation 3.

$$Total\ OTTV = ((A_1*OTTV_1) + (A_2*OTTV_2) + \ldots + (A_n*OTTV_n)) / (A_1+A_2+\ldots+A_n) \quad (3)$$

Qw = heat conduction through solid walls (w)
Qg = heat conduction through openings (w)
Qsol = heat radiation through openings (w)
Ai = total area of openings and solid walls (m^2)
Aw = area of solid walls (m^2)
Uw = thermal transmittance of solid walls (w/m^{2o}C)
TDeq = temperature difference equivalent (°C)
Af = area of openings (m^2)
Uf = thermal transmittance of openings (w/m^{2o}C)
ΔT = inside-outside temperature difference (°C)
SC = shading coefficient of openings (unitless)
SF = solar factors (w/m^2)

To attain the efficient energy use of GOP 9 Building, the OTTV of its original design is calculated. The steps of façade design modification will be changing the material type of openings, or reducing the area of openings, re-constructing the solid walls, or alternating the shading design. The modifications of its original façade design are applied to minimize the total heat gain and get the value of final OTTV to less than 35 w/m^2.

4. Results

To know the overall thermal value transferred through the original design, it is important to calculate the basic technical parameters for gaining the OTTV. To calculate the OTTV, it needs to get the total area of exterior wall and its openings. Since the building design has facades on all orientation, the total area of façade is calculated from eight parts of the external wall: north, northeast, east, southeast, south, southwest, west, and northwest. The total area of external walls and openings is shown in Table 1.

Table 1. Total area of original external walls and openings

Orientation	Wall (m^2)	Openings (m^2)	Total Area (m^2)
North	115.78	1,456.85	1,572.62
Northeast	48.80	614.04	662.84
East	317.25	0.00	317.25
Southeast	67.79	853.00	920.79
South	168.52	2,120.49	2,289.01
Southwest	37.76	475.20	512.96
West	163.30	81.65	244.95
Northwest	34.52	434.43	468.95
		Total area	6,989.37

Based on the original design, it is specified that the wall is made of concrete at 30 cm thick covered by external and internal plastering finished with white emulsion painting. From its wall composition, it is calculated that the heat absorbance (α) is 0.27 and the heat transmittance (Uw) is 2.723 $W/m^{2o}C$. The openings of the building use clear glass cladding with prefabricated shading coefficient (SC_k) of 0.93 and its heat transmittance (Uf) is 5.7 $W/m^{2o}C$. From its specification, heat conduction through wall (Qw), heat conduction through openings or glass (Qg), solar heat gain through openings (Qsol), and the total OTTV of building's original design are shown in Table 2.

Table 2. Total OTTV of original design

Orientation	Qw (W)	Qg (W)	Qsol (W)	Qw+Qg+Qsol (W)	Area (m^2)	OTTV (W/m^2)
North	0.55	26.40	112.00	138.95	1,572.62	218,513.89
Northeast	0.55	26.40	97.35	124.30	662.84	82,392.62
East	7.43	0.00	0.00	7.43	317.25	2,358.75
Southeast	0.55	26.40	83.57	110.52	920.79	101,763.51
South	0.55	26.40	83.57	110.52	2,289.01	252,976.64
Southwest	0.55	26.40	151.63	178.58	512.96	91,604.13
West	4.96	9.50	75.33	89.79	244.95	21,993.24
Northwest	0.55	26.40	181.78	208.73	468.95	97,885.47
				Total	6,989.37	869,488.26
				TOTAL OTTV		124.40 W/m^2

The result of OTTV calculation based on original design shows that the OTTV is much higher than the threshold stated in the Indonesia's National Standard through SNI 03-6389-2011 for energy conservation. The steps of façade modification below are conducted to lower the OTTV by modifying possible façade elements in order to reduce the penetrated heat gain. First modification is directed to reduce the heat transfer through openings. The original clear glass cladding will be replaced by double glass cladding that has air cavity in between. The thickness of outside and inside glass is 8 mm and 6 mm respectively. This new modification of cladding construction is applied to all side of the façade and reduces the heat transmittance through openings (Uf) from 5.7 to 2.96 $W/m^{2o}C$. As the remain components remain the same, the overall thermal transfer value after first modification is shown on Table 3.

Table 3. Total OTTV after 1st modification

Orientation	Qw (W)	Qg (W)	Qsol (W)	Qw+Qg+Qsol (W)	Area (m^2)	OTTV (W/m^2)
North	0.55	13.73	112.00	126.28	1572.62	198,589.75
Northeast	0.55	13.73	97.35	111.63	662.84	73,994.86
East	7.43	0.00	0.00	7.43	317.25	2,358.75
Southeast	0.55	13.73	83.57	97.85	920.79	90,097.73
South	0.55	13.73	83.57	97.85	2,289.01	223,976.36
Southwest	0.55	13.73	151.63	165.91	512.96	85,105.25
West	4.96	4.94	75.33	85.23	244.95	20,876.58
Northwest	0.55	13.73	181.78	196.06	468.95	91,944.16
				Total	6,989.37	786,943.45
				TOTAL OTTV		112.59 W/m^2

Second modification is based on design after applying new double glass as constructed on 1st modification, and it is still aimed on reducing the heat transfer through openings. The original area of glass cladding will be reduced by adding parapet construction along the façade (Figure 2). It will be constructed 1 m high from each floor elevation with 15 cm thick of brick-plastering material, which has 12 cm of brick thickness, and 1.5 cm for each outside and inside plastering thickness. Based on the conductivity calculation results, this new façade design has reduced the area of openings and the heat transfer on each façade orientation. The OTTV value after second modification is displayed on Table 4.

Table 4. Total OTTV after 2nd modification

Orientation	Qw (W)	Qg (W)	Qsol (W)	Qw+Qg+Qsol (W)	Area (m^2)	OTTV (W/m^2)
North	2.33	10.09	82.33	94.76	1,572.62	149,020.52
Northeast	2.33	10.09	71.56	83.99	662.84	55,673.77
East	7.32	0.00	0.00	7.32	317.25	2,320.81
Southeast	2.33	10.09	61.43	73.86	920.79	68,009.17
South	2.33	10.09	61.43	73.86	2,289.01	169,065.82
Southwest	2.33	10.09	111.46	123.89	512.96	63,551.53
West	5.15	4.38	66.84	76.38	231.89	17,711.06
Northwest	2.33	10.09	133.63	146.06	468.95	68,493.81
				Total	6,976.31	593,846.49
				TOTAL OTTV		85.12 W/m^2

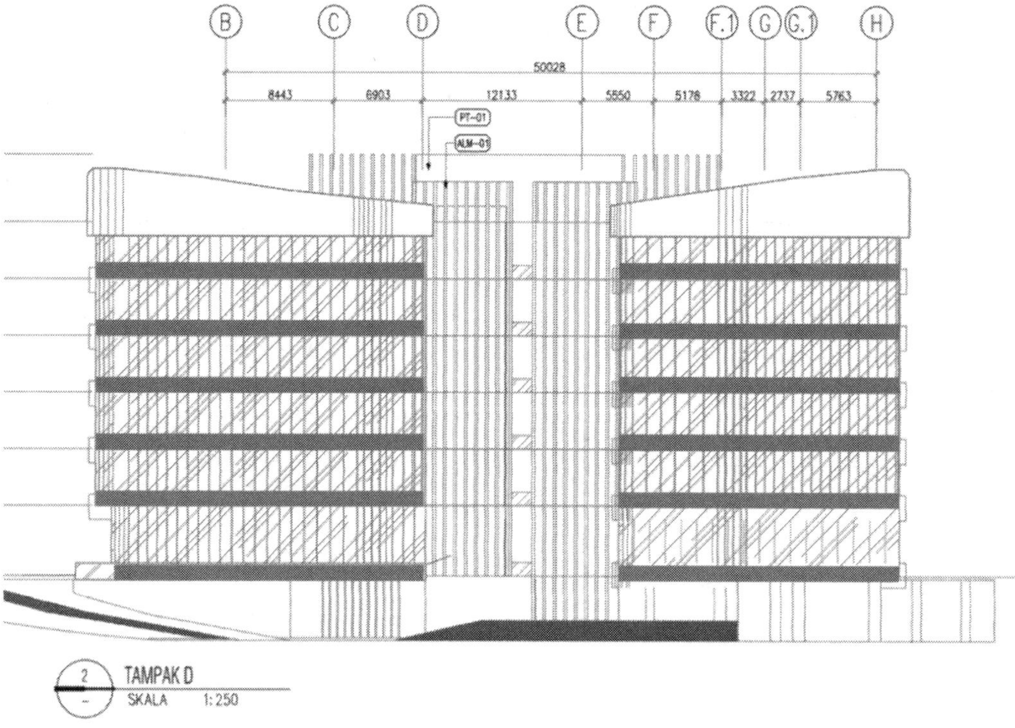

Figure 2. Building façade after additional parapet

In third modification, a 12 mm thick gypsum board layer will be added to the interior side of brick wall constructed in 2nd modification. This gypsum board has 0.17 W/m°C conductivity and will contribute to the new heat transmittance of wall (Uw) of 2.296 $W/m^{2o}C$. Although it is designed to layer all interior side of the wall, this additional gypsum board layer has only reduced the OTTV by less than 1 W/m2, from 85.12 W/m^2 on 2nd design modification to 84.69 W/m^2). The total OTTV obtained from third modification is displayed in Table 5.

Table 5. Total OTTV after 3rd modification

Orientation	Qw (W)	Qg (W)	Qsol (W)	Qw+Qg+Qsol (W)	Area (m^2)	OTTV (W/m^2)
North	1.96	10.09	82.33	94.38	1,572.62	148,425.75
Northeast	1.96	10.09	71.56	83.61	662.84	55,423.08
East	6.13	0.00	0.00	6.13	317.25	1,944.70
Southeast	1.96	10.09	61.43	73.48	920.79	67,660.93
South	1.96	10.09	61.43	73.48	2289.01	168,200.11
Southwest	1.96	10.09	111.46	123.51	512.96	63,357.53
West	4.32	4.38	66.84	75.54	231.89	17,517.46
Northwest	1.96	10.09	133.63	145.68	468.95	68,316.45
				Total	6,976.31	590,846.01
				TOTAL OTTV		84.69 W/m^2

Lowering the shading coefficient (SC) is the main goal for fourth modification. Additional metal-vertical and horizontal shading (SC_{eff}) will be constructed on northwest, west, and southwest façade, since those three elevations have the most exposed surface to solar radiation (highest solar factor). As shown on Figure 3, the vertical shading will have 40 cm of width (X), 12 mm of thickness, and constructed every 100 cm (Y) along the façade. Meanwhile, the horizontal shading will be designed to have 80 cm of length (X) for covering overall height (220 cm) of the glass cladding.

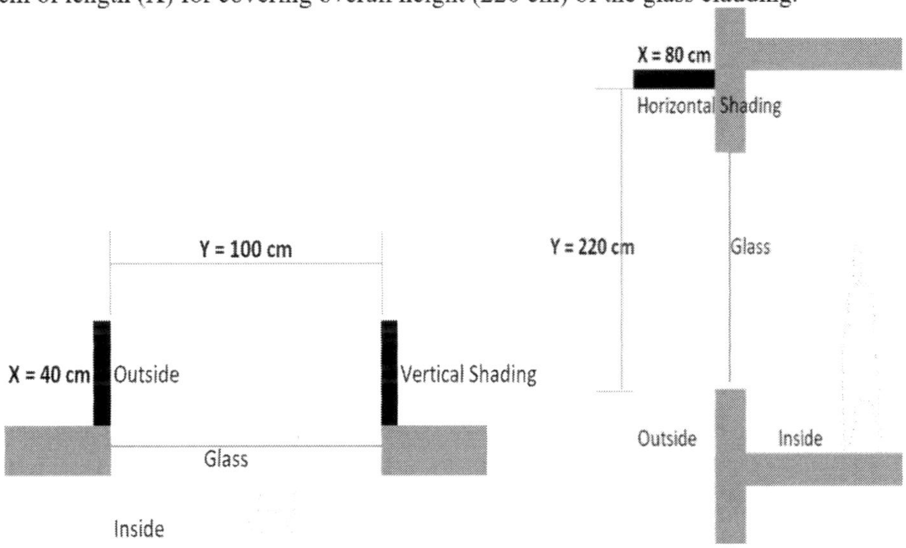

Figure 3. Design of vertical and horizontal shading

Both vertical and horizontal shading have the ratio (R) between its horizontal projection and the span of 0.40 and 0.36 with prefabricated shading coefficient (SC_k)of 0.93. The combination of SC_k and shading construction (SC_{eff}) gives the total shading coefficient (SC) for the openings on northwest, west, and southwest façade of 0.581, 0.661, and 0.569 respectively. This design of physical shading has successfully reduced the solar heat gain through the openings (Qsol) on northwest façade from 133.63 W to 83.50 W, on west from 66.84 W to 47.49 W, and on southwest from 111.46 to 68.23 W. Following the solar heat gain reduction, the OTTV has also been decreased to 77.50 W/m^2 (Table 6).

Other than constructing physical shading, reducing the solar heat gain through openings will also be conducted by modifying the prefabricated shading coefficient (SC_k). In fifth design modification, the original clear-glass cladding (SC_k of 0.93) will be replaced by 8 mm dark-blue glass with SC_k of 0.4. Since the value of new SC is much lower than the original, this modification gives almost half of solar heat gain (Qsol) of original clear glass usage for all façade orientation. The OTTV is also significantly reduced from 77.50 W/m^2 to 39.98 W/m^2 (Table 7).

Table 6. Total OTTV after 4th modification

Orientation	Qw (W)	Qg (W)	Qsol (W)	Qw+Qg+Qsol (W)	Area (m^2)	OTTV (W/m^2)
North	1.96	10.09	82.33	94.38	1,572.62	148,425.75
Northeast	1.96	10.09	71.56	83.61	662.84	55,423.08
East	6.13	0.00	0.00	6.13	317.25	1,944.70
Southeast	1.96	10.09	61.43	73.48	920.79	67,660.93
South	1.96	10.09	61.43	73.48	2,289.01	168,200.11
Southwest	1.96	10.09	68.23	80.28	512.96	41,178.39
West	4.32	4.38	47.49	56.19	231.89	13,029.47
Northwest	1.96	10.09	83.50	95.55	468.95	44,807.82
				Total	6,976.31	540,670.25
				TOTAL OTTV		77.50 W/m^2

Table 7. Total OTTV after 5th modification

Orientation	Qw (W)	Qg (W)	Qsol (W)	Qw+Qg+Qsol (W)	Area (m^2)	OTTV (W/m^2)
North	1.96	10.09	35.41	47.46	1,572.62	74,638.81
Northeast	1.96	10.09	30.78	42.83	662.84	28,389.84
East	6.13	0.00	0.00	6.13	317.25	1,944.70
Southeast	1.96	10.09	26.42	38.47	920.79	35,424.85
South	1.96	10.09	26.42	38.47	2,289.01	88,063.59
Southwest	1.96	10.09	29.34	41.39	512.96	21,233.82
West	4.32	4.38	20.42	29.13	231.89	6,753.95
Northwest	1.96	10.09	35.91	47.96	468.95	22,492.64
				Total	6,976.31	278,942.21
				TOTAL OTTV		39.98 W/m^2

Table 8. New total area of external walls and openings after 6th design modification

Orientation	Wall (m2)	Openings (m2)	Total Area (m2)
North	711.29	861.33	1,572.62
Northeast	421.06	241.78	662.84
East	317.25	0.00	317.25
Southeast	503.35	417.44	920.79
South	939.84	1,349.17	2,289.01
Southwest	373.24	139.72	512.96
West	163.30	68.59	231.89
Northwest	359.20	109.75	468.95
		Total area	6,976.31

The dramatic change of façade design is conducted in sixth modification. Without changing both structural and architectural layout plan, in this modification there will be additional all-floors vertical walls constructed on each curved corner of the building. The span of this new walls will be 4 m stretched from the middle of each corner, replacing 4 m span of glass cladding at the same stretch. As shown in Figure 4 this replacement will reduce the area of openings, therefore reduces the total solar heat gain. The changes of total area of walls and openings after this modification is displayed in Table 8.

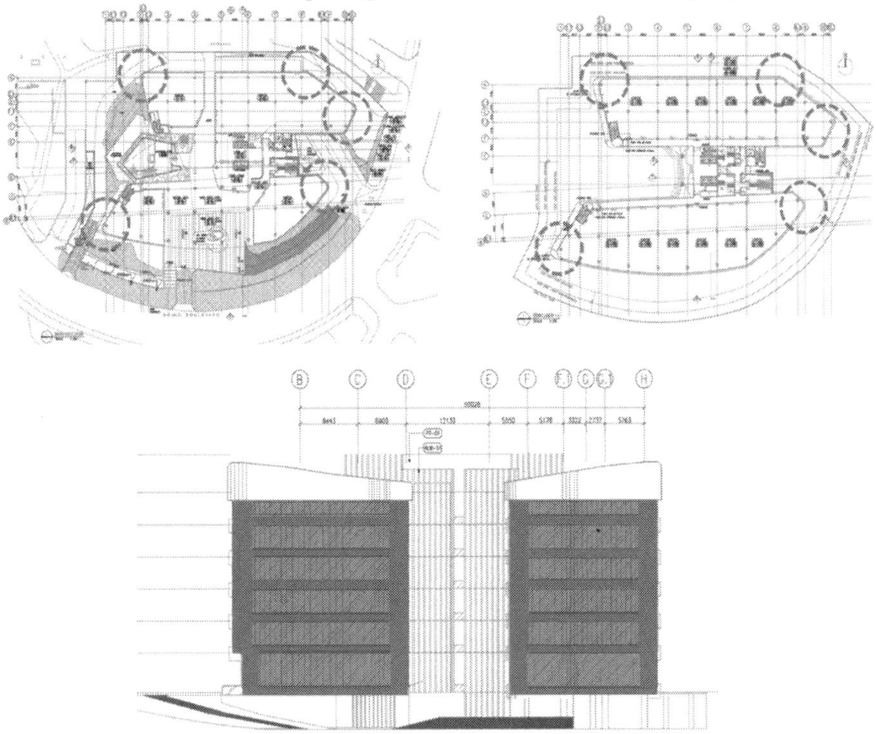

Figure 4. Additional vertical walls on building's curved corners

The area changes of walls and openings after sixth design modification has lead to dramatic changes of heat conduction both through walls and openings, and solar heat gain through the openings. The values of Qw at all façade orientation have increased, meanwhile the values of Qg and Qsol have decreased. As for the solar heat gain through openings (Qsol), it is significantly much lower compared to those on original openings area, particularly on northwest and southwest façade. 35.91 W and 29.34 W on original openings area has decreased to 12.34 W and 11.74 W on new façade design. The total OTTV has also been reduced accordingly, and it is shown in Table 9.

Table 9. Total OTTV after 6th modification

Orientation	Qw (W)	Qg (W)	Qsol (W)	Qw+Qg+Qsol (W)	Area (m^2)	OTTV (W/m^2)
North	2.77	8.12	17.54	28.43	1,572.62	44,704.32
Northeast	3.89	5.41	10.30	19.60	662.84	12,993.89
East	6.13	0.00	0.00	6.13	317.25	1,944.70
Southeast	3.35	6.72	17.59	27.66	920.79	25,470.12
South	2.52	8.74	22.87	34.12	2,289.01	78,108.86
Southwest	4.46	4.04	11.74	20.23	512.96	10,379.65
West	4.32	4.38	20.42	29.13	231.89	6,753.95
Northwest	4.70	3.47	12.34	20.51	468.95	9,616.55
				Total	6,976.31	189,972.05
				TOTAL OTTV		27.23 W/m^2

5. Conclusion

The results show that the modification on façade design can reduce the heat gain and improve thermal performance of the building. There is a dramatic change of OTTV from 124.40 w/m^2 on original design to 27.73 w/m^2 after several steps of façade modification. The biggest reductions of OTTV is obtained when modifications on openings are applied. Since the heat transmission is mainly coming into buildings through openings, modifications on glass and windows are the keys. The change of glass material, from clear glass to those with lower shading coefficient (SC) gives reduction of 37.52 w/m^2. Another significant modification that can lead to general façade design strategies is reducing the area of openings by replacing them with solid wall. In the case of GOP 9 Building, constructing new parapet and vertical solid wall can reduce OTTV by 27.47 w/m^2 and 12.25 w/m^2 respectively. Some good options of modification that will not substantially affect the overall architectural appearance are on glass and shading construction. Adding shading construction on openings and changing the double glazing can reduce the OTTV by 7.19 w/m^2 and 11.81 w/m^2 respectively.

Acknowledgement

The completion of this study could not have been possible without the assistance of Green Building Council Indonesia (GBCI) in Jakarta for giving the opportunity to have the case study and hands-on guidance during Greenship Professional (GP) Training. Their contributions are sincerely appreciated and gratefully acknowledged.

References

[1] Pramesti P U, Ramandhika M, Hasan M I and Werdiningsih H 2021 The influence of building envelope design in energy efficiency: OTTV calculation of multi storey building *IOP Conference Series: Earth and Environmental Science* **623** 012075

[2] Zhao H and Magoulès F 2012 A review on the prediction of building energy consumption *Renewable and Sustainable Energy Reviews* **16** 3586–92

[3] J. Litardo, R. Hidalgo-León, J. Macías, K. Delgado and G. Soriano 2019 Estimating energy consumption and conservation measures for ESPOL Campus main building model using EnergyPlus *2019 IEEE 39th Central America and Panama Convention (CONCAPAN XXXIX)* 2019 IEEE 39th Central America and Panama Convention (CONCAPAN XXXIX) pp 1–6

[4] Kusumawati L, Setyowati E and Purnomo A B 2021 Rule of thumb for the calculation of overall transfer thermal value in Bandung *IOP Conference Series: Earth and Environmental Science* **780** 012014

[5] Abdul Nasir M H and Hassan A S 2020 Thermal performance of double brick wall construction on the building envelope of high-rise hotel in Malaysia *Journal of Building Engineering* **31** 101389

[6] Machdijar L K, Setyowati E and Purnomo A B 2019 Anticipation of solar radiation through the building C envelope of campus a Universitas Trisakti *AIP Conference Proceedings* **2114** 040009

[7] Riniardi K, Priyomarsono N W and Rilatupa J 2019 Re-designing facade of kadin tower building (application of retrofit programme with OTTV) *IOP Conference Series: Materials Science and Engineering* **508** 012028

[8] Loekita S and Priatman J 2015 OTTV (SNI 03-6389-2011) and ETTV (BCA 2008) Calculation for Various Building's Shapes, Orientations, Envelope Building Materials: Comparison and Analysis *ced* **17** 108–16

[9] Natephra W, Yabuki N and Fukuda T 2018 Optimizing the evaluation of building envelope design for thermal performance using a BIM-based overall thermal transfer value calculation *Building and Environment* **136** 128–45

[10] Lim Y-W, Seghier T E, Harun M F, Ahmad M H, Samah A A and Majid H A 2019 Computational BIM for Building Envelope Sustainability Optimization *MATEC Web Conf.* **278**

[11] Bhatia A, Sangireddy S A R and Garg V 2019 An approach to calculate the equivalent solar heat

gain coefficient of glass windows with fixed and dynamic shading in tropical climates *Journal of Building Engineering* **22** 90–100

[12] Rabbani B A 2019 Comparative Study of Energy Consumption between OTTV and Sefaira in a House *IOP Conference Series: Earth and Environmental Science* **248** 012089

[13] Sandi R Y and Ismail M 2019 Optimizing Building Planning and Design to Control External Energy Loads *IOP Conference Series: Earth and Environmental Science* **328** 012028

[14] Kusumawati L, Setyowati E and Purnomo A B 2021 Practical-Empirical Modeling on Envelope Design towards Sustainability in Tropical Architecture *Sustainability* **13**

[15] Jia J and Lee W L 2018 The rising energy efficiency of office buildings in Hong Kong *Energy and Buildings* **166** 296–304

[16] Muhfizaturrahmah, Yuniarti N, Sukisno T and Urdifat Y 2021 The influence of Overall Thermal Transfer Value (OTTV) on building energy consumption *Journal of Physics: Conference Series* **1833** 012046

[17] Chan L S 2019 Investigating the environmental effectiveness of Overall Thermal Transfer Value code and its implication to energy regulation development *Energy Policy* **130** 172–80

[18] Hwang R-L, Shih W-M, Lin T-P and Huang K-T 2018 Simplification and adjustment of the energy consumption indices of office building envelopes in response to climate change *Applied Energy* **230** 460–70

IOP Conf. Series: Earth and Environmental Science **847** (2021) 012029 doi:10.1088/1755-1315/847/1/012029

The forecasting model of discharge at Brantas sub-basin using autoregressive integrated moving average (ARIMA) and decomposition methods

G Idfi[1], A Yulistyorini[1], T Rahayuningsih[1], V A K Dewi[1] and E Setyawan[1]

[1]Department of Civil Engineering, Faculty of Engineering, Universitas Negeri Malang, Malang, Jawa Timur, Indonesia.

gilang.idfi.ft@um.ac.id

Abstract. A watershed is a combination of several rivers and tributaries with certain boundaries that function to drain rainwater into a lake or a sea. One of the hydrological data contained in the watershed is discharge data. If there is incomplete discharge data, it must be extended based on historical data. ARIMA and decomposition are methods that can predict time series data. The purposes of this research are to determine the historical discharge patterns of Brantas Sub-basin, to know the discharge forecasting model of Brantas Sub-basin, to know the results of forecasting Brantas Sub-basin discharge, and to compare the accuracy between ARIMA and decomposition methods. The accuracy is obtained by calculating MSE and RMSE values. The best method is a method that has the smallest MSE and RMSE values. The results of the research showed that Brantas Sub-basin discharge data in 2007-2017 has a seasonal pattern. The best ARIMA model is ARIMA (0,0,3)(1,0,1)12 model, while the best decomposition model is the additive decomposition model. The Decomposition method has better accuracy than the ARIMA method in predicting discharge of Brantas Sub-basin.

1. Introduction

Some rivers and tributaries will form a watershed which has certain boundaries on land and sea that are used as rainwater container and then flow into lake or sea [1]. The watershed can consist of one or more sub-basin that function to receive and drain rainwater through tributaries to the main river. One of the hydrological data contained in watersheds is discharge data [2]. According to Limantara, the discharge is a volume of water that passes through a river cross-section which is calculated per unit of time [3].

Some rivers and tributaries will form a watershed which has certain boundaries on land and sea that are used as rainwater container and then flow into lake or sea [4]. The watershed can consist of one or more sub-basin that function to receive and drain rainwater through tributaries to the main river [5]. One of the hydrological data contained in watersheds is discharge data. According to Zhang, the discharge is a volume of water that passes through a river cross-section which is calculated per unit of time forecasting can be done according to right method [6]. These patterns include seasonal, cyclical, trend, and irregular patterns.

Auto-Regressive Integrated Moving Average (ARIMA) is a method that can be used to extend data using time series data [7]. According to Juwono in Pramujo, time-series data used in the ARIMA model must be stationary [8]. This makes the ARIMA model forecasting results better than the ARMA model. There is also a decomposition method that can be used to forecast data. According to Meilano state that components of the decomposition method include trend factor, seasonal factor, cycle factor, and a random factor [9]. The advantage of this method is that patterns or components can be calculated separately, so that more accurate forecasting results are obtained.

Discharge data is an example of time series data in the field of hydrology, so it can be extended using stochastic methods. Some previous research state that the ARIMA method can be used to predict discharge data and provide good results [10]. Based on this, the research tries to compare the accuracy of the ARIMA method with the decomposition method. ARIMA and decomposition methods are used to forecast discharge data with the help of Minitab 18 software.

The research was conducted at Brantas Sub-Basin to predict the availability of water resources in the future, this is because the communities around the site need large amounts of water that need to be used in industry, food crop agriculture, plantation, and tourism. This is in accordance with the Decree of Minister of Settlement and Regional Infrastructure Number: 360/KPTS/M/2004 in 2004 which states that river discharge data forecasting is needed to estimate water availability needed. The results of forecasting are used when planning, designing, building, operating and maintaining water resources. Data commonly used are annual, seasonal, monthly, semi-monthly, and ten-daily discharge data. Discharge forecasting can also be used to predict flooding.

The purpose of this research is to conduct a comparative analysis of ARIMA and decomposition methods for the extension of discharge data. The results of this research are expected to obtain information on the reliability level of both methods in predicting the extension of discharge data.

2. Research method

Discharge forecasting of Sumber Brantas Sub Watershed uses ARIMA and Decomposition Methods which is modeled with Minitab 18 software. The historical data used is discharge data of Brantas Sub-basin with a discharge gauge at Sengkaling Dam. The data began in 2007 until 2018 which is divided into two parts, namely in-sample data (data processed for model determination) and out sample data (data used to calibrate forecast results for the next period). Discharge data in 2007-2017 are used as in sample data, while discharge data in 2018 are used as out sample data.

2.1. Deletion of outlier data and pattern analysis

Outlier data is data that has value too far from other data. The data needs to be deleted so that forecast results have good accuracy [11]. Analysis of historical data pattern needs to be done before making time series forecasting. This is important because data pattern will affect type of forecasting method that will be used

2.2. Forecasting with arima method

The ARIMA method consists of several processes with different models. The models is :

2.2.1. Auto regressive (AR) process

The equation for auto regressive (AR) is

ARIMA (p,0,0)

$$Xt = \mu + \phi 1Xt{-}1 + \phi 2Xt{-}2 + \cdots + \phi pXt{-}p + et \tag{1}$$

Where:

μ = constanta

ϕp = autoregresif parameters

et = margin errors

2.2.2. Moving average (MA) process

The equation for moving average (MA) is

ARIMA (0,0,q) or MA (q)

$$Xt = \mu + et - \theta 1et{-}1 - \theta 2et{-}2 - \cdots - \theta qet{-}q \tag{2}$$

Where:

$\theta 1$ until θq = moving average parameters
et−q = margin errors
μ = constanta

2.2.3. Steps forecasting
The stages of forecasting with ARIMA methods are.

- Input in sample data in Minitab 18 software.
- Test stationarity of data on variant and average.
- Determine ARIMA models based on ACF and PACF patterns.
- Test parameters of model, white noise assumption and residual normality.
- Determine the best ARIMA model.
- Forecast based on the best ARIMA model.

2.3. Forecasting with decomposition method
Decomposition is a method for analyzing time series data by calculating each component separately [12]. Forecasting results from the method are obtained by combining the estimates of each component. This method depend on 4 component :

- Trend factor : the trend factor shows that data has increased, decreased or been constant over a long period of time. Trends can be represented by straight lines, exponential S curves and other long-term patterns.
- Cycle factor : the cyclical factor shows that data has a varying pattern where the data moves from large values to small values and returns to large values in a longer period of time than seasonal factors.
- Seasonal factor : seasonal factors show that data has a steady increase and decrease over a long period of time, such as monthly, quarterly and yearly.
- Random factor : the random factor is assumed to be the difference between the combined effect of the three decomposition factors and the actual data

2.3.1. Decomposition equation model
The general model of the decomposition method can be seen in equation below

$$Xt = f(It, Tt, Ct, Et) \quad (3)$$

where
Xt = time series value in t period,
It = season indeks in t period
Tt = trend component in t period
Ct = cycle component in t period
Et = random component in t period

2.3.2. Decomposition aditif model
The aditif model of the decomposition method can be seen in equation below

$$Xt = (It + Tt + Ct) + Et \quad (4)$$

2.3.3. Decomposition multiplikatif model
The multiplikatif model of the decomposition method can be seen in equation below

$$Xt = (It * Tt * Ct) * Et \quad (5)$$

2.3.4. Steps forecasting
The stages of forecasting with decomposition methods are:

- Input in sample data in Minitab 18 software.
- Analyze data using additive and multiplicative decomposition models.
- Determine the best decomposition model.

IOP Conf. Series: Earth and Environmental Science **847** (2021) 012029 doi:10.1088/1755-1315/847/1/012029

- Forecast based on the best decomposition model.

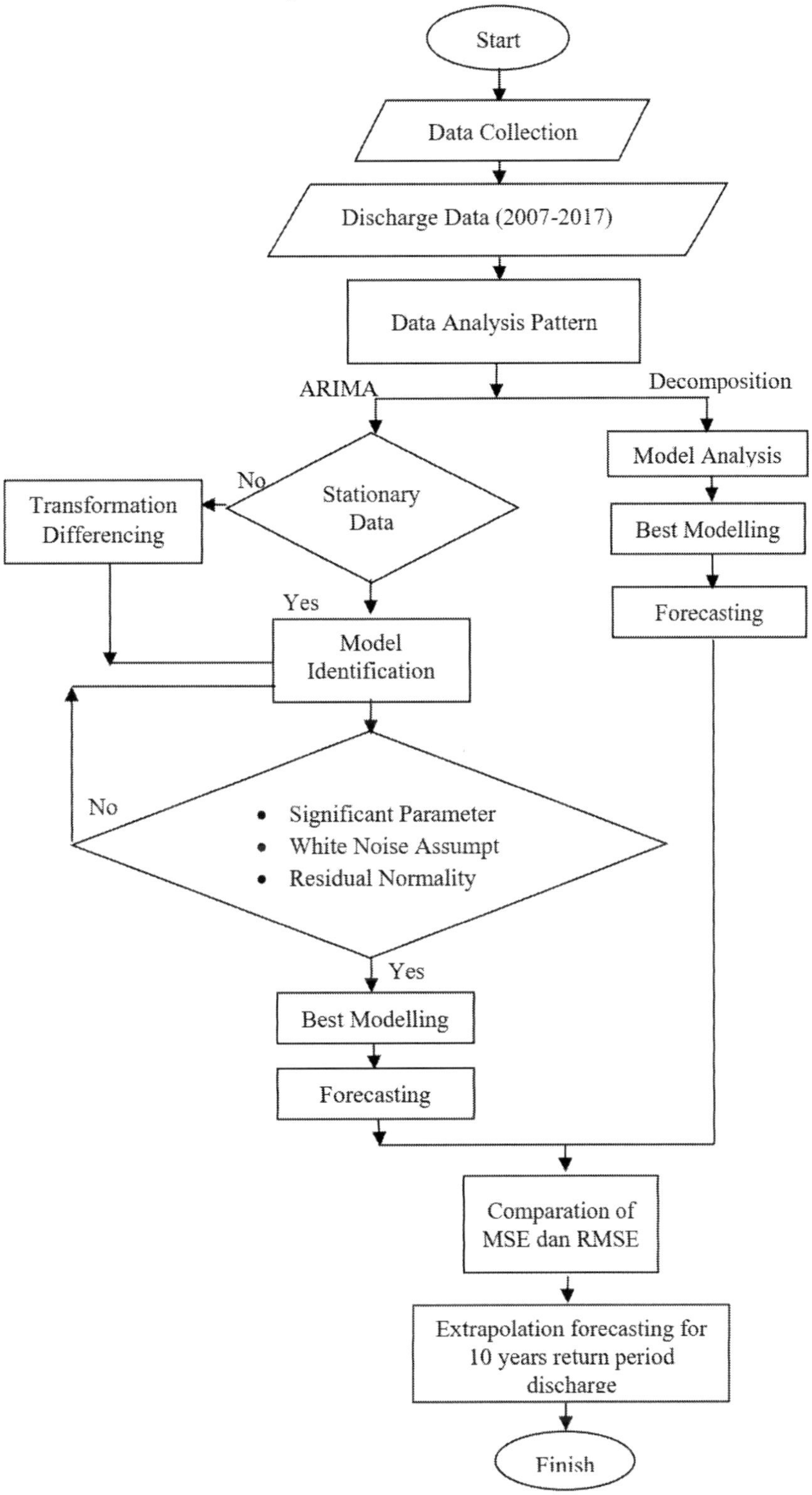

Figure 1 Flow chart of research

2.4. Fitting test

Data accuracy testing must be done in every forecasting process. This can be done by minimizing the error rate, which is the difference between the actual value and the forecast value [13]. There are several ways that can be used to test the accuracy of forecasting results, including MSE and RMSE [14].

Mean Square Error (MSE) is the result of dividing the error value squared by the amount of data [15]. The MSE calculation formula can be seen in equation below

$$MSE = \frac{\varepsilon(X_t - F_t)^2}{n} \tag{6}$$

where:
Xt = actual value in t period
.Ft = forecast value in t period
..n = forecast period

Root Mean Square Error (RMSE) is the root of the MSE value[16]. The RMSE calculation formula can be seen in equation below.

$$RMSE = \sqrt{\frac{\varepsilon(X_t - F_t)^2}{n}} \tag{7}$$

where:
Xt = actual value in t period
.Ft = forecast value in t period
..n = forecast period

To make the implementation of the research does not deviate from the formulation of the problem and the purpose, a flow diagram is used as in Figure 1. The diagram is compiled from the data which were obtained from the relevant institutions or survey research locations.

3. Results and discussion

Pattern of Brantas Sub-basin discharge data in 2007-2017 can be seen in Figure 2 below

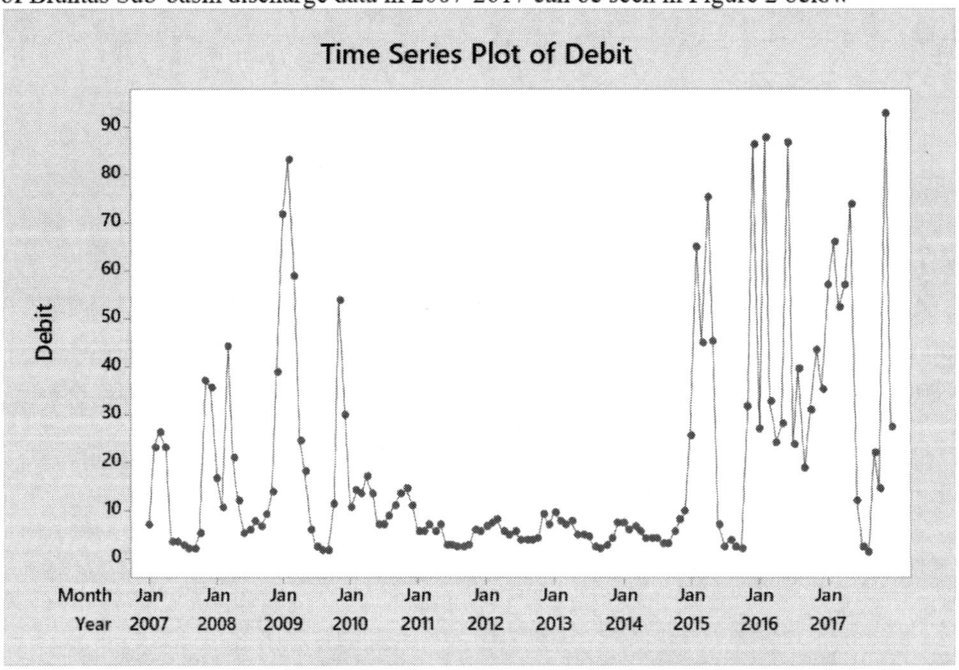

Figure 2 Pattern of Brantas sub watershed discharge data in 2007-2017

The first step is to enter in sample data in Minitab 18 software. Then test stationarity of data on variant and average. Data stationarity test on variant is done by looking at rounded value on Box-Cox transformation graph [11]. Data stationarity test on average is done by looking at trend pattern of data Result of data stationary test on variant and test for average can be seen in Figure 3 and 4 below

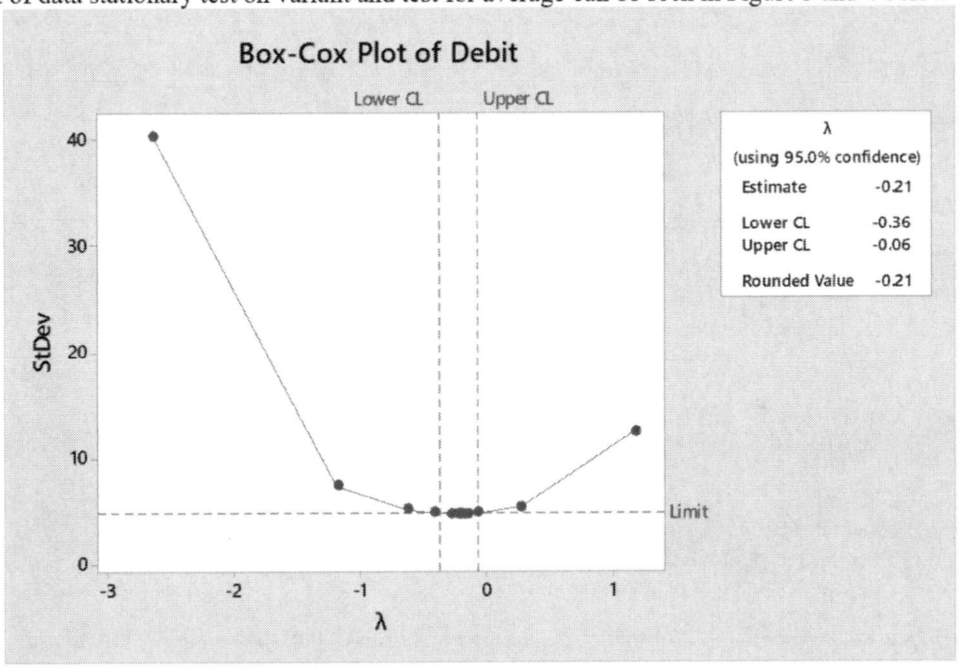

Figure 3 Result of data stationary test on variant

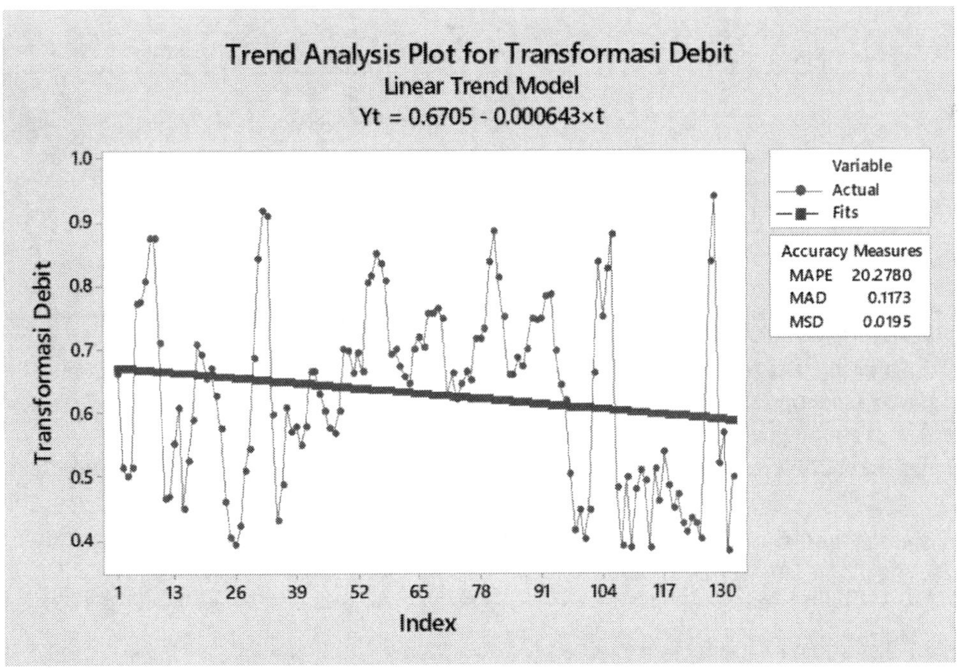

Figure 4 Result of data stationary test for average

The best comparison of the graph of the actual discharge value and the ARIMA fit value $(0,0,3)(1,1,3)^{12}$ can be seen in Figure 5 below

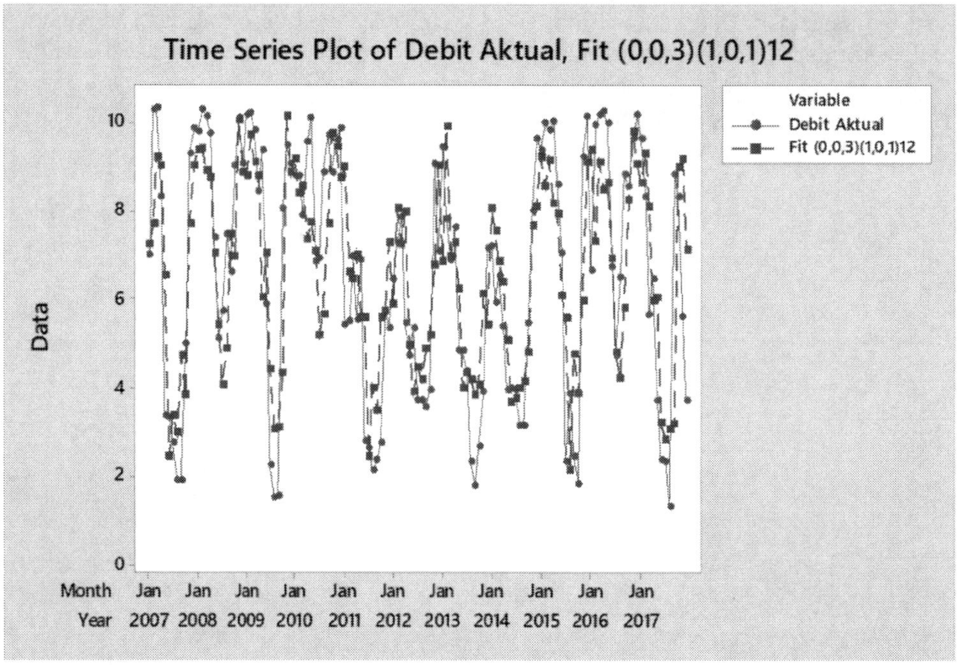

Figure 5. Comparison of observed discharge compared to the results of ARIMA model

The tests are also carried out on other ARIMA models and then calculate MSE and RMSE values of each model [13]. Recapitulation of significance tests, MSE and RMSE calculation results can be seen in Table 1 below

Table 1 Recapitulation of Significance Tests, MSE and RMSE Calculation Results

No	Models	Parameter Estimation	Diagnostic Test		MSE	RMSE
			White Noise Test	Residual Normality Test		
1	ARIMA $(0,0,3)(0,1,1)^{12}$	Fulfilled	Fulfilled	Fulfilled	7,095	2,664
2	ARIMA $(0,0,3)(1,0,0)^{12}$	Fulfilled	Fulfilled	Fulfilled	3,093	1,759
3	ARIMA $(0,0,3)(1,0,1)^{12}$	Fulfilled	Fulfilled	Fulfilled	2,865	1,693
4	ARIMA $(0,0,3)(1,1,3)^{12}$	Fulfilled	Fulfilled	Fulfilled	6,933	2,633
5	ARIMA $(0,1,1)(0,1,3)^{12}$	Fulfilled	Fulfilled	Fulfilled	7,680	2,771
6	ARIMA $(0,1,2)(1,0,1)^{12}$	Fulfilled	Fulfilled	Fulfilled	3,410	1,847
7	ARIMA $(1,0,0)(0,1,1)^{12}$	Fulfilled	Fulfilled	Fulfilled	6,987	2,643

Table 1 Recapitulation of Significance Tests, MSE and RMSE Calculation Results (contd.)

No	Models	Parameter Estimation	Diagnostic Test		MSE	RMSE
			White Noise Test	Residual Normality Test		
8	ARIMA $(1,0,0)(0,1,3)^{12}$	Fulfilled	Fulfilled	Fulfilled	6,650	2,579
9	ARIMA $(1,0,0)(1,0,1)^{12}$	Fulfilled	Fulfilled	Fulfilled	2,912	1,707
10	ARIMA $(1,1,1)(0,1,1)^{12}$	Fulfilled	Fulfilled	Fulfilled	7,843	2,800

Results of additive model analysis include time series, component analysis graph and seasonal analysis graph. Time series plot of additive decomposition model can be seen in Figure 6 below.

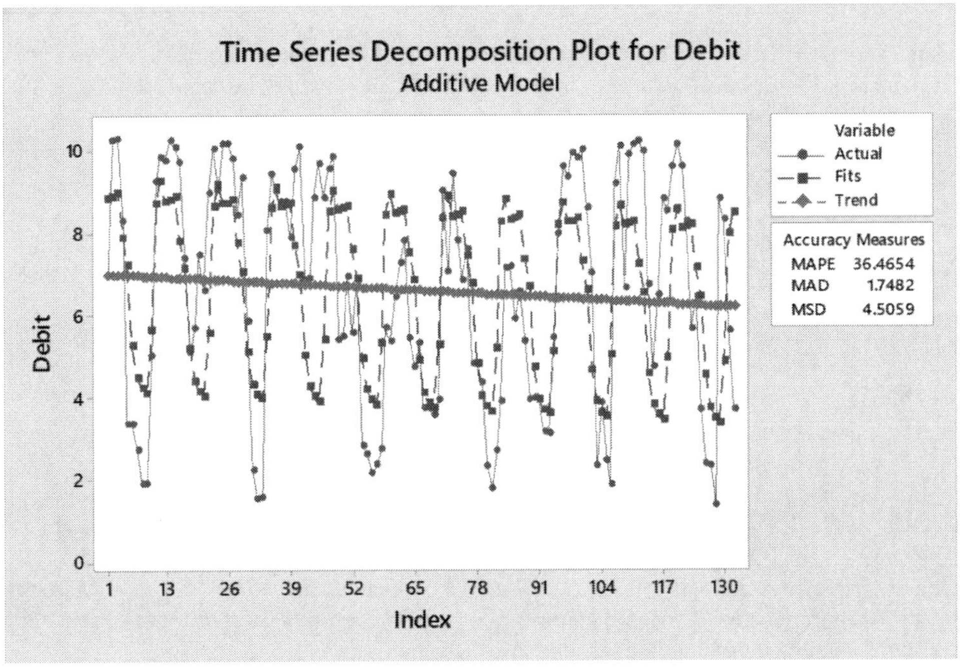

Figure 6. Comparison of observed discharge compared to the results of decomposition model

The analysis is also carried out on multiplicative model and then calculate MSE and RMSE values of each model. Recapitulation of MSE and RMSE calculation results can be seen in Table 2 below

Table 2 Recapitulation of MSE and RMSE Calculation Results

No	Models	MSE	RMSE
1	Additive	4,506	2,123
2	Multiplicative	4,621	2,150

Comparison of accuracy between ARIMA and decomposition methods can be seen in Table 3 below

Tabel 3 Comparison of accuracy between ARIMA and decomposition methods

No	Methods	In Sample		Out Sample	
		MSE	RMSE	MSE	RMSE
1	ARIMA	2,865	1,693	11,167	3,342
2	Decomposition	4,506	2,123	7,970	2,823

4. Conclusions

Discharge data of Brantas Sub-basin in 2007-2017 has seasonal pattern. It can be seen from existence of monthly seasonal factor that affect data. Therefore, forecasting process can be carried out using ARIMA and decomposition methods. Discharge forecasting model of Brantas Sub-basin are made using ARIMA $(0,0,3)(1,0,1)^{12}$ and additive decomposition model. Both models are chosen as forecasting models because they have the best accuracy among other ARIMA and decomposition models.

The results of forecasting Brantas Sub-basin discharge data in 2018 using decomposition method have MSE and RMSE values that are smaller than ARIMA method. It shows that the results forecasting using decomposition method has smaller error values and data variance that is closer to actual data variance. Decomposition method has better accuracy than ARIMA method in predicting discharge data of Brantas Sub-basin. That is because MSE and RMSE values of decomposition method is smaller than ARIMA method for out sample criteria.

References

[1] Harto S 1993 *Analisis Hidrologi (Hydrological Analysis)* (Jakarta: Gramedia Pustaka)
[2] Limantara L M 2010 *Hidrologi Praktis (Practical Hydrology)* (Bandung: Lubuk Agung)
[3] Viessman W, Knapp J W, Lewis G L and Harbaugh T E 1977 *Introduction to Hydrology* (Newyork: Harper and Row)
[4] Nathan R J and McMahon T A 1990 *Water Resources Research* **26** 1465–1473
[5] Stewart M K 2015 Hydrol. *Earth Syst. Sci.* **19** 2587–2603
[6] Zhang R, Li Q, Chow T L, Li S and Danielescu S 2013 *Hydrol. Process* **27** 2659–2665
[7] Xu C 2002 *Hydrologic Models* (Uppsala: Uppsala University Department of Earth Sciences Hydrology)
[8] Pramujo B, Juwono P T and Soetopo W 2014 *Discharge Modeling Using ARIMA Method to Determine Selorejo Reservoir Operation Pattern (Journal of Water Resources Engineering)* (Malang : Brawijaya University)
[9] Meilano A 2016 *Forecasting Number of Regional Library Visitors at Central Java Provincial Using Decomposition Method* Undergraduate thesis published (Semarang: Faculty of Mathematics and Natural Sciences, State University of Semarang)
[10] Danish Hydraulic Institute 1982 *NAM Model Users Guide* (Denmark: Danish Hydraulic Institute)
[11] Tsutsumi A, Jinno K and Berndtsson R 2004 *Hydrological Sciences* **49(2)** 205-226
[12] Shakti D W, Hadiani R and Setiono 2014 Discharge Analysis Based on 15 Daily Cumulative Rain by Backpropagation Artificial Neutral Network Method at Tirtomoyo Watershed for Sulingi Point *e-Journal of Civil Engineering Matrix* 768-775
[13] Utami T N, Hoyyi A and Rusgiyono A 2017 Penggunaan Metode Peramalan Kombinasi Trend Deterministik dan Stokastik pada Data Jumlah Penumpang Kereta Api (Studi Kasus: KA Argo Muria) (The Use of Forecasting Methods Combination of Deterministic and Stochastic Trends on the Number of Train Passengers Data (Case Study: Argo Muria Train)) *Jurnal Gaussian* **6(1)** 131-140
[14] Kuok K K, Harun S, and Chiu PC 2011 *Int. J. Advance. Soft Comput. Appl.* **3(3)** 1-17
[15] Jain S K 1993 *Sciences -Journal- des Sciences Hydrologiques* **38(5)** 431-441
[16] Yuni S, Talakua M W, Lesnussa Y A 2015. Forecasting Number of Pattimura University Library Visitors at Ambon Using Decomposition Method *Journal of Mathematical Sciences and Applied* **9(1)** 41-50

IOP Conf. Series: Earth and Environmental Science **847** (2021) 012030 doi:10.1088/1755-1315/847/1/012030

Duri Kosambi Sponge Park: The application of Sponge City concept on green open space in reducing flood intensity in Jakarta

C Hadhinata[1], N Firdaus[2] and M M A Pratama[3]

[1]Civil Engineering Education Study Program, Department of Civil Engineering, Faculty of Engineering, Universitas Negeri Malang, Malang, Indonesia
[2]English Language Education Study Program, Department of English, Faculty of Letters, Universitas Negeri Malang, Malang, Indonesia
[3]Department of Civil Engineering, Faculty of Engineering, Universitas Negeri Malang, Malang, Indonesia

mirza.abdillah.ft@um.ac.id

Abstract. Flood in Jakarta has been a big issue nowadays. It soaked a thousand hundred hectares of resident area. There are many factors generating flood in that area, such as the dysfunctional of green open space as flood control. Furthermore, the large of provided green open space has not been complied the requirement, as only 9.9% of pre-requirement large of 30% of large area. Sponge City Concept is on of ideas in the form water controlling succeeded to be applied in China having 4 main concepts, which are water ecology, water safety, water environment, and water resources. Therefore, this article takes green open space Duri Kosambi as the specific research object by redesigning it into Duri Kosambi Sponge Park. This paper utilizes 3 (three) steps in redesigning green open space Duri Kosambi, pre-design, design, and post-design. The result of this paper Duri Kosambi Sponge Park design consisting of a plan-layout design, an aesthetic bioswale, a multifunction "O'Area", and an absorption poll visualized in 3D picture. Additionally, a plan and analysis of Duri Kosambi implementation are discussed in this paper. By stating this idea, it is expected to be able to give the positive effect for environment, economics, and social culture in the society.

1. Introduction

Flooding in Jakarta has been a big issue recently. It soaked 1,397 neighbourhoods with a total population of 454,158 people and 114,305.69 ha of drown area [1]. The Meteorology, Climatology, and Geophysics Council (BMKG) stated that the flood in Jakarta in 2021 is generated by extreme weather, impacting the increase of rain cloud growth and rainfall intensity [2]. Besides, the land subsidence in Jakarta occurring every year impacts the increase of risk of being flooded by high tide and flood. The Directorate General of Water Resources Ministry of Public Works and Housing (PUPR) [3] stated that the land in Jakarta undergo subsidence with the range 0-12 cm/year due to the escalation of groundwater usage for household consumption and industry. The decrease of the soil is caused by the widespread of ground water for household consumption and industry in Jakarta. Pratiwi, et.al [4] also stated that Jakarta area is a downstream area located under the sea level. Therefore, if the overflow of water on the surface occurs, the flood will occur.

Flood occurring in Jakarta is the annual duty for government of Jakarta. On 2021, government collaborating with Regional People's Representative Assembly (DPRD) of Jakarta have been allocated the fund by IDR 1.5 trillion for solving the flood [5]. The fund has been allocated for revitalisation of pump house, river normalization, and even for alerting Public Facilities and Infrastructure Handling workers [5,6]. However, that have not been able to reduce flood intensity coming annually.

In fact, the existing green open space has been unable to resist high puddles. It is only planted with trees without any water absorption into the soil before being channelled to the river concept. This condition worse with the fact that the total amount of green open space in Jakarta is only 9.9% of 30% of pre-requirement [7].

The concept of flood control in urban areas has been developed recently, one of them is the "Sponge City Concept". It is an idea discussing ecological rain management developed in China [8]. It is also applied in some countries in the world before with different terms as shown by Figure 1.

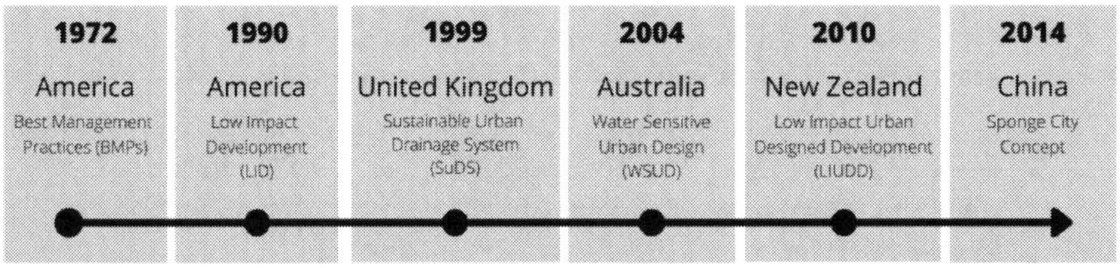

Figure 1. The development of *Sponge City* concept discovery [9]

For instance, Best Management Practices (BMPs) in America, Sustainable Urban Drainage System (SuDS) in England, and Water Sensitive Urban Design (WSUD) in Australia succeeded in being implemented [6]. There are 4 (four) main principles in planning and implementation of the sponge city concept, such as: water ecology; water safety; water environment; and water resources [10]. However, the application of the concept has not been applied massively in Jakarta, especially in the green open space area. The government only focuses on the addition of biopore holes around the street [11,12] and less attention regarding the value of green open space as the media to reduce the flood. Based on the explanation, the writer has the idea in the form of green open space implementing sponge city concept to reduce flood intensity, especially in Jakarta.

2. Methods

The methods used in designing Duri Kosambi Sponge Park shown as Figure 2.

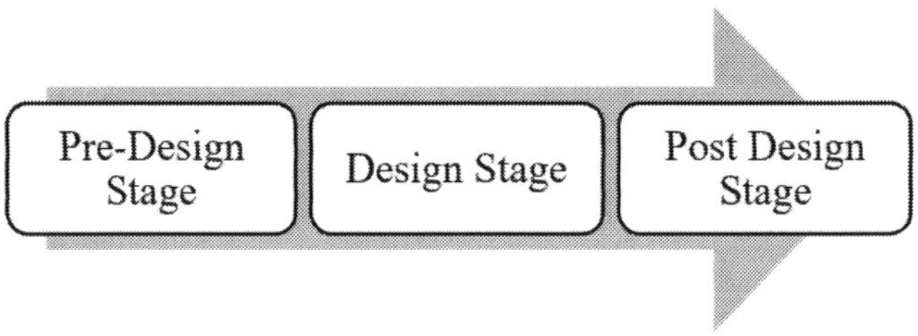

Figure 2. The Method Stages of Designing Duri Kosambi Sponge Park

2.1. Pre-design stage

Pre-design is conducted by collecting the data required in designing Duri Kosambi Sponge Park in the form of secondary data. The data is obtained by reviewing numerous literature sources, including from books, news articles, national and international journals related to the discussed topic.

2.2. Design stage

The collected data is processed and presented into systematic results later. This stage is undergone by creating the preliminary design of Duri Kosambi Sponge Park using AutoCAD, SketchUp, and Lumion. Afterwards, the design is deliberated with experts to get constructive inputs, suggestions, and final design.

2.3. Post design stage
Later, the approved final design will be analyzed regarding the steps of implementation the design and how the effect if the design is implemented. The result is shown in the form of the chart and table.

3. Results and discussion

3.1. Duri Kosambi sponge park layout plan
The concept of Duri Kosambi Sponge Park is shown Figure 3. The layout setting is supposed to form an artistic, appropriate, and multifunction design of the planned element. The layout design of this park is divided into 4 (four) zones (Figure 4), including.

- Zone 1 is an absorption pool area.
- Zone 2 is a playground area.
- Zone 3 is a *Multifunction* O'Area.
- Zone 4 is a *Tree Canopy Expansion* by adding the variety of trees in the garden.

Figure 3. The existing condition of green open space of Duri Kosambi

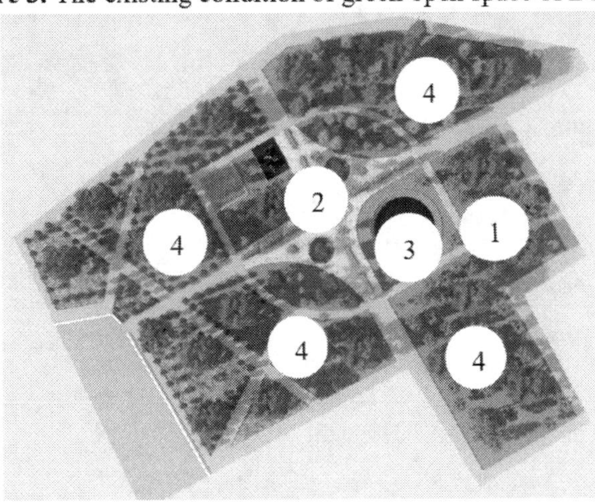

Figure 4. Layout plan of "Duri Kosambi Sponge Park"

The road area plan is designed by utilizing paving blocks with raw material and mix compositions which can absorb rainwater (porous concrete). It is expected to accelerate rainwater into the soil. Putri et al. [13] revealed that porous concrete has rapid infiltration acceleration at a rate of 2,804.71 inches/hour. There are two paving blocks employed in the road area. They are rectangle shape (conventional) and hollow shape. Its combination will also add to the aesthetics of the garden.

IOP Conf. Series: Earth and Environmental Science **847** (2021) 012030 doi:10.1088/1755-1315/847/1/012030

3.1.1. Aesthetic bioswale

The next design is providing a building aesthetic bioswale in the median and garden roadside as revealed by Figure 5. Bioswale is one of the engineering design infrastructures ensuring that the water flow is directed to a reservoir [14]. The Aesthetic bioswale design is in the form of elongated basin with a certain depth and it is planted with ornamental plants vegetation.

Figure 5. Aesthetic Bioswale design

Bioswale has a simple system in contributing water absorption into the soil. Later, the road runoff water getting into bioswale, then filtered and deposited by the variety of plants and rocks. Furthermore, the water is infused into the soil sequentially through the hollow pipes shown in the Figure 6.

Figure 6. The Detail of Aesthetic Bioswale

The production of bioswale brings a lot of benefits, such as being able to reduce the content of mud, heavy metals, and rainwater pollutants; enhancing the ability of water absorption into the soil; and being able to add aesthetics and recreation for the local community [15]. It has been proven in study conducted by Xiao, et.al. [16] identifying the existence of nitrogen, phosphate, and total organic loading by 99.1%, 99.5%, and 99.4% and surface runoff by 99.4% decrease, compared to the after bioswale construction control. Herein in Purvis, et.al. [17] also identified the existence of enterococci concentration, fecal coliform concentration, and total suspended sediment decrease by 55%, 75%, and 10% to bioswale overflow and by 59%, 65%, and 88% to underdrain bioswale.

3.1.2. Multifunction O'area

Multifunction O'Area is an additional facility in the form of outdoor seating in the middle of Duri Kosambi Sponge Park. It is designed as multilevel design with a grass layer cushion, as being shown in Figure 7. It is integrated with an absorption well (Figure 8), functioning as rainwater reservoir falling to the Multifunction O'Area. Rina, et al. [18] stated that the existence of the absorption well can lower flood rate surface up to 0.01020 m^3/s of 0.02211 m^3/s flood rate without water reservoir. Putri and Sudarmadji [19] identified that there is the effectiveness of absorption well construction by 18.7% up to 26.1% in ability to let the rainfall be absorbed into the soil. Taufik and Anggraeni [20] also identified the effectiveness of absorption well construction by 34.408% up to 99.96% in reducing water rate. Therefore, it can be concluded that the absorption well construction has some advantages, such as [21]: to fulfil soil pores resisting the land degradation occurs; to increase the amount of water entering the soils; and to reduce surface runoff which can prevent flooding.

Figure 7. Multifunction O'Area design

Figure 8. The detail of absorption well system

3.1.3. Water absorption pool

Water absorption pool or retention pool is a pool made for accommodating water runoff surface water that is reabsorbed later into the soil [14]. It is designed as multilevel design by utilizing the cut and fill process for land surface formation as shown by Figure 9. The existence of an absorption pool can add the aesthetic value of Duri Kosambi Sponge Park through multilevel water absorption contour. Besila [14] stated that the existence of the absorption pool is more effective in accommodating and absorbing runoff water than a sloping runoff niche.

Figure 9. The water absorption pool design (self-made by the authors)

3.2. The stages of Duri Kosambi Sponge Park implementation

The stages of Duri Kosambi Sponge Park implementation can be seen on Figure 10 and explained below.

- Socialization. This step is supposed to educate the local community of Duri Kosambi regarding the flood, the effect, and how to overcome with flood. This activity also discusses the grand design of Duri Kosambi Sponge Park built in Duri Kosambi green open space.
- Land study. In this step, there will be land study supposing to implement the first design of Duri Kosambi Sponge Park in the whole of green open space area.
- Permission. This step is done to get the permission regarding the construction of Duri Kosambi Sponge Park from the related agencies.
- Agreement. This step is done by making an agreement by signing *Memorandum of Understanding* (MoU) done by green open space manager, related agencies, and the local community regarding the use and the management of green open space area together.
- The DED (*Detailed Engineering Design*) making is a shop drawing of design component of Duri Kosambi Sponge Park. It is done by project tender system referring the Presidential Regulation No. 16 Year 2008 about The Government Procurement of Goods/Services [22].
- Inauguration. This step is the last step of Duri Kosambi Sponge Park implementation. It is undergone symbolically by green open space manager, related agency, and the local community.

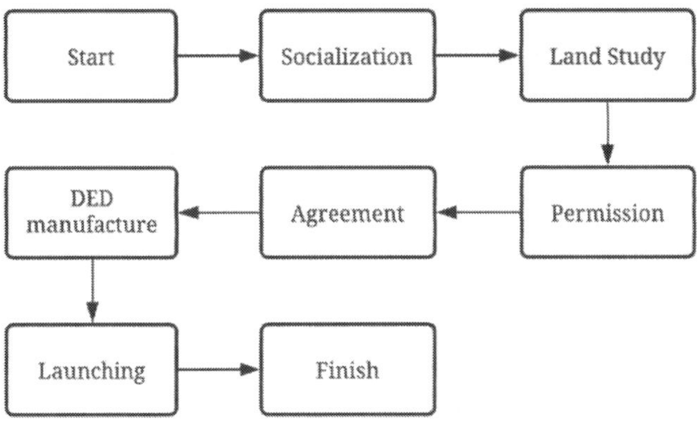

Figure 10. The Duri Kosambi Sponge Park implementation stages

3.3. The analysis of effect from implementing Duri Kosambi Sponge Park

Every construction conducted in a region let the society get the impact. Its impact of conducting a construction may be positive or even negative. Moreover, there are three fields having a strong impact caused by construction, such as environment, economics, and social culture [23]. The analysis of effect from Duri Kosambi Sponge Park construction towards those three aspects is shown by Table 1.

Table 1. The Analysis of Effect from Implementing Duri Kosambi Sponge Park

No	Impacted Area	Explanation
1	Environment	The environment becomes more beautiful, cozy, neat, and comfortable due to the increase of plant variety. In addition to that, the water goes to a water reservoir and absorption first before flowing to drainage.
2	Economics	The employment opportunity and business in the community increases, impacting the increase and acceleration of the community's income.
3	Social Culture	The corporation among the local community increases through some activities, such as taking care of the facilities and joint activities, like exercise and community service done in Duri Kosambi Sponge Park.

4. Conclusion

To sum up, flood occurring in Jakarta is a natural disaster that can be anticipated by conducting some infrastructure engineering, which one of them done by maximizing the function of green open space as flood control. Duri Kosambi Sponge Park is one of the alternatives taken by the Jakarta government in preventing floods by implementing the sponge city concept. Furthermore, the construction of Duri Kosambi Sponge Park is designed to require 4 (four) principles including water ecology; water safety; water environment; and water resources. Besides, the implementation of the design is believed to have positive effects for the environment, economics, and social culture.

5. References

[1] Setiaji H 2021 Solusi Banjir Jakarta: Banjir Wacana, Kering Realita
[2] Dewi R K 2021 Banjir Jakarta, Waspada Hujan Lebat hingga 25 Februari 2021
[3] Hutabarat L E 2017 Studi Penurunan Muka Tanah (Land Subsidence) Akibat Pengambilan Air Tanah Berlebihan di DKI Jakarta (Jakarta: UKI Press)
[4] Pratiwi V, Yakti B P and Widyanto B E 2020 Flood Control Reduction Analysis using HEC-RAS due to Local Floods in Central Jakarta IOP Conf. Ser. Mater. Sci. Eng. 879
[5] Antara and Arjanto D 2021 Anggota DPRD DKI Ungkap Penyebab Anggaran Banjir Jakarta Belum Digunakan
[6] Jakarta Smart City 2020 A Flood-Free Future of Jakarta
[7] Subagio J 2019 Jakarta Masih Kekuragan Ruang Teerbuka Hijau, Ini Penjelasan Ahli
[8] Sun Y, Deng L, Pan S-Y, Chiang P-C, Sable S S and Shah K J 2020 Integration of green and gray infrastructures for sponge city: Water and energy nexus Water-Energy Nexus 3 29–40
[9] Ulku M, Xinxin S, van der Lans M, Peynado T D, Zheng J and Fong C 2018 Sponge City Multidisciplinary Project: ERQI sponge city final report,China
[10] Yin D, Chen Y, Jia H, Wang Q, Chen Z, Xu C, Li Q, Wang W, Yang Y, Fu G and Chen A S 2021 Sponge city practice in China: A review of construction, assessment, operational and maintenance J. Clean. Prod. 280 124963
[11] Inggita N 2019 Baru Urban Village to Have 1,500 Biopore Infiltration Holes
[12] Jakarta Smart City 2018 Biopore: A Solution to Prevent Flood and Drought

IOP Conf. Series: Earth and Environmental Science **847** (2021) 012030 doi:10.1088/1755-1315/847/1/012030

[13] Putri T S, Ngii E, Mustarif R A, Sriyani R, Sukri A S, Talanipa R and Balaka R 2018 BETON KRIB POROUS SEBAGAI KONSTRUKSI PERLINDUNGAN DI BELOKAN SUNGAI Teknol. Terap. Berbas. KEARIFAN Lokal 1 1–9
[14] Besila Q A and Kusumadewi A W 2018 Penerapan Konsep Ekologis untuk Pendidikan Lingkungan pada " Taman Pintar " di Kelurahan Kayu Putih , Jakarta Timur Biology Education Conference vol 15 pp 775–82
[15] UNEP 2017 Climate Change Adaptation Technologies for Water
[16] Xiao Q, Gregory McPherson E, Zhang Q, Ge X and Dahlgren R 2017 Performance of two bioswales on urban runoff management Infrastructures 2 1–14
[17] Purvis R A, Winston R J, Hunt W F, Lipscomb B, Narayanaswamy K, McDaniel A, Lauffer M S and Libes S 2018 Evaluating the water quality benefits of a bioswale in Brunswick County, North Carolina (NC), USA Water (Switzerland) 10
[18] Rina E, Sujatmoko B, Fauzi M, Jurusan M, Sipil T, Teknik F and Riau U 2018 Efektifitas Pemanfaatan Sumur Resapan Untuk Mereduksi Limpasan Permukaan Jom FTEKNIK 5 1–7
[19] Putri M A and Sudarmadji 2018 EFEKTIVITAS SUMUR RESAPAN AIR HUJAN SEBAGAI UPAYA KONSERVASI AIRTANAH DI SUB-URBAN BANTENG, SLEMAN (Universitas Gadjah Mada)
[20] Taufik M and Anggraeni D 2019 VARIASI SUMUR RESAPAN UNTUK MENGURANGI DEBIT BANJIR THE 9TH UNIVERSITY RESEARCH COLLOQIUM (URECOL)
[21] BADAN PENGKAJIAN DAN PENERAPAN TEKNOLOGI 2019 KONSERVASI AIR TANAH DENGAN SUMUR RESAPAN
[22] Presiden Republik Indonesia 2018 PERATURAN PRESIDEN REPUBLIK INDONESIA NOMOR 16 TAHUN 2018 TENTANG PENGADAAN BARANG/JASA PEMERINTAH (Indonesia)
[23] Dewi L A and Susanti R 2019 Dampak Pembangunan Taman Kasmaran Pada Perubahan Kondisi Ekonomi, Sosial-Budaya Masyarakat, Dan Lingkungan Kampung Wonosari Tek. PWK (Perencanaan Wil. Kota) 8 226–38

IOP Conf. Series: Earth and Environmental Science **847** (2021) 012031 doi:10.1088/1755-1315/847/1/012031

Geopolymer hybrid fly ash concrete for construction and conservation in peat environment: A review

R Yanuari[1], D Septari[1], J A Rindy[1] and M Olivia[1]
[1]Department of Civil Engineering, Faculty of Engineering, Universitas Riau, Pekanbaru 28293, Indonesia

monita.olivia@lecturer.unri.ac.id

Abstract. Fly ash is industrial residue from the coal combustion in electric and steam generating plants and recognized as an environmental pollutant. Fly ash contains a high concentration of silica and alumina and can be used as geopolymer concrete. Studies the use of fly ash as a geopolymer concrete mixture has been developing in recent decades. This review focuses on the characteristics of fly ash as hybrid geopolymer concrete, and its resistance is peat environment. The hybrid geopolymer is made by activating fly ash using an alkaline activator and Portland cement. The addition of fly ash geopolymer concrete with Portland cement forms N-A-S-H bond structure from fly ash, and C-S-H hydration bond from Portland cement forms C-A-S-H, which is more solid and stronger against aggressive environment such as peat. In addition, the geopolymer concrete is proven to be able to mobilize heavy metals in fly ash so the concrete has low leaching rate and not harmful to the peat ecosystem. This review proves that hybrid geopolymer concrete using fly ash is a potentially innovative and promising material in the future. The geopolymer hybrid not only has its resistance to peat and acts as conservative material in an environmentally friendly peat environment.

1. Introduction

Indonesia's coal production is continuously increasing over time. In 2017, the realization of coal production reached 456 million tons and increased to 548 million tons in 2018; according to Dewi [1], fly ash waste resulting from coal combustion is classified as a toxic and hazardous substance to the environment and living creatures. Fly ash contains heavy metals such as Pb, Cu, Cd, Cr and Zn. These five elements are hazardous if dissolved in water and potential to disrupt the peat ecosystem. Therefore, fly ash waste needs to be transformed using stabilization and solidification methods such as geopolymer material. Geopolymer is an effective method to mobilize heavy metals in fly ash. The heavy metals in fly ash used in geopolymer concrete remain physically bonded solidly in concrete and will not cause long-term environmental pollution [2]. Therefore, geopolymer can be a construction material and can be conservative material that will minimize pollution potential from heavy metals in fly ash waste.

The peatland map compiled by BB Litbang SDLP [3] shows that the total area of peatland in three major islands is around 14.9 million ha. The details of the peatland area are in Sumatra 6.40 million ha, Kalimantan is 4.70 million ha and Papua 3.60 million ha. The largest peatland area in Sumatra is located in Riau Province, with an area of 3.861.401 ha or around 24% of the total peatland area in Indonesia [4]. The majority of construction in Riau Province is carried out on peatland, and around

57% of the total peatland area in Riau has been converted to infrastructure, agriculture, industry and even human settlement.

Peat environment generally has poor physical properties to concrete construction because it has high organic contents, high acidity (pH 3-5) and low bearing capacity [5]. Zivica & Bajza [6] explained that acid ions in peat would attack the cement paste bond in concrete and cause interference to the concrete bonding system. As a result, the concrete is easily porous, with reduced serviceability and decreased concrete performance. The study result [7] showed that OPC (Ordinary Portland Cement) experience a significant compressive strength decrease starting from 28 days to 150 days in peat water soaking.

Geopolymer is a polymerization material synthesized from silica and alumina, resistant to high temperatures [8]. The silica and alumina can be obtained from industrial residue such as fly ash. Geopolymer concrete is made from the aggregate mixture, water and binder material resulted from polymerization reaction between activator alkaline with silica and alumina in fly ash material [9]. Fly ash geopolymer almost has the same and even better physical and mechanical properties than OPC concrete, especially in an aggressive environment such as peat [10-13]. Bakharev [47] showed that fly ash geopolymer concrete compressive strength did not experience a significant decrease in the sulfuric acid environment.

The study by Mejía et al. [14], Suwan & Fan [15], Nath & Sarker [16] and Wijaya et al. [17] showed that the addition of OPC could increase setting time and increase the fly ash geopolymer strength without high temperature. Yanuari et al. [18] conducted a study on concrete geopolymer using palm ash, and it also showed the same trend. Therefore, the addition of PPC to geopolymer concrete can increase geopolymer strength without high temperature. Although the effect of PCC and OPC addition to hybrid geopolymer concrete has been studied, the relationship between the durability and the environmental performance of fly ash geopolymer concrete with Portland cement such as PCC and OPC has not been studied. Therefore, a literature review is conducted, which is purposed to discover the relationship between the durability and the environmental performance of hybrid geopolymer concrete in a peat environment

2. Methods

The study method used was a literature review using a tracing of secondary data, which is indirectly obtained from laboratory tests (primary data) or the primary study data. The flowchart of this review is presented in Figure 1.

The stages of data collection, data analysis and conclusion drawing are explained as follows:

1. A literature review is searching for information and data related and relevant to the keyword of fly ash, geopolymer concrete, hybrid, heavy metals and peat, through a searching database such as Google Scholar, ResearchGate and Scopus of 2010-2020. Two thousand articles related to the keywords were found, which will be selected and reviewed to be 50 full-text articles most relevant to the research topic and article review.
2. We are reducing and summarizing all research data obtained during data collection and focusing on the reviewed aspect. This study examines the strength performance (compressive strength and the change of compressive strength) and environmental performance (heavy metals leaching).
3. The data was analysed by tabulating, mapping the quantitative data, deriving them into other equations and comparing them to various comparable data and related regulations.
4. We present the analyzed data into narrative text, graphs, or tables for ease of understanding.
5. Conclusion drawing in the secondary data-based study is obtained from synthesizing preceding data, normalizing data, and making data from various sources as equal as possible to conclude the best recommendations.

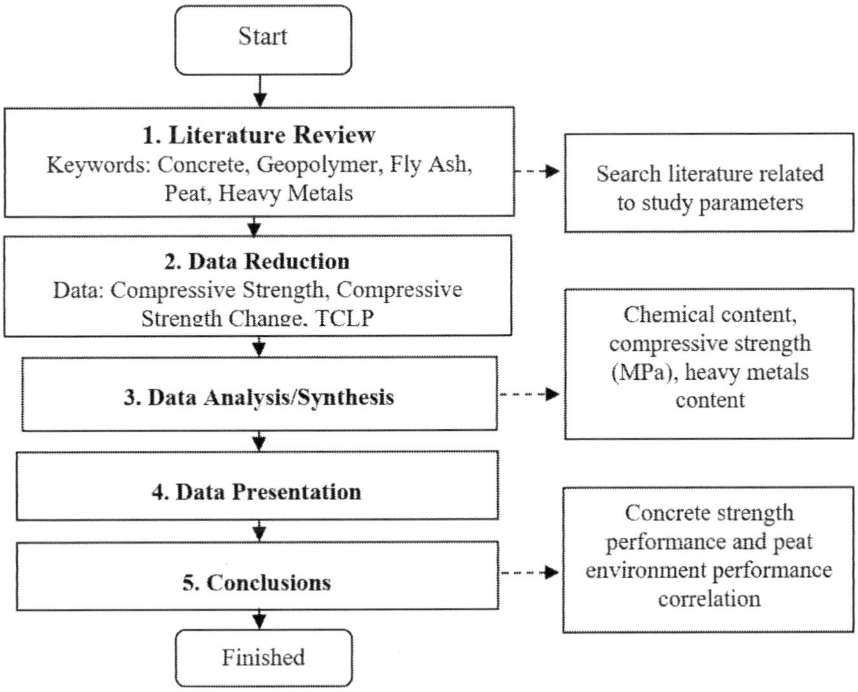

Figure 1. Flowchart of research method

3. Results and discussion

3.1. Result of Coal Fly Ash Chemical Composition Test

Fly ash is an industrial by-product of coal combustion. Coal combustion produces approximately 5% solid pollutants, 80-90% fly ash, and 10-20% is bottom ash. Fly ash is formed from inorganic mineral material and fine particles passing No. 200 sieves [19]. Fly ash is pozzolanic because it contains silica (Si) and alumina (Al) but little calcium (Ca). Fly ash handling is currently limited as an ameliorant material and mostly hoarded on empty spaces around steam power plants (Figure 2).

(a) (b)

Figure 2. (a) Fly ash particles and (b) Fly ash disposal [19]

Fly ash can be used both as addition or partial replacement material of cement in concrete mixture. The reaction between fly ash and calcium will form adhesive properties similar to cement and has a better density compared to conventional concrete. Fly ash can be used as filler to reduce porosity in concrete thereby increasing the strength of the concrete. Based on the test result by Research and Development Center of Coal and Mineral Technology [20] it was found the heavy metal elements in fly ash as presented in Table 1. The heavy metal elements are not hazardous in open-air. However, they will be poisonous and deadly in water. Likewise in peat water, the presence of these elements indirectly causes living creatures around the peat water poisoned.

Table 1. Heavy metal content in coal fly ash [20]

Type of Coal	Heavy Metal Content (mg/L)				
	Cu	Pb	Zn	Cr	Cd
Asam-Asam fly ash	298	19	391	224	11
Ombilin fly ash	87	15	153	120	tt

ASTM C 618 [21] classifies fly ash into three classes: C, N and F. The C class has a minimum of 50% of SiO_2+ Al_2O_3+Fe_2O_3 and more than 10% CaO (low-calcium fly ash) content requirement. Meanwhile, the F class has a minimum of 70% of SiO_2+ Al_2O_3+Fe_2O_3 and a maximum 10% CaO (low-calcium fly ash) content requirement. Therefore, F class fly ash is more suitable for a concrete mixture in an acid environment such as peat water because it contains low calcium that is not easily dissolved by the peat water acidity.

Table 2. Chemical composition of fly ash in several steam power plants in Indonesia

SiO_2	Al_2O_3	Fe_2O_3	CaO	MgO	K_2O	Na_2O	LOI	Origin	References
48.51	15.3	8.68	3.76	2.21	1.71	1.05	-	Mpanau, Sulawesi	[22]
52.2	38.6	2.90	0.70	0.50	0.40	0.50	1.40	Paiton, East Java	[23]
36.23	6.25	4.34	2.85	0.49	0.14	0.93	-	Amurang, North Sulawesi	[9]
48.06	34.76	3.91	2.51	14.53	28.56	0.75	-	North Sulawesi	[24]
37.89	20.52	2.17	2.1	23.76	0.58	-	-	Pelalawan, Riau	[25]
57.3	16.94	9.73	5.57	5.43	0.79	0.73	0.96	Suralaya 2, Banten	[26]
59.25	29.25	5.45	1.54	0.31	2.23	0.68	18.98	Ombilin, West Sumatera	[27]
61.63	17.71	9.30	4.24	1.98	1.28	1.31	4.27	Sibolga	[28]
35.74	12.01	19.32	8.49	1.75	0.69	0.25	0.68	Pangkal Pinang	[28]
55.32	22.43	5.12	1.67	0.49	-	-	12.96	Ombilin Unit 1	[28]
50.29	24.66	6.41	0.78	0.78	-	-	13.64	Ombilin Unit 2	[28]

From Table 2, it can be seen the majority of fly ash from steam power plants in Indonesia contains approximately 70%-90% of SiO_2+Al_2O_3+Fe_2O_3 and is included in the F class pozzolan has a low content of calcium. Therefore, according to Hardjito & Rangan [29], this fly ash is suitable for being applied as a geopolymer material because it has a high silica and alumina content and low calcium. Thereby, the geopolymer can react to alkaline to form a polymerization bond, having good performance and not quickly disintegrate in an acid environment.

3.2. Characteristics of Peat Water

Based on Wetland International calculation, the peatland area worldwide is 400 million m^2 where it covers approximately 2% of the whole land area on earth. Meanwhile, Indonesia is the world's fourth-largest peatland area after Canada, Russia, and United States [30]. Indonesia is estimated to have 20.6 million hectares of peatland or 10.8% of the land area. Peat has a significant role in hydrology and the environment for the life of human beings and other living creatures. However, peat is a prone ecosystem to damage due to toxic substances. The damage occurring in peatland will result in disaster around the environment and disrupt biota living in it. Therefore, peatland must be protected and conserved.

Peat is formed from the accumulation of organic material from weathered plants decomposition found mostly in waterlogged conditions, swamps, and river basins [31]. Continuous stagnant water on the soil causes a change of water characteristics such as turbid, brownish red in colour, high content of mineral salt and low pH number between 3-5 resulting in acid peat water [32]. For example, peat water in Riau Province is highly acid, pH 3-5, contains high organic substances and low hardness levels. Generally, the characteristics of peat water in several locations in Indonesia can be seen in Table 3.

Table 3. Characteristics of peat water in several locations in Indonesia [32]

Parameter	Unit	Peat Water			
		South Kalimantan	West Kalimantan	South Sumatera	Riau
Color	PtCo	753	527	1315	1125
Turbidity	mg/l SiO2	32	0	5	9
DHL	μ mho/cm	-	30	78	75
pH	-	4.10	3.90	5	4
Organic substances	mg/l KMnO4	278	194	290	243
Hardness	°D	2.05	0.48	5.50	1.40
Calcium	mg/l	-	-	4.5	-
Magnesium	mg/l	8.83	2.10	20.90	6.20
Iron	mg/l	-	-	-	-
Manganese	mg/l	-	-	-	-
Chloride	mg/l	11.11	5.48	162	18
SO4	mg/l	-	-	11.20	-
HCO3	mg/l	-	51.40	-	-

3.3. Performance of Conventional Concrete in Peat Water Acid Environment

Peat environment is acid that will impair the mechanism and performance of concrete. (Neville & Brooks [33] explained that Portland cement is not resistant to acidity. The acid ion in peat water will unravel calcium (Ca) in cement paste so that the concrete is easily porous. The higher the degree of acidity, the easier the concrete to suffer from structural damage and serviceability decrease and remain residual sediment in peat environment [6].

The compressive strength of concrete in peat water environments tends to decrease until the day of 180, by 40% [12]. This is because of the unwell cement hydration process and cement matrix decomposition with acid ion, thereby decreasing the density and compressive strength of the concrete. The formation of gypsum and ettringite in concrete reacting with acid will damage concrete durability. Excessive calcium in cement paste will unravel, resulting in the concrete experiencing cracks and pores, decreasing the compressive strength of the concrete. In addition, gypsum and ettringite are hazardous elements if released and decomposed in peat water. The acid endemic animals and even plants are no longer able to survive due to peat environment damage.

3.4. Fly Ash Geopolymer Concrete

Davidovits [8] first proposed the geopolymer in the 1990s. Initially, geopolymer was used as an anti-fire building material. The material then was developed as an alternative substitution of cement using the silica and alumina activator from industrial waste such as fly ash, palm oil fuel ash, and rice husk ash. The activation process of silica and alumina uses an activator solution of sodium hyroxide and sodium silicate [8].

Hybrid fly ash geopolymer concrete is the mixture of sand, gravel, water, and binder material from polymerization reaction between alkaline with silica and alumina from fly ash and Portland cement as a mixture. However, the pozzolanic reaction from fly ash reacts adequately slowly that the initial strength of fly ash geopolymer concrete is lower. Therefore, it is necessary to use the combination of PCC and alkaline activated fly ash geopolymer [34]. Figure 3 illustrates the composition of Portland cement and fly ash in hybrid geopolymer mortar.

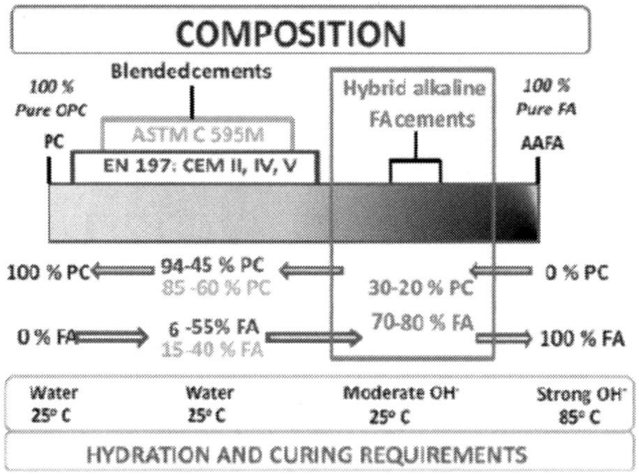

Figure 3. Composition of fly ash and Portland cement in hybrid geopolymer mortar

Several geopolymer development studies with OPC addition can be seen in Table 4. Based on the study results, generally, Portland cement can function as an agent to accelerate the reaction, curing time, increase mechanical properties and reduce porosity due to the formation of a stronger hydration product than the pure geopolymer bond.

Table 4. Summary of hybrid geopolymer concrete studies

References	Findings
[18]	Replacement of PCC to POFA geopolymer mortar mixture increases compressive strength effectively under room temperature curing.
[17]	Hybrid fly ash geopolymer mortar using OPC has higher compressive strength than PCC from the age of 7 to 28 days under room temperature curing.
[35]	Replacement of fly ash with partial Portland cement can increase the compressive strength under room temperature curing.
[34]	The addition of fly ash geopolymer concrete with OPC forms C-S-H hydration structure from OPC and N-A-S-H from fly ash resulting in stronger C-A-S-H.
[14]	The addition of OPC or GBFS increases compressive strength (at 28 days of drying) up to 127% compared to fly ash-based geopolymer system.
[16]	The addition of OPC helps heat process from fly ash geopolymer concrete; thereby, the concrete no longer requires high temperature (oven) curing but adequate room temperature.
[36]	The replacement of fly ash with cement in geopolymer will decrease bonding time, porosity and increase compressive strength and modulus of elasticity.

3.5. Resilience Performance of Hybrid Fly Ash Geopolymer Concrete in Peat Environment

Based on Table 5, several tests present the relationship between the type and composition of concrete with the compressive strength of concrete. The study result by [37,38] showed decreased compressive strength of OPC and PCC concrete with peat water immersion kept increasing along with the age of the concrete. The study by Olivia et al. [37], Nath et al. [39] showed a similar decrease trend with OPC concrete by pure fly ash addition. Meanwhile, the fly ash geopolymer concrete with the addition of OPC showed a different result. The compressive strength in peat water immersion increased along with the concrete age.

Table 5. Summary of earlier compressive strength studies in peat water immersion [38] [37] [39]

7		Concrete Mixture (kg/m³)				Alkaline Activator		Compressive Test (MPa)					Ref.
OPC	PCC	Fly Ash	Water (kg)	Coarse agg (kg)	Fine agg (kg)	NaOH	Na_2SiO_3	0 day	7 days	28 days	90 days	120 days	
507.895	-	-	194.78	987.425	684.138	-	-	-	-	35	34	30.85	[38]
-	507.895	-	194.78	987.425	684.138	-	-	-	-	35.05	31.75	31.72	
350.88	-	-	224.01	930.83	829.96	-	-	21.75	22.71	20.21	-	-	[37]
76.83	-	512.2	120	1014	546	109.27	218.54	15.59	16.75	20.6	-	-	
165.44	-	116.4	225.36	932.73	814.81	-	-	11.16	12.90	14.63	-	-	
263.16	-	814.81	225.36	932.73	814.81	-	-	19.64	21.17	21.37	-	-	
-	-	100% (730)	-	-	1178	83.4	208.60	-	28	55	65	-	[39]
8% (58.4)	-	90% (671.6)	-	-	1178	83.4	208.60	-	31	56.50	62.50	-	
10% (73)	-	90% (657)	-	-	1178	83.4	208.60	-	26	58	71	-	

Furthermore, [40] conducted study on the change of compressive strength of hybrid geopolymer concrete in peat environment. The formula used to calculate the compressive strength change is based on ASTM C-267 [41].

$$\text{Compressive Strength Change} = \frac{S2-S1}{S1} \times 100 \quad (1)$$

Where S1 = the average compressive strength of concrete after 28 days curing (initial strength) period, S2 = the average compressive strength of concrete after a test period of peat water exposure.

A change of compressive force with a positive number shows that the strength (durability) increases after peat water exposure from its initial strength. Otherwise, the change of compressive strength with a negative number indicates that the strength decreases after peat water exposure.

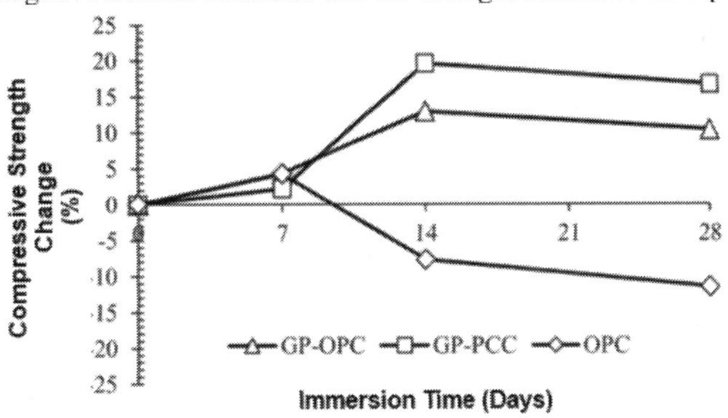

Figure 4. The graph of compressive strength change of concrete in peat water immersion [40]

Figure 4 presents that the change of compressive strength of OPC concrete (control concrete) decreases until the age of 28 days in peat water immersion. This is because the OPC concrete contains high calcium, which is susceptible to the acid ion of peat water. Meanwhile, the change of compressive strength of GP OPC (Geopolymer with the addition of OPC cement) and GP PCC

(Geopolymer with the addition of PCC cement) increases at the age of 14 days in peat water immersion. GP PCC has the most significant change of compressive strength compared to the other concretes because PCC contains high pozzolanic material and less Ca (calcium) than OPC. After 28 days of immersion, GP OPC and GP PCC concrete decrease slightly in strength.

3.6. Environmental Performance of Hybrid Fly Ash Geopolymer Concrete in Peat

In this study of environmental performance in peat environment, earlier research data of heavy metals leaching of fly ash hybrid geopolymer concrete in peat environment is conducted and compared to other concretes. Fly ash has leachate potential with a high concentration of heavy metals, resulting in terrible and severe environmental impacts. Heavy metals are metals having high toxicity if released into the environment. The toxic substances in heavy metals pollute soil, air and water but will also cause damage to human health through the food chain [42]. One of the tests to determine heavy metals leaching level is TCLP (Toxicity Characteristic Leaching Procedure) test. This test is used to identify the toxic concentration of waste.

Table 6 presents the leaching concentration of heavy metals in various types of concrete mixtures. Based on the study of [40], fly ash geopolymer concrete with the substitution of OPC 15% and PCC 15% by fly ash did not experience a significant change of heavy metals leaching after 28 days of immersion. However, if compared to the study of Prasanda et al. [43] and Aziz & Azhari [44], the concrete with the addition of fly ash has a higher leaching concentration.

Table 6. Summary of TCLP test result data [40] [43] [44] [45] [46]

Concrete Mixtures (Immersion Water)	Immersion Time	Metals (mg/l)					References
		Pb	Cd	Cu	Cr	Zn	
Fly Ash (Peat)	-	5.55	7.56	14.3	5.22	23.1	[40]
Geopolymer FA + OPC (Peat)	0	0.396	0.114	0.29	0.397	0.582	
	28	0.472	0.185	0.382	0.432	0.687	
Geopolymer FA + PPC (Peat)	0	0.491	0.157	0.461	0.5	0.731	
	28	0.717	0.19	0.553	0.69	0.836	
Fly Ash (Water)	28	-	-	-	0.23	0.29	[43]
Bottom Ash (Water)	28	-	-	-	0.24	1.25	
SCBA + 20% Steam (Water)	28	-	-	-	0.36	1.27	
SCBA + 20% Non-Steam (Water)	28	-	-	-	0.34	1.28	
Bauxite Residue Geopolymer (Water)	-	0.021	0.0002	0.021	0.002	0.043	
Fly Ash (Water)	-	8.9	-	7.978	5.735	-	[44]
Fly Ash 25% Carbide Waste 2,5 % Cement 72,5% (Water)	-	1.31	-	0.245	0.087	-	
Fly Ash 25% Carbide Waste 5 % Cement 70% (Water)	-	1.232	-	0.245	0.055	-	
Fly Ash 25% Carbide Waste 10 % Cement 65% (Water)	-	0.969	-	0.261	0.043	-	
U.S.EPA	-	5	0.5	-	5	200	[46]
PP 101/2014	-	5	0.15	10	2.5	50	[45]

The table above also describes the concentration limit of heavy metals, such as Cu (copper), Pb (lead), Cd (cadmium), Cr (chromium) and Zn (zinc) allowed by Government Regulation No. 101 [45] and the United States Environmental Protection Agency (USEPA) [46]. Based on Table 6, it can be concluded that the concentration of all heavy metals in hybrid geopolymer concrete is allowed according to [45] and [46]. In contrast to regular fly ash concrete without geopolymer, the heavy metals concentration in concrete using fly ash mixture exceeds the limit allowed by [45], except for Zn (zinc), which is still allowed. Furthermore, the concentration of heavy metals decreases significantly before and after the solidification of fly ash to be geopolymer concrete. This evidence shows that geopolymer concrete is effective to mobilize heavy metals in fly ash.

4. Conclusions

This study investigated the mechanical properties and shrinkage of FABA-GP (fly ash bottom ash) This literature study concludes that most fly ash in Indonesia has a high concentration of Silica and Alumina in chemical composition and have hazardous heavy metals content for peat environments such as Cu, Pb, Cd, Cr and Zn. The solidification technique of fly ash to be hybrid geopolymer material effectively increases concrete strength in an acid environment and mobilises heavy metals in fly ash. Thereby, it is suitable for conservative construction material to have superior resistance in a peat environment.

Acknowledgement

This research was funded through PKM-Research activities, Program Kreativitas Mahasiswa, Direktorat Pembelajaran dan Kemahasiswaan, Direktorat Jenderal Pendidikan Tinggi Kementerian Pendidikan dan Kebudayaan Republic Indonesia year 2020.

References

[1] Dewi N R, Dermawan D, and Ashari M L 2016 Studi Pemanfaatan Limbah B3 Karbit dan Fly Ash Sebagai Bahan Campuran Beton Siap Pakai (Bsp) (Studi Kasus : Pt. Varia Usaha Beton) Media Komunikasi Dan Pengembangan Teknik Lingkungan **13(1)** 34-43

[2] Arioz E, Arioz O, and Kockar O M 2012 Leaching of F-type fly ash based geopolymers Procedia Engineering 42 1114–1120.

[3] Ritung S, Wahyunto N K, Sukarman H and Suparto TC 2011 Peta Lahan Gambut Indonesia Skala 1: 250.000 *BB Litbang SDLP* Bogor

[4] Agus F, Anda M and Jamil A 2016 Lahan gambut Indonesia: pembentukan, karakteristik, dan potensi mendukung ketahanan pangan *IAARD Press*

[5] Moayedi H, Kassim K A, Kazemian S, Raftari M, and Mokhberi M 2014 Improvement of Peat Using Portland Cement and Electrokinetic Injection Technique *Arabian Journal for Science and Engineering* **39(10)** 6851–6862

[6] Zivica V, and Bajza A 2001 Acidic attack of cement based materials - A review. Part 1. Principle of acidic attack *Construction and Building Materials* **15(8)** 331–340.

[7] Olivia M 2015 Kuat Tekan Beton dengan Semen Campuran Limbah Agro-Industri di Lingkungan Asam. *Andalas Civil Engineering National Conference* 306–312.

[8] Davidovits J 1994 Properties of Geopolymer Cements *First International Conference on Alkaline Cements and Concretes* 131–149.

[9] Manuahe R, Sumajouw M DJ, and Windah R S 2014 Kuat Tekan Beton Geopolymer Berbahan Dasar *Jurnal Sipil Statik* **2(6)** 277–282.

[10] Bakharev T 2005 b Resistance of geopolymer materials to acid attack. *Cement and Concrete Research* **35(4)** 658–670

[11] Olivia M 2011 Durability Related Properties of Low Calcium Fly Ash Based Geopolymer Concrete *Issue Doctoral dissertation, Curtin University*

[12] Olivia M, Wulandari C, Sitompul I R, Darmayanti L, and Djauhari Z 2016 Study of fly ash (FA) and palm oil fuel ash (POFA) geopolymer mortar resistance in acidic peat environment *Materials Science Forum* **841** 126–132.

[13] Part W K, Ramli M, and Cheah C B 2015 An overview on the influence of various factors on the properties of geopolymer concrete derived from industrial by-products *Construction and Building Materials* **77** 370-395

[14] Mejía J M, Rodríguez E, Mejía De Gutiérrez R, and Gallego N 2015 Preparation and characterization of a hybrid alkaline binder based on a fly ash with no commercial value. *Journal of Cleaner Production* **104** 346–352.

[15] Suwan T, and Fan M 2014 Influence of OPC replacement and manufacturing procedures on the properties of self-cured geopolymer. *Construction and Building Materials* **73** 551–561.

[16] Nath P, and Sarker P K 2015 Use of OPC to improve setting and early strength properties of low calcium fly ash geopolymer concrete cured at room temperature *Cement and Concrete Composites* **55** 205–214

[17] Wijaya M F, Olivia M, and Saputra E 2019 Kuat Tekan Mortar Geopolimer Abu Terbang Hybrid menggunakan Semen Portland *Jurnal Teknik* **13(1)** 60–68

[18] Yanuari R, Ikrammulah M, Septari D, Wijaya M F, and Olivia M 2020 Studi parametrik mortar geopolimer hybrid abu sawit (Palm Oil Fuel Ash/POFA) *Rekayasa Sipil* **14** 83-90

[19] Munir M 2008 Pemanfaatan Abu Batubara (Fly Ash) untuk Hollow Block yang Bermutu dan Aman Bagi Lingkungan *Doctoral dissertation, Program Pascasarjana Universitas Diponegoro*

[20] Damayanti R 2018 Abu batubara dan pemanfaatannya: Tinjauan teknis karakteristik secara kimia dan toksikologinya *Jurnal Teknologi Mineral dan Batubara* **14(3)** 213-231

[21] ASTM C 618 2014 Standard Specification for Coal Fly Ash and Raw or Calcined Natural Pozzolan for Use in Concrete *ASTM International* West Conshohocken PA

[22] Suarnita I W 2011 Kuat tekan beton dengan aditif fly ash ex. PLTU Mpanau Tavaeli *Jurnal Smartek* **9(1)** 1–10.

[23] Ekaputri J J, and Triwulan T 2013 Sodium sebagai Aktivator Fly Ash , Trass dan Lumpur Sidoarjo dalam Beton Geopolimer *Jurnal Teoritis Dan Terapan Bidang Rekayasa Sipil* **20(1)** 1–10.

[24] Philip A, Marthin M, Sumajouw D J, and Windah RS 2015 Pengaruh Penambahan Abu Terbang (Fly Ash) Terhadap Kuat Tarik Belah Beton *Jurnal Sipil Statik* **3(11)** 729–736.

[25] Islami A N, Saputra E, and Olivia M 2015 Kajian Parameter Mortar Geopolimer Menggunakan Campuran Abu Terbang (Fly Ash) dan Abu Sawit (Palm Oil Fuel Ash) *Prosiding 2nd Andalas Civil Engineering National Conference* 160–167.

[26] Utami R I A, Herbudiman B, and Irawan R R 2017 Efek Tipe Superplasticizer terhadap Sifat Beton Segar dan Beton Keras pada Beton Geopolimer Berbasis Fly Ash *Reka Racana* **3(1)** 1–12.

[27] Olivia M, Wibisono G, and Saputra E 2019 Early strength of various fly ash based concrete in peat environment *MATEC Web of Conferences* **276** 01022.

[28] Gunawan G, and Nono 2019 Potensi Pemanfaatan Bahan Limbah Fly Ash dan Bottom Ash untuk Lapisan Fondasi Jalan Semen (Potential for Utilization of Fly Ash and Bottom Ash Waste Materials for Cement Road Foundation) *Jurnal Jalan-Jembatan* **36**.

[29] Hardjito D, and Rangan B V 2005 Development and properties of low-calcium fly ash-based geopolymer concrete *Research Report GC* **94**

[30] Tjahjono E 2006 Kajian Potensi Endapan Gambut Indonesia Berdasarkan Aspek Lingkungan *Pusat Sumber Daya Geologi* 1–8.

[31] Arioz E, Arioz O, and Kockar O M 2012 Leaching of F-type fly ash based geopolymers **42** 1114–1120

[32] Kusnaedi 2006 Mengolah Air Gambut Dan Air Kotor Untuk Air Minum. *Penebar Swadaya* Jakarta

[33] Neville A M, and Brooks J J 1987 Concrete technology (2nd ed.) *Longman Scientific & Technical.*

[34] Garcia-Lodeiro I, Donatello S, Fernández-Jiménez A, and Palomo Á 2016 Hydration of hybrid alkaline cement containing a very large proportion of fly ash: A descriptive model *Materials* **9(8)**

[35] Assi L N, Eddie E, Elbatanouny MK, and Ziehl P 2016 Investigation of early compressive strength of fly ash-based geopolymer concrete. *Construction & Building Materials* **112** 807–815

[36] Phoo-Ngernkham T, Chindaprasirt P, Sata V, and Sinsiri, T 2013 High calcium fly ash geopolymer containing diatomite as additive *Indian Journal of Engineering and Materials Sciences* **20(4)** 310–318.

[37] Olivia M, Ismeddiyanto, Wibisono G, and Sitompul IR 2017 Early age compressive strength, porosity, and sorptivity of concrete using peat water to produce and cure concrete. *AIP Conference Proceedings*

[38] Olivia M, Pradana T, and Sitompul I R 2017 Properties of Plain and Blended Cement Concrete Immersed in Acidic Peat Water Canal *Procedia Engineering* **171** 557–563

[39] Nath P, Sarker P K, and Rangan V B 2015 Early age properties of low-calcium fly ash geopolymer concrete suitable for ambient curing *Procedia Engineering* **125** 601–607.

[40] Wijaya MF, Olivia M, Wibisono G, Saputra E, and Wang S 2019 Characteristics of geopolymer hybrid concrete in peat water. *IOP Conference Series: Materials Science and Engineering* **615(1)**

[41] ASTM C-267 1998 Standard Test Methods for Chemical Resistance of Mortars, Grouts, and Monolithic. *ASTM International* West Conshohocken PA

[42] Damayanti R 2018 Abu batubara dan pemanfaatannya: Tinjauan teknis karakteristik secara kimia dan toksikologinya *Jurnal Teknologi Mineral Dan Batubara* **14(3)** 213–231.

[43] Prasanda A F E, Triwulan and Ekaputri JJ 2015 Paving Geopolimer Berbahan Dasar Bottom Ash dan Sugar Cane Bagasse Ash (SCBA) *JURNAL TEKNIK ITS* **4(2)** D110-D115.

[44] Aziz M and Azhari 2014 Pembuatan Bahan Geopolimer Berbasis Residu Bauksit untuk Bahan Bangunan *Jurnal Teknologi Mineral dan Batubara* **10(1)** 32–43.

[45] U.S Environmental Protection Agency 1993 Clean water act. Section 503 58 32.

[46] Peraturan Pemerintah No 101 2014 Pengelolaan Limbah Bahan Berbahaya dan Beracun, Republik Indonesia **7** 219–232

[47] Bakharev, T 2005 a. Durability of geopolymer materials in sodium and magnesium sulfate solutions *Cement and Concrete Research* **35(6)** 1233–1246.

IOP Conf. Series: Earth and Environmental Science **847** (2021) 012032 doi:10.1088/1755-1315/847/1/012032

The effect of forest area change in tropical islands towards baseflow and streamflow

N L Khomsiati [1], N Suryoputro [1], A Yulistyorini [1], G Idfi [1] and N E B Alias [2]

[1]Department of Civil Engineering, Faculty of Engineering, Universitas Negeri Malang, Jawa Timur, Indonesia.
[2]Department of Hydraulics and Hydrology, Faculty of Civil Engineering, Universiti Teknologi Malaysia, Johor, Malaysia

nurlatifah.khomsiati@gmail.com

Abstract. Baseflow is a very important component in maintaining river flow. Baseflow is generated from groundwater flow. Baseflow is very important role in water conservation. Indonesia is located in a tropical climate. Based on the Koppen climate classification, Indonesia is included in the tropical rainforest climate (type Af), so that the characteristics of the climate, soil, flora and fauna are unique. About 70% of Indonesia's territory consists of water. Baseflow in the tropical island region has a unique characteristic. This research aimed to obtain the knowledge on the effect of forest area change in tropical islands towards the characteristics of baseflow and streamflow that involved RDF filter parameter calibration, baseflow contribution towards streamflow based on the Baseflow Index (BFI) and Q90 on Q50 ratio, rain contribution on the total baseflow, baseflow stream domination, and discharge stability of streamflow. This research was conducted in three sub watersheds with forest area percentage of ±18% Bango Sub watershed, ±23% Brantas Hulu Sub watershed, and ±59% Cobanrondo Sub watershed. The results showed that One Parameter Algorhitm method satisfied the application of baseflow separation. There was a positive correlation between rain and baseflow. Baseflow dominated the streamflow in the research location. The greater the forest area, the more stable the flow condition from year to year.

1. Introduction

Areas with tropical wet and dry climates are characterized by two main seasons: wet and dry. Baseflow can be a great predictor to display the low-flow condition. The variations of baseflows depend on the changes in groundwater abstraction, changes in land use and land cover patterns, climate change, and other factors from time to time [17]. Therefore, understanding the baseflow is vital in predicting the drought. Baseflow prediction is beneficial to understand the climate change that influences the resilience of ecosystem components and the allocation of drinking water and irrigation.

Baseflow is an essential component in maintaining river flow. Baseflow is the result of groundwater flow. According to [4], baseflow is the flow of the river that occurs during rainless period. This definition is rather tricky to understand because, during the rainless period, the river flow also contains interflow element. In his research, Harto proposed three watershed parameters that affecting the baseflow such as source frequency (the ratio between segment 1 river on the total streams in all segments), drainage density (the sum of the river lengths of all segment per unit area of the watershed), and drainage frequency (total rivers in all segments per unit area of the basin).

The technique to separate baseflow from the hydrograph of river flow has a long history in hydrology field and is used to predict the low-flow and water quality. Multiple methods were developed to separate the baseflow. One of the low-pass filter methods is the Recursive Digital Flter (RDF) method which was adapted from signal processing theory [18]. In RDF method, surface runoff is viewed as a high-frequency signal while baseflow is seen as a low-frequency signal. By filtering the high-frequency signal (surface runoff) from the flow, low-frequency signal (baseflow) can be identified [8]. The filtered results mostly depend on some specific watershed parameters. These parameters can be calibrated based on the measured data in the baseflow. However, the measured data are rare, and in practice, these parameters are often determined arbitrarily [9].

The change in forest land use and others are related to the changes in evapotranspiration, infiltration, and groundwater recharge in a watershed, which all influence the baseflow (Price, 2011; Dou et al., 2015). Planning and managing watershed requires practical knowledge in its correlation between the change in forest land towards other land uses and baseflow process. The most significant negative impact of baseflow is the changes in forest land followed by agriculture and settlement. Meadows, badlands and shrubs did not have a significant impact [5].

This research aimed to find the effect of forest area in tropical islands towards baseflow characteristics including the RDF filter parameter calibration, baseflow contribution on total streamflow based on the Baseflow Index (BFI) and ratio of Q90 on Q50, rain contribution on total baseflow, baseflow domination, and streamflow stability.

2. Location and Research Method

2.1. Research location

This research was conducted in Brantas Hulu Watershed, Malang Regency, East Java Province. Three sub-watersheds were taken as research locations to represent differences in forest area in the watershed: Bango Sub watershed, Sumber Brantas Sub watershed, and Cobanrondo Sub watershed. The coordinate for this research were 112.44976^0 - 112,94181^0 East Longitude and 7.74690^0-7.94661^0 South Latitude with +500 up to +1,800 elevation. The characteristics from each sub watershed are: 1) Bango Sub watershed has 239.71 km^2 area and main river length of 12.5 km, 2) Sumber Brantas Sub watershed has 174.98 km^2 area and main river length of 15.5 km, and 3) Cobanrondo Sub watershed has 18.14 km^2 area and main river length of 12.16 km. The average land slope in Bango Sub watershed is ±2.0% (slope), Sumber Brantas Sub watershed is ±14.0% (rather steep), and Cobanrondo Sub watershed is ±18.0% (steep).

The forest area percentage from each sub watershed in 2018 were: 1) ±18% in Bango Sub watershed, 2) ±23% in Brantas Hulu Sub watershed, and 3) ±59% in Cobanrondo Sub watershed. The soil fraction at the study site generally consisted of 26% sand, 51% silt and 23% clay so that it was classified as silt loam texture with an infiltration rate of 7.5-15 mm / hour included in moderate slow classification [14]. This fraction caused a lower tendency of surface runoff.

2.2. Hydroclimatological data

The rain data was obtained from 11 rain stations. The rain data and rate were recorded from 2004 to 2018 (15 years). Before analyzed, the data was screened [13]. The analysis was conducted using the Polygon Thiesen method [3].

2.3. Baseflow separation method

This research used the Recursive Digital Filter (RDF) with One Parameter Algorithm, and Flow Duration Curve (FDC) filters to separate the baseflow. The RDF method used Hydro Office BFI+ 3.0 software [2]. The RDF equation with One Parameter Algorithm filter is presented in Equation 1:

$$q_{b(t)} = \frac{k}{2-k} q_{b(t-1)} + \frac{1-k}{2-k} q_t \quad (1)$$

Where:

q(t) : The actual river discharge value on the t-day

qb(t) : The actual baseflow value on the t-day
qb(t-1) : The baseflow value before t-day
k : Filter parameters given by recession constants
t : Daily time interval

This research used the Equation 2 Flow Duration Curve (FDC) [16].

$$P = 100\left(\frac{m}{n+1}\right) \quad (2)$$

Where:
P: The probability of specifics flows to be equal or exceeded
m: Ranking for daily or monthly flow in descending order
n: The total number of observations (e.g., 365 for daily data in a year)

The flow rate at P=50% (Q50) was taken as the river flow median. The flow higher than Q50 was picked as the low-flow rate. Also, the slope of the curve outside the Q50 mark illustrates the dominant flow type of river flow. The low slope represents the baseflow domination, whereas the steep slope after Q50 shows the surface runoff domination. The ratio of Q90 on Q50 displays the proportion of groundwater flow contributions or percentage of baseflow components [6].

2.4. Assessment of river flow stability

The river flow stability assessment in the watershed used the Coefficient of River Regime and Coefficient of Variation (CV) parameters.

2.4.1. A subsubsection Coefficient of River Regime (KRS).

KRS is the comparison between maximum rate (Qmax) and minimum rate (Qmin) of a watershed.

$$\text{Coefficient of River Regime (KRS)} = (Q\ max)/(Q\ min) \quad (3)$$

Note:
Q max (m3/sec) = highest average daily discharge (Q)
Q min (m3/sec) = lowest average daily discharge (Q)

Table 1. Coefficient of River Regime Classification [4]

No	Value	Classification
1	< 50	Good
2	50 – 120	Medium
3	> 120	Poor

2.4.2. Coefficient of Variation (CV).

The coefficient of variance (CV) is a description of the variation conditions of the annual discharge (Q) from a watershed.

$$CV = \frac{\textbf{Sd}}{\textbf{Qaverage}} \times 100\ \% \quad (4)$$

With:
Sd = annual standard discharge data deviation (Q)
Qaverage = annual average discharge data

Table 2. CV Classification [4]

No	CV Value	Classification
1	< 10 %	Good
2	10% – 30%	Medium
3	> 30%	Poor

2.5. Baseflow characteristics evaluation

The baseflow characteristic evaluation consisted of the below steps:

- Calibration evaluation of the RDF method filter parameters. This research used the coefficient of the determinant method (r^2).

$$r^2 = \left(\frac{\sum_{i=1}^{n}(O_i - \bar{O})(P_i - \bar{P})}{\sqrt{\sum_{i=1}^{n}(O_i - \bar{O})^2}\sqrt{\sum_{i=1}^{n}(P_i - \bar{P})^2}} \right)^2 \tag{5}$$

Where n is the number of observation data during the period under review, Oi is the observed value of the ith model, $\bar{O}$ is the average observed value, Pi is the output value of the ith model, $\bar{P}$ is the average output value. And the Nash-Sutcliffe Efficiency (NSE)

$$\text{NSE} = 1 - \frac{\sum_{i=1}^{n}(O_i - P_i)^2}{\sum_{i=1}^{n}(O_i - \bar{O})^2} \tag{6}$$

The range of NSE lies between 1.0 (perfect fit) and - ∞

- Proportion evaluation of baseflow contribution on total river flow in a year was analyzed based on the Baseflow Index (BFI) and Q90 ratio on Q50
- Proportion evaluation of rain contribution on total baseflow in a year was analyzed based on the average rain total ratio on baseflow in a year
- Evaluation of baseflow dominance based on the slope curve of FDC outside of Q50 mark
- Evaluation of the effect of changes in forest area to the stability of river flow based on the coefficient of variance (CV) and river regime coefficient (KRS).

3. Result and Discussion

3.1. Calibration evaluation of the filter parameter

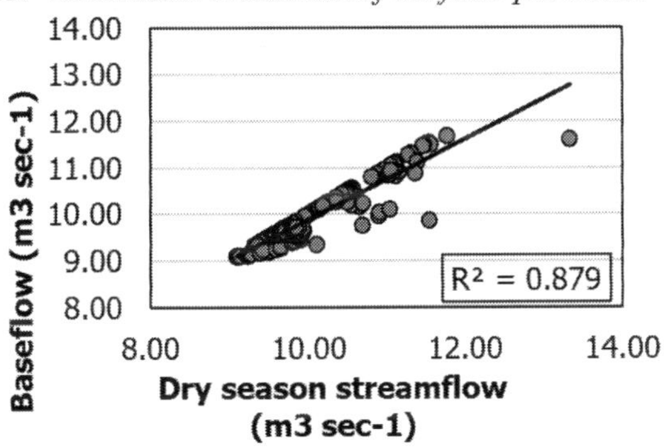

Figure 1. Comparison of discharge and baseflow in the dry season (June–November) Bango Sub watershed

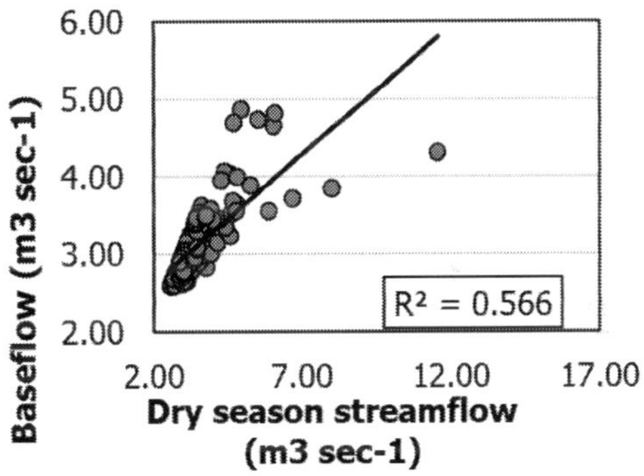

Figure 2. Comparison of discharge and baseflow in the dry season (June–November) Sumber Brantas Sub watershed

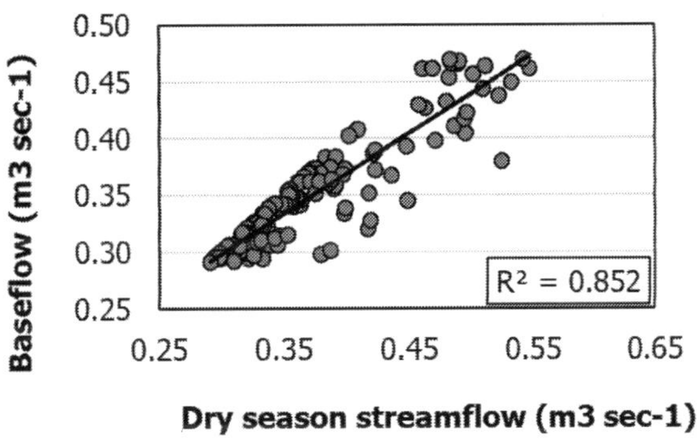

Figure 3. Comparison of discharge and baseflow in the dry season (June–November) Cobanrondo Sub watershed

The parameter filter value in this research was k=0.925 [8]. According to them, the acceptable filter parameter value is between 0.90–0.95. Figure 1 shows the results of the RDF filter parameter calibration during the dry season (June–November). The coefficient of determinant in Figure 1-3 is $r^2>0.5$, meaning, the RDF filter with One Parameter Algorithm filter result (k=0.925) was satisfying.

3.2. Proportion evaluation of baseflow contribution on total river flow

Figure 4-6 and Table 3 show the baseflow analysis results on overall river flow

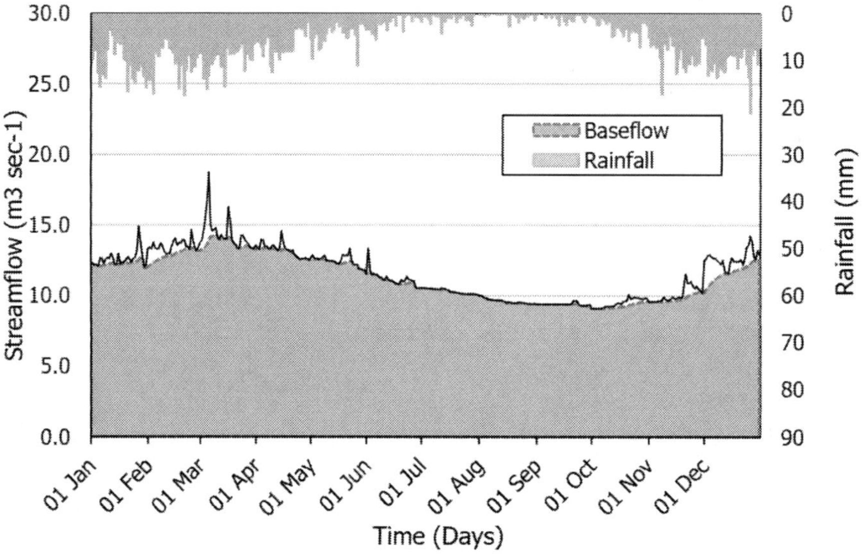

Figure 4. Streamflow, baseflow, and average rain (2004–2018) in Bango-Sari Sub watershed,

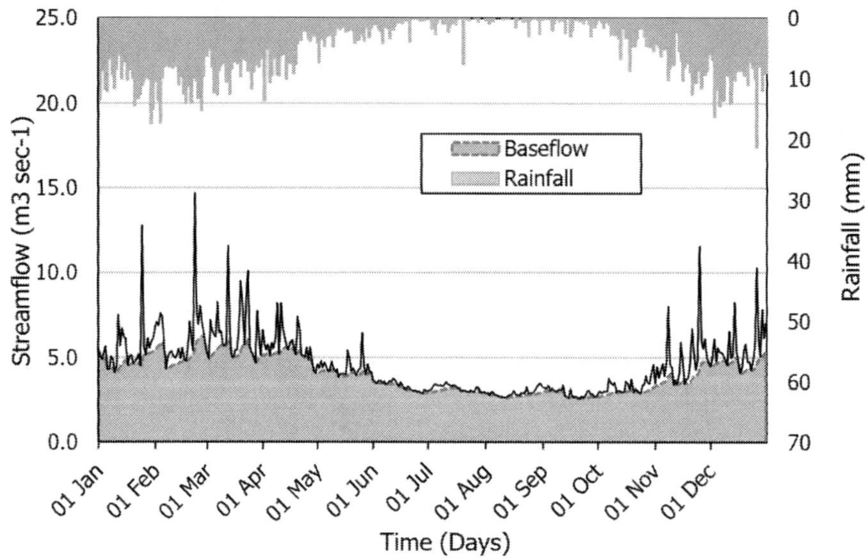

Figure 5. Streamflow, baseflow, and average rain (2004–2018) in Brantas Hulu Sub watershed

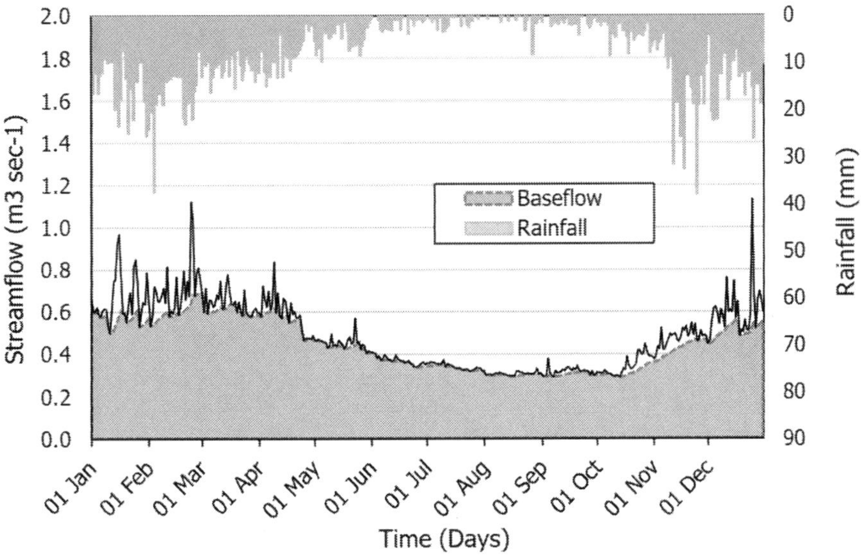

Figure 6. Streamflow, baseflow, and average rain (2004–2018) in Cobanrondo Sub watershed.

Table 3. The parameter contribution of baseflow to river flow in each sub watershed

Sub watershed	Baseflow Index (BFI)	Q90/Q50
Bango-Sari Sub watershed	0.97	0.80
Brantas Hulu Sub watershed	0.87	0.68
Cobanrondo Sub watershed	0.92	0.67

3.3. Proportion Evaluation of Baseflow Contribution on Total River Flow

The correlation between gross rainfall on baseflow can be seen in Figure 3. Figure 3 shows that the rain had a 30% contribution to the baseflow, and the remaining portion was influenced by other factors. Figure 3 also presents the positive correlation between rain and baseflow, meaning the more significant the rain, the higher the baseflow created. A similar pattern was achieved by several researchers such as Rumseya et al. and Ouyang et al. The increase in baseflow due to precipitation occurred when precipitation becomes the primary source for groundwater recharge [7]. Silt loams are grouped at infiltration rates and moderately rapid recharge classification.

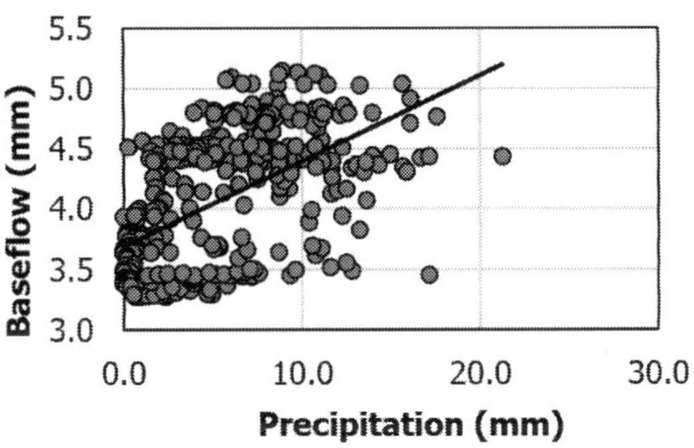

Figure 7. Correlation between annual rainfall (precipitation) on baseflow in Bango Sub watershed

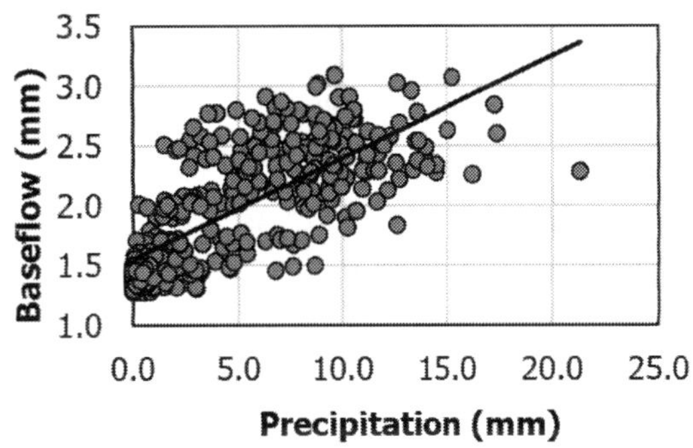

Figure 8. Correlation between annual rainfall (precipitation) on baseflow in Sumber Brantas Sub watershed

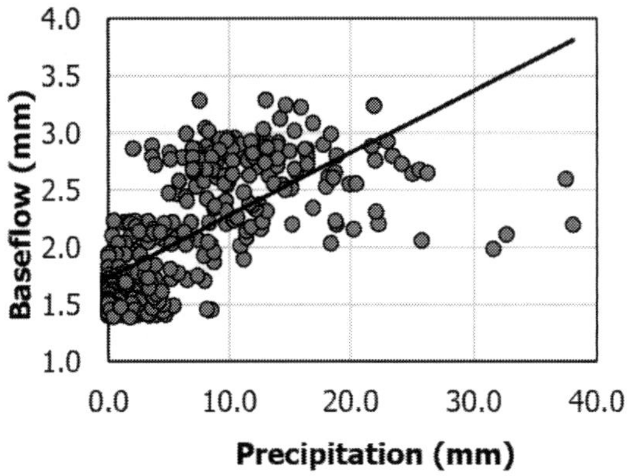

Figure 9. Correlation between annual rainfall (precipitation) on baseflow in Cobanrondo Sub watershed

The negative and essential effect on the change of baseflow is influenced by factors such as changes in the forest area, followed by agriculture and settlement. Meadows, badlands, and shrubs did not produce a significant impact [5]. Changes in the forest land use and its relation with changes in evapotranspiration, infiltration, and groundwater recharge in a watershed could influence baseflow [11]. The decrease of rain contribution on the baseflow with large forest area percentage is predicted due to the canopy density which causes increased loss of rain due to interception, stemflow [15], and evapotranspiration so that the infiltration and recharge of groundwater decrease.

3.4. Baseflow dominance evaluation

The FDC method resulted in a curve outside Q50 mark with relatively flat curve that shows a significant contribution from baseflow. In other words, baseflow dominates the river in each sub watershed. This occurrence was caused by the land characteristics in research location that was silt loam and is a part of the land with a moderately rapid infiltration rate and recharge.

3.5. Stability evaluation of streamflow

If the variation in annual flow (Q) is small, the flow condition (Q) year after year does not change much. In another side, if the yearly flow (Q) is high, the flow condition (Q) year after year will experience changes, which shows the unstable watershed/sub watershed. The cause is the change in land use and or water use in watersheds, such as El Nino and La Nina. Table 4 shows the value of CV in Bango Sub watershed (18% area) that is 64.83%, Brantas Hulu Sub watershed (23% area) that is 31.98%, and Cobanrondo Sub watershed (59% area) that is 29.67%. Hence, the more percentage of forest area generates smaller CV, which means that the flow (Q) year after year in watershed with forest area percentage above 60% is more stable compared to a basin with forest area below 30%.

The high Coefficient of River Regime represents the high Qmax and Qmin; or, the range of runoff values in the rainy season (flood) that occurs is high, while in the dry season the flow of water that happens is minimal or shows drought. Indirectly, this condition indicates that the ground absorption ability in watershed unable to save or store the rain and the runoff water continues to enter the river and is discharged into the sea so that the availability of water in the watershed during the dry season is little. Based on Table 4, the KRS value is less than 50, which means the stability.

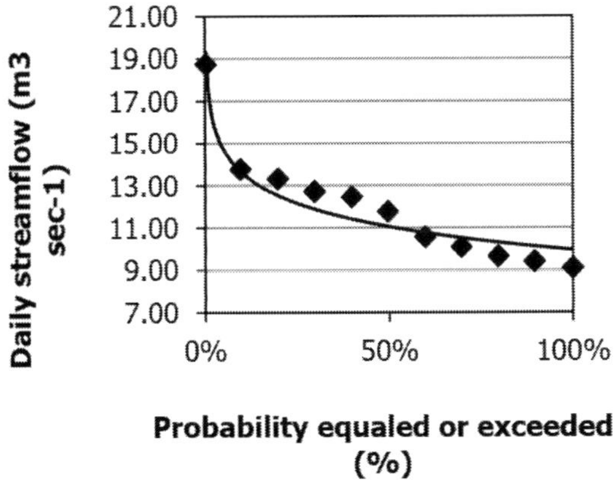

Figure 10. Daily Flow Duration Curve (FDC) in Bango Sub watershed

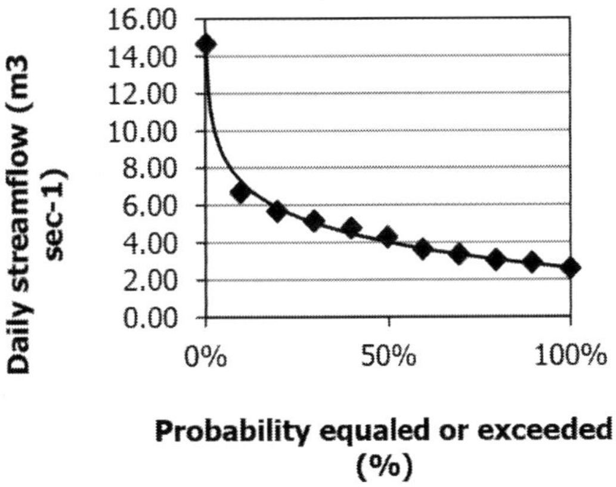

Figure 11. Daily Flow Duration Curve (FDC) in Bango Sub watershed

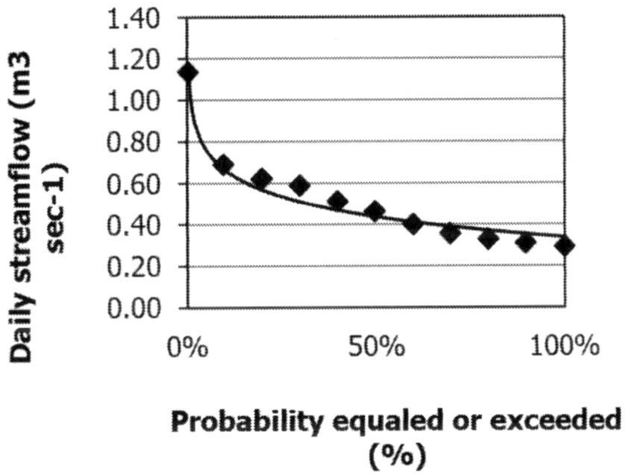

Figure 12. Daily Flow Duration Curve (FDC in Cobanrondo Sub watershed

Table 4. River flow stability

Sub watershed	Coefficient of River Regime (KRS)	Coefficient of Variant (CV)
Bango-Sari Sub watershed	1.53	64.83%
Brantas Hulu Sub watershed	2.24	31.98%
Cobanrondo Sub watershed	2.28	29.67%

4. Conclusion

- This research conducted a study in the change of forest area in tropical island towards baseflow and streamflow characteristics including the RDF filter parameter calibration, baseflow contribution on total river flow based on the Baseflow Index (BFI) and Q90 ratio on Q50, rain contribution on total baseflow, baseflow domination, and streamflow stability.
- Based on the RDF filter parameter calibration with One Parameter Algorithm method, the result showed 0.925 filter parameter value or satisfying calibration.
- The baseflow contribution proportion on streamflow (BFI value) was above 0.80 with ratio Q90/Q50 showed higher than 0.60, which meant that the baseflow dominated the streamflow in the research location.

- There was a positive correlation between rain and baseflow, meaning the higher rain flow generated higher baseflow.
- The result from the FDC method showed the curve outside of Q50 with relatively flat slope, which indicated a substantial contribution from baseflow; or in other words, baseflow dominated the streamflow. This occurrence was caused by the land characteristic in the research location that was silt loam or land with moderately rapid infiltration and recharged.
- The higher the percentage of large forest area, the more stable the condition of streamflow year after year. The absorption ability of watershed area in the research location could withstand and store the rainfall; hence, water availability during the dry season was quite large, which meant river flow stability was excellent, or there was only small gap between streamflow discharge in rain season and dry season.

References

[1] Dou X., Chen B., Black T.A., Jassal R.S., and Che M. 2015 *Impact of Nitrogen Fertilization on Forest Carbon Sequestration and Water Loss in a Chronosequence of Three Douglas-fir Stands In The Pacific Northwest* (MDPI Forests) **6** 1897–1921.

[2] Gregor M., 2010 *BFI +3.0 User's Manual* (Bratislava: Hydro Office)

[3] Hand D, and Bray M., 2006 *Automated Thiessen Polygon Generation* (Water Resources Research) **42** 1-5.

[4] Harto S. 1993 *Analisis Hidrologi* (Jakarta: Gramedia Pustaka Utama)

[5] Huang X.D, Shi Z H, Fang N.F, and Li X 2016 *Influences of Land Use Change on Baseflow in Mountainous Watersheds* (MDPI Forests) 7 **1** 16-26.

[6] Lyne V.D and Hollick M 1979 *Stochastic Time-Variable Rainfall-Runoff Modelling* (Hydrology and Water Resources Symposium, Institution of Engineers Australia, Perth) 89–92.

[7] Memon B.A., 1995 *Quantitative Analysis of Springs* (Environmental Geology) **26** 111-120

[8] Nathan R.J, and McMahon T.A., 1990 *Evaluation of Automated Techniques for Baseflow and Recession Analyses* (Water Resources Research) **26** 1465–1473

[9] Nejadhashemi A.P., Shirmohammadi A., and Montas H.J. 2003 *Evaluation of Streamflow Partitioning Methods* (Pap. No. 032183 in ASAE Annual International Meeting, edited by St. Joseph, M., Las Vegas, Nevada, USA.)

[10] Ouyang L., Liu S., Ye J., Liu Z., Sheng F., Wang R., and Lu Z. 2018 *Quantitative Assessment of Surface Runoff and Baseflow Response to Multiple Factors in Pengchongjian Small Watershed* (MDPI Forests) **9** 1-16.

[11] Price K. 2011 *Effects of Watershed Topography, Soils, Land Use, and Climate on Baseflow Hydrology in Humid Regions* (A review. *Prog. Phys. Geogr*) **35** 465–492.

[12] Sheng H., and Cai T., 2019 *Influence of Rainfall on Canopy Interception in Mixed Broad-Leaved—Korean Pine Forest in Xiaoxing'an Mountains, Northeastern China.* (MDPI Forests) **10** 1-11

[13] Soewarno S 1995 *Hidrologi: Aplikasi Metode Statistik untuk Analisis Data* (Bandung: Nova)

[14] Vargas T.B., Uribe C.M., Aristizábal E.F.G., and Martínez J.F.E 2017 *Identification and Characterization of Regional Water Flows Contributing to The Recharge of Anunconfined Aquifer* (Revista Facultad de Ingeniería) **85** 70-85

[15] Venkatraman K, and Ashwath N., 2016 *Canopy Rainfall Intercepted by Nineteen Tree Species Grown on a Phytocapped Landfill* (Int J Waste Resour) **6** 202

[16] Welderufael W.A., Woyessa Y. E 2010 *Streamflow Analysis and Comparison of Baseflow Separation Methods. Case Study of The Modder River Basin in Central South Africa* (European Water)**31** 3-12.

[17] Zelenhasic E., and Salvai A., 1987 *A Method of Streamflow Drought Analysis* (Water resources research) **23** 156-168.

[18] Zhang Y., Ahiablame L., Engel B., and Liu L 2013 *Regression Modeling of Baseflow and Baseflow Index for Michigan* (USA : Water) **5** 1797-1815.

IOP Conf. Series: Earth and Environmental Science **847** (2021) 012033 doi:10.1088/1755-1315/847/1/012033

Elektroof: Smart roof as an energy independent solution for Indonesia in the future

A W Azhar[1], M H Al-Fikri[1], R I Sulasmono[1], N Sholikhah[1], N A Wafiyah[1] and C P Dewi[1]

[1]Department of Civil Engineering, Faculty of Engineering, Universitas Negeri Malang, Indonesia

cynthia.dewi.ft@um.ac.id

Abstract. The Electric Steam Power Plant is still being one of the most commonly used power plants in Indonesia. The use of coal has so many bad impacts on humans and also environmental health. The existence of coal reserves is decreased each year caused of non-renewable energy. Therefore, an innovation for alternative power plants sourced from renewable energy is needed to reduce pollution and be more environmentally friendly. This study examines the possibility of using solar energy and raindrops as the main energy source for elektroof systems. The method used for this study is by comparing the literature about the utilization of piezoelectric and solar cells as the new source of power generators in typical house buildings in Indonesia. The results of the literature study showed that 1 elektroof can produce electrical energy of 183 wh when the weather is hot and when the weather rains elektroof can produce 156 wh so that the elektroof can be used as a tile on the roof of the house that can simultaneously become a self-contained power plant. Elektroof can generate energy by utilizing renewable energy in the form of sunlight and rainwater. Thus, elektroof is possible for use in Indonesia which is a tropical country.

1. Introduction

The Economic growth in Indonesia has grown significantly over the last few decades. Until now, the Indonesian economy is the largest in Southeast Asia and the 7th largest in the world based on the gross domestic product (GDP) based on purchasing power parity (PPP) [1]. The rapid economic growth has led to an increase in electricity demand in Indonesia. Indonesia is the largest energy consumer in the Association of Southeast Asian Nations (ASEAN), accounting for more than 36% of the region's energy demand [2]. According to World Bank data, Indonesia's electricity consumption in 2019 reached 1084 kWh per capita, which has increased significantly in recent years. To meet this electricity growth, Indonesia is still very dependent on fossil fuel power plants [1]. Data from the world bank said that Indonesia's electricity production in 2015 was contributed by oil, gas, and coal amounting to 89.3% of the total energy mix.

Most of the country's electricity sources, both developed and still under development, still use fossil fuels [3]. The use of fossil fuels is still very popular because of their capability in delivering energy and its price [4]. Moreover, fossil fuels as no renewable energy will be run out in the future. The use of fossil fuels is also one of the causes of various environmental problems such as global warming and air pollution which causes health problems and affects the quality of life of the population [3]. In addition, the increase in population and people's living standards along with the increasing demand for electrical energy can have an impact on the high rate of growth of CO_2 emissions. From the results of the final energy demand projection, CO_2 emissions produced in 2030 will reach 912 million tons of CO_2 eq (BaU) [5]. Increased CO_2 emissions will have an impact on increasing temperatures on earth so that it can cause climate change. Many impacts caused by the use of fossil fuels have led to the need for the use of new, renewable energy sources that are more environmentally friendly.

We will review and discuss the possibility of using *Elektroof* as a standalone power plant in every home so that it can become an environmentally friendly clean energy producer. The *Elektroof* is a development of solar power plants that are limited to utilizing the dry season only, while Indonesia has one more season, namely the rainy season which is also a source of renewable energy. Electrophysical components consist of solar and piezoelectric panels that serve as electricity producers. This source of electricity is sourced from renewable energy that utilizes 2 seasons in Indonesia, namely sunlight and rainwater so that this tool does not produce pollution that negatively impacts people and the environment.

2. Coal: the main fuel for power plant

Data from outlook energy 2019 show that the coal power plant is the largest electricity contributor in Indonesia with a percentage of 56% of all existing power plants [5]. The use of coal is still the dominant and main used as a power plant material compared to other energy sources. Even though Indonesia only has 3% of the world's coal reserves [6]. If coal is used continuously as the main source of electricity generation in Indonesia, it is estimated that coal reserves in Indonesia will be used up in 56 years [7]. In addition, the use of coal as the main ingredient in the operation of electric steam power plants also brings various impacts to humans and nature.

Coal mining causes irreversible damage to land, air, and water sources [6]. The International Energy Agency (IEA) revealed that fossil fuels Coal accounted for 44% of total global CO_2 emissions. Coal burning is the largest source of greenhouse gas (GHG *emissions*), which trigger climate change. Serious smog in the form of 90% of SO_2 emissions, 70% of dust emissions, and 67% of NOx emissions is produced during the coal combustion process [8]. In addition, coal combustion also produces several pollutants in the form of SOx, NOx, CO, and particles in the form of fly ash which are released through the chimney so that they can be carried by the wind to the surrounding community who live near the power plant [9]. Pollutants produced from coal combustion are one of the contributors to air pollution in Indonesia. Air pollution is a silent killer causing 3 million premature deaths worldwide and can lead to an increased risk of lung cancer, stroke, heart disease, and respiratory disease [6].

The results of the GEOS-Chem atmospheric modeling conducted by the Harvard University – Atmospheric Chemistry Modeling Group (ACMG) research team show that air pollution from the current operation of the Indonesian coal-fired power plant has caused around 6,500 premature deaths per year. The main causes are stroke (2,700), ischemic heart disease (2,300), chronic obstructive pulmonary disease (400), lung cancer (300), and other cardiovascular and respiratory diseases (800). Air pollution caused by coal-fired power plants is not only limited to the area where the power plant is located but also affects other areas in Indonesia [6]. If the use of this coal continues in the future, it can cause more serious impacts on humans and the environment.

3. The possibility of using renewable energy in Indonesia

Indonesia is one of the growing countries due to several factors, namely economic fast growth, increasing urbanization, and stable population growth in Indonesia. Some of these factors resulting in Indonesia are included in the category of countries that have a high level of energy consumption. Indonesia uses 40% of the total energy in ASEAN (Association of Southeast Asian Nations). Between 2000 and 2014, Indonesia's energy consumption increased by around 65% and is predicted to reach 80% by 2030 [10]. With high energy consumption, Indonesia must develop all renewable energy in Indonesia. So that all fossil resources will become energy reserves, and electricity needs in Indonesia in 2050 can be achieved by clean and environmentally friendly renewable domestic energy sources. The maximum share of renewable energy that can be achieved in 2050 is 33% and 80% without technology diffusion with the least cost optimization resulting in lower generation costs than CO_2 emissions. Total CO_2 emissions in 2050 are predicted to be five to six times greater than current emissions.

Indonesia is on the equator so it has a lot of potential renewable energy such as solar, water, biomass, geothermal and tidal waves [10]. Geothermal energy is one of the renewable energy that can be found in Indonesia and has a potential of 40%. Around 252 geothermal locations in Indonesia are spread out following volcanic formation pathways that stretch from Sumatra, Java, Nusa Tenggara, Sulawesi, to Maluku. This has a total potential of about 27 GWe. Geothermal energy is renewable and environmentally friendly energy, this huge potential needs to be increased in its contribution to meet domestic energy needs which can reduce Indonesia's dependence on dwindling fossil energy sources. The potential of geothermal energy is expected to meet the target of developing geothermal energy to generate electricity through a Geothermal Power Plant of 6000 MWe by 2020 [11]

On the other side, wind power generation is no longer a new problem in the world. However, in Indonesia, wind energy is not developed optimally, until now the installed capacity of wind power plants is only about 1.6 MW of the total predicted 9.29GW capacity, due to several factors, and one of them is the low level of confidence. on the availability of wind in Indonesia [12]. Ocean wave energy also has great potential in Indonesia, it is predicted that located south of the Java Sea has the most promising location for wave power potential, with the highest energy resources available from January to December. The maximum annual average wave power in the Java Sea can reach 22 kW/m [13]. However, the utilization of renewable energy is still hasn't maximized until now it is only limited to the use of solar energy. This is very unfortunate because if another renewable energy is also implemented optimally, it can make Indonesia independent of clean and environmentally friendly energy.

Solar energy has great potential in Indonesia because it is located at the equator, so it has a relatively high level of solar radiation and rainfall throughout the year. Most areas of Indonesia get enough solar radiation intensity with average daily radiation of about 4 kWh/m^2 [14]. Even the intensity of rainfall in Indonesia reaches 2000 – 3000 mm/year. So this can be utilized properly to support the development of new renewable energy in Indonesia by utilizing both seasons simultaneously. Optimally and bring up an idea or ideas, namely *Elektroof*: Multi-weather Electricity Generating Tile as an Environmentally Friendly Clean Energy Alternative Solution.

4. *Elektroof* Details and system

Elektroof is a multi-weather electric tile that utilizes 2 seasons in Indonesia. Elektroof is capable of generating electrical energy in hot and rainy weather. The design of the elektroof is described using SketchUp software as shown in figure 1 below. The elektroof consists of Monocrystalline Solar Panel, Piezoelectric PVDF, and Elektroof Connector.

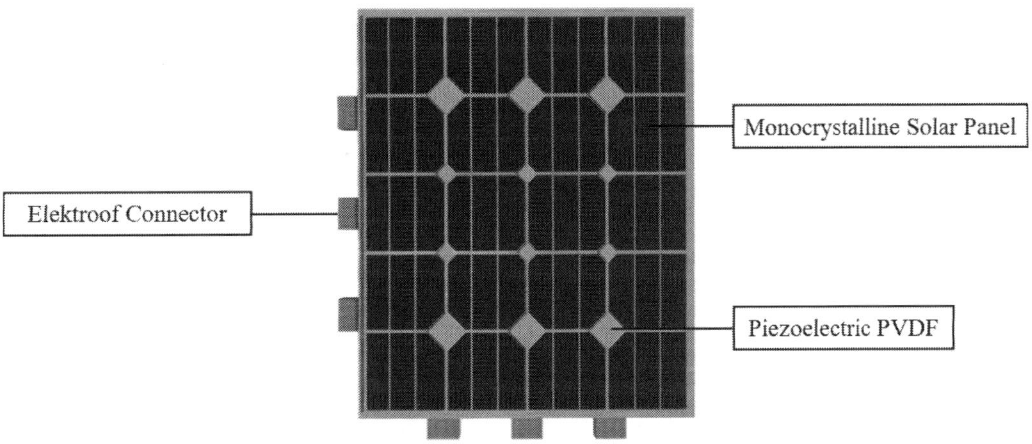

Figure 1. Elektroof design and component

4.1. Monocrystalline Solar Panel

Monocrystalline Solar Panel made from single molded silicone. In the course of making, the crystal is sliced and formed into bars. The excess of this single silicone has more space to stream the electron that generates the electricity. Moreover, this type of solar panel resistance could use in high temperatures and also in shady weather. The resulting efficiency panel reaches 15-20%; vertical solar panel (20,4818%) and flexible solar panel (16,4044%). The flexible reach higher output than the vertical. Depending on the radiation intensity to the panel surface differences are observed between energy productions. The average energy performances of monocrystalline and polycrystalline are 546.82 Wh and 517.52 Wh, respectively. Maximum and minimum energy production from panels were actualized on 19.08.2014 and 24.08.2014, respectively. The results obtained from the measurement show that monocrystalline solar panels performed better compared to polycrystalline on every experiment day [15].

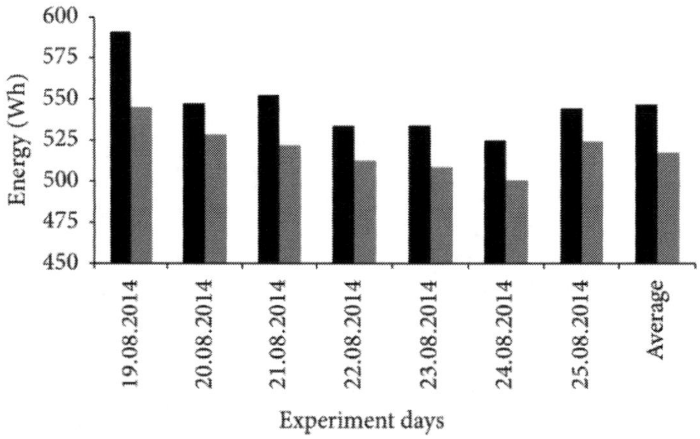

Figure 2. Comparison bar of energy output between monocrystalline and polycrystalline daily [16]

The bar in figure 2 shows from the daily experiment for a week that the output energy of the monocrystalline is always more than the polycrystalline. The highest energy output was on Monday and the lowest was on Saturday caused of the different sun exposure times. Besides, the comparison between monocrystalline and polycrystalline is that monocrystalline solar panels produce higher hourly power output performance than polycrystalline.

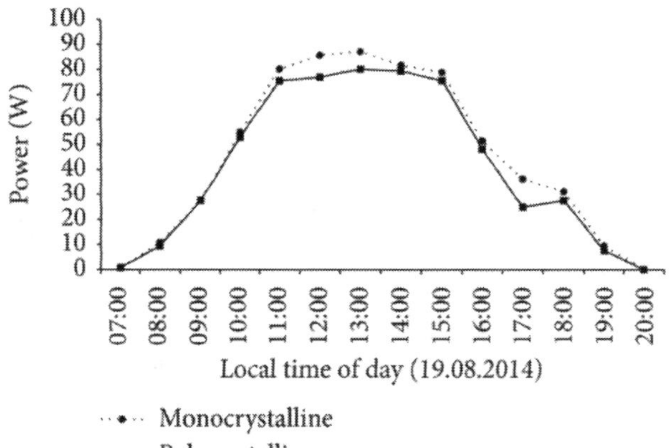

Figure 3. Hourly power output graphics [16]

The Graphics in figure 3 shows that Solar Panel Monocrystalline resulting higher output than polycrystalline with the highest at 1 pm and also has a significant increase and decrease of power output graphics.

4.2. Piezoelectric PVDF

Piezoelectric is a material that produces electricity based on the pressure effect on it. The composition of *Polyvinylidene Difluoride* (PVDF) is CH2-CF2 or it's also called 1,1 *difluoro–ethylene*. The pre-eminence of PVDF is flexible, easy handling, and also undemanding. As its molecular structure, PVDF material has three forms which are; phase α, β, γ. The most usage of the form is the "*phase β*". Piezoelectric PVDF is used for producing electricity from rainfall pressure. The research shows that rainfall resulting much more energy than gas pressure and also disc loop. PVDF polymer piezo films have many advantages, mainly due to their mechanical properties: flexibility (adapted to the non-planar surface), high mechanical resistance, dimensional stability, homogeneous piezo activity within the plane of the film, high piezoelectric coefficients without any aging effect for temperatures up to 808°C (PVDF) or 1108°C (*copolymer*), high dielectric constant, chemically inert material with a low acoustic impedance close to that of water[17].

Figure 4. PVDF crystal [18]

The PVDF crystal is easy to be shaped and also flexible cause the PVDF crystal has a high content of yield strain. The Piezoelectric PVDF works by the collision into the crystal and the crystal excrete the electric field that produces electricity. PVDF crystals as shown in figure 4 above.

4.3. Elektroof Connector

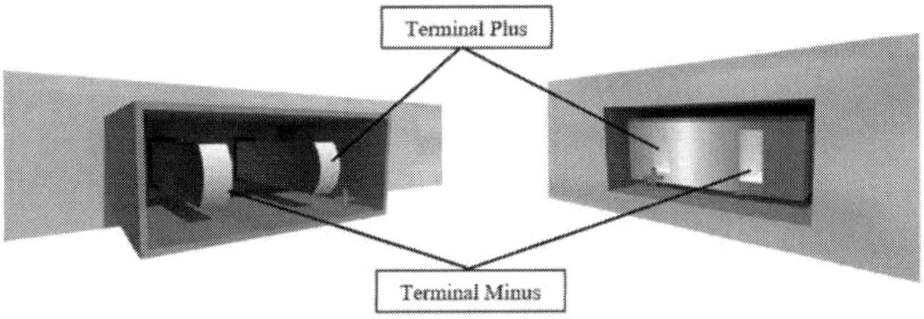

Figure 5. *Elektroof* connector plug **Figure 6.** *Elektroof* connector plug hole

The Elektroof Connector serves as the connector between one to other elektroof. Furthermore, the connector is used to transfer the energy from one connector to another. The advantage of the connector is to reduce the human error in Solar panel wiring which can give a big impact on energy distribution. installation of solar panels generally using cables plus and minus to cause a human error during installation. therefore, the elektroof connector will be able to reduce human error at the time of installation. On the connector elektroof, the plus and minus parts of the cable have been adjusted to facilitate the user in installation. On the installation of the user just need to install the elektroof connector plug in figure 5 with the elektroof connector plug hole in figure 6 that has been adjusted the pluses and minuses. In figure 5 and figure 6 are also shown the plus terminal and minus terminal parts on the elektroof connector. The connector was made from zinc alloy. Clearly, as the nickel content in the electrodeposit increases the protective nature of the coating will change too, the alloy becoming electrochemically nobler and hence being less sacrificial to the substrate steel [19].

5. Measurement energy

5.1. Energy conversion

Elektroof tile is formed from 2 components that produce electricity which have different sources. Because it has 2 electricity generators, the power produced by the elektroof has 2 variations. When the weather is hot, solar panels work to generate electrical energy. Elektroof using monocrystalline solar panels 50 wp size 54×64×3 cm which can provide an output power of 50 watts. The input power received by solar modules from solar radiation does not 100% enter the inverter because it is affected by the loss of components and systems, along with the calculation of power to the energy generated [20].

Table 1. Losses of components and systems of solar panels [20]

Type of losses	Value of losses (%)
Losses manufacture (power tolerance)	3
Losses dirt	5
Losses temperature module	5,7
Loses cable	5

Based on the data loss component and solar panel system in table 1 above then for solar panels 50 wp will produce an output power of 40.65 watts. So if it is known that Indonesia's solar irradiation 4.5 kWh day/m^2 [21], it will be obtained energy produced by solar panels as follows :

$$\frac{40,65 \text{ wp}}{1000 \text{ W/m}^2} \times 4500 \text{ Wh day/m}^2 = 183 \text{ Wh} \tag{1}$$

From the calculation above (1) it is known that one elektroof tile can produce 183 Wh of power. Then when the weather is rainy, the solar panels on the elektroof will automatically stop working and their role is replaced by piezoelectric which converts the boom caused by rain into electrical energy. The power generated by the piezoelectric is as follows:

$$\text{Piezoelectric PVDF size 1 x 1 cm [22]} \tag{2}$$

$$\text{Piezoelectric voltage} = 0.15 \text{ V [22]} \tag{3}$$

$$\text{Panel voltage} = 18 \text{ V} \tag{4}$$

Because the panel and piezoelectric voltages are not the same (3) and (4), they must be equalized first. This voltage equalization is important so that the piezo can be in sync with the panel and not become a burden that ends up reducing the power generated by the panel. Voltage equalization is done by increasing the number of piezoelectrics and stringing them together.

$$\frac{18 \text{ V}}{0.15 \text{ V}} = 120 \text{ piezo} \tag{5}$$

To get the same voltage, 120 piezoelectric PVDF pieces are needed in series with a size of 1 x 1 cm. The 120 piezos are divided into two groups, namely group 1 and group 2. Group 1 is a rhombus-shaped area with a size of 4 x 4 cm, while group 2 is a rhombus-shaped area with a size of 2 x 2 cm. each grub is placed in 6 different points. As figure 6 shows below.

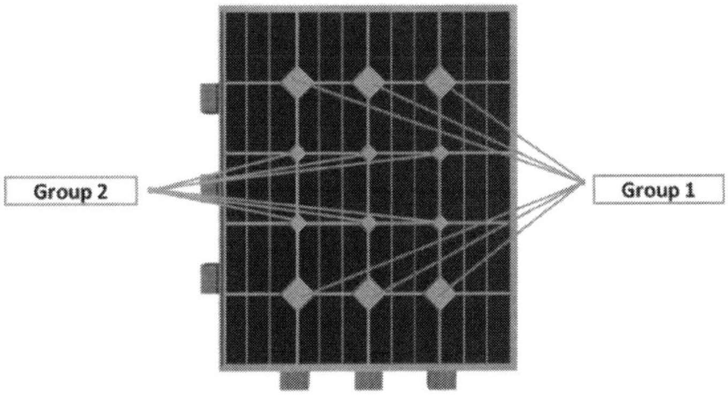

Figure 7. Piezoelectric group division

For group 1 there are 6 groups each measuring 4 x 4 cm and because for 1 piezo it has a size of 1 x 1 cm then group 1 can accommodate 16 piezos for 1 point. Then the power generated by group 1 is as follows:

$$P = \frac{V^2}{R} = \frac{(16\times0.15)^2}{1} = 5.76 \text{ Watt} \quad (6)$$

Because there are 6 groups (6), then:

$$5.76 \times 6 = 34.56 \text{ watt} \quad (7)$$

For group 2 there are 6 groups each measuring 2 x 2 and because 1 piezo has a size of 1 x 1 cm, group 2 can accommodate 4 piezos for 1 point. Then the power produced by group 2 is as follows:

$$P = \frac{V^2}{R} = \frac{(4\times0.15)^2}{1} = 0.36 \text{ Watt} \quad (8)$$

Because there are 6 groups (8), then:

$$0.36 \times 6 = 2.16 \text{ watt} \quad (9)$$

So, for a total of 120 piezos, it can produce a power of (7) and (9) 36.72 watt this assuming torrential rain intensity of 50 mm-100 mm for 1 hour. If within 1 day of heavy rain with an intensity of 50 mm-100 mm for 5 hours then the resulting power is 183.6 watts. The power generated from piezoelectric cannot be directly supplied to home devices because it is still in the form of DC. Therefore, the power must first be converted into AC. If the DC is converted to AC then the power must be reduced for the calculation of losses of 15% of its DC power [23]. So that for piezoelectric energy output obtained as follows:

$$183.6\times15\% = 27.54 \text{ Watt} \quad (10)$$

$$183.6 - 27.54 = 156.06 \approx 156 \text{ Watt} \quad (11)$$

So, for one elektroof module can be generated electrical energy of 183 wh when the weather is hot and when it is raining then the elektroof can generate electrical energy of 156 wh in one day..

5.2. Energy Distribution

The energy produced by the elektroof goes through several stages as shown in figure 8. Elektroof connector functions as a transfer of the resulting power to flow power from one elektroof to another elektroof. This power supply has a tip, namely the elektroof which is at the lower left end of the roof.

The elektroof which is at the left end of the roof is then in charge of flowing to the charge controller. The charge controller functions to protect, control the process of charging, discharging the battery, and determining the photovoltaic operating point with its voltage level and thereby affecting the yield of photovoltaic energy [24]. Then from the charge controller, the power is stored in the battery. Batteries function to support fluctuating solar resources and provide energy at night and during the day [24]. The battery here also has a very important role in the power generated by the piezoelectric on the elektroof. Because piezoelectric power is inconsistent when directly supplied to home appliances, a battery is needed to overcome this limitation [25] This inconsistency is due to the nature of the rain having fluctuating and erratic intensity. The capacity of the battery used as a place to accommodate power depends on the amount of power needed by a house. Please note, the power stored in the battery is still in DC and must first go through an inverter to be converted into AC before being used for home appliances. This inverter will later play a role in controlling the interaction with the voltage and frequency of the electric utility at home [26].

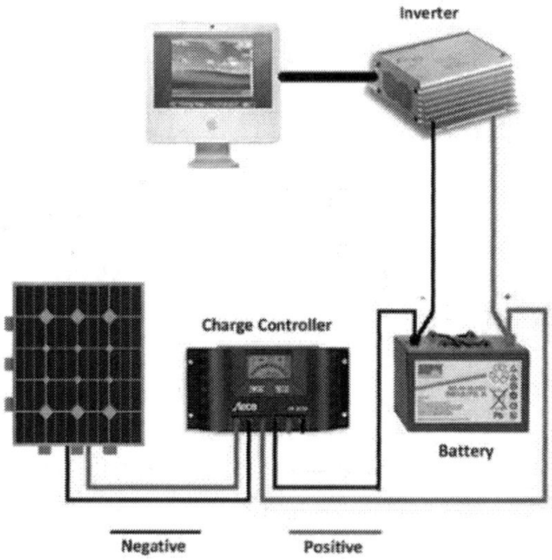

Figure 8. Elektroof electricity distribution

Figure 9. Left end elektroof which is connected to charge controller

Power can be supplied to home appliances after the current is converted to AC. From the calculations that have been done previously (1), it is known that the elektroof can produce 183 watts of energy per hour when the solar panel is active or the weather is sunny. If it is calculated, the energy produced by just one panel can be distributed to turn on a 34-watt fan for 2 hours, 65-watt television for 1 hour, and

Led lamps with 10-watts of power for 5 hours. In addition, from previous calculations, it is also known that when the piezoelectric power generated in the active elektroof reaches 156 Watts per hour (11). Although the power produced is less than when the solar panel is active, this power can already be used to power home appliances, such as a 34-watt fan for 2 hours, 65-watt television for 1 hour, and Led lamps with 10-watts of power for 2 hours.

6. Application of *elektroof* in a building

Elektroof is a tile producing electricity that consists of renewable energy. In its application, elektroof can be applied as a roof cover as well as a power plant in the house. Elektroof is used on part of the roof of the house with an amount tailored to the electricity needs of the house. For example, if used type 36 house with daily electricity consumption of 2876 watts, it will need 30 elektroof modules to meet all the electricity needs. For its application as a tile on the roof of the house can be seen in figure 9 below.

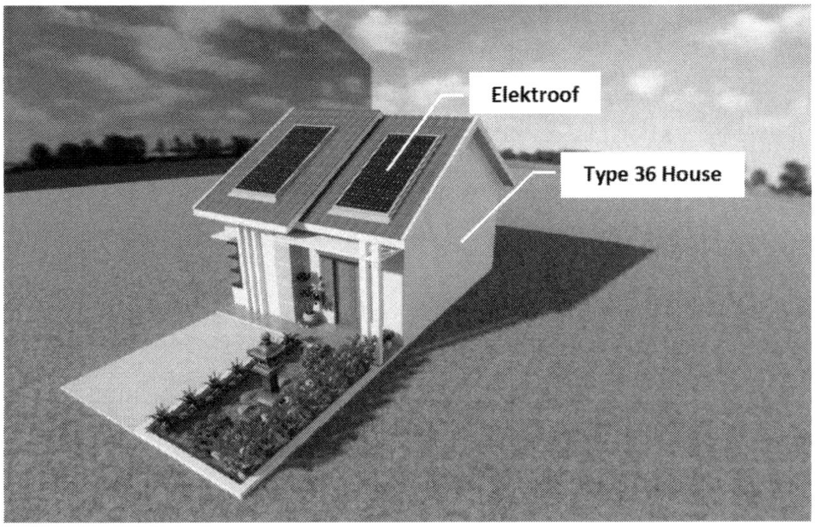

Figure 10. Type 36 house with elektroof roof

6.1. Cost Analysis of the installation

The unit cost of a 50 wp monocrystalline solar panel is IDR 325,000 [15]. Photovoltaic system power generation has a higher capital cost per unit compared to fossil-based power sources. However, the operating costs of photovoltaic systems are much lower than fossil-based power sources because photovoltaic systems have safe efficiency and can last for 25 years [16]. Piezoelectric PVDF has received wide attention due to its good flexibility, strength, easy processing, low cost, and other advantages [17]. While the cost for a PVDF type piezoelectric unit with dimensions of 1cm × 1 cm is IDR 10,000 [18]. In one PVDF type piezoelectric electrode used as many as 120 pieces so that the overall cost incurred for the use of piezoelectric is IDR 10,000 × 120 = IDR 120,000. So the total cost for 1 unit of elektroof is around IDR 500,000. The use of elektroofs does require a large investment cost, but for the next 25 years, if these costs are accumulated with steam power plants, of course, the use of elektroofs will be cheaper. On the other hand, several years later the development of solar panels and piezoelectric shows a decrease in costs caused by the accelerated market growth and economies of scale. Thus, an annual cost reduction of about 4% can be made in the coming years [27].

6.2. Cost Analysis of the installation

The roof of the house can be used as an independent power plant location because it is a strategic location to get sunlight and rainwater. If every house in Indonesia can supply its own electricity needs, then Indonesia can become an energy-independent country. The use of PLN electricity in each house will be reduced because the community can be energy independent. The reduction in the use of PLN electricity will naturally stop the operation of coal-fired power plants and later reduce the negative impact on the community and the surrounding environment.

Application elektroof will take place in several stages. The first stage is to conduct further research on the material and the effectiveness of the power generated by the elektroof. In the second year, an experiment will be conducted with elektroof or sampling. Sampling is an experimental stage of a settlement in Indonesia and is evaluated on an ongoing basis. The last stage of the application of elektroofs will be to replace the roof of all houses in Indonesia using elektroofs with the aim that Indonesia can become an energy independent country and can achieve SDGs point 7 of clean and affordable energy. SDGs indicator point 7 targets to substantially increase the share of renewable energy in the global energy mix to take urgent action to combat climate change and its impacts and ensure access to affordable, reliable, sustainable, modern, and clean energy services to increase efficiency [28].

7. Conclusion

As has been explained and clearly stated, sunlight and raindrops can be utilized on a large scale to generate electrical energy. Sunlight is converted into electrical energy by utilizing solar panels and the collision of rainwater can be converted into electrical energy using piezoelectric. Elektroof is a combination of solar panels and piezoelectric which are used to generate electrical energy. Elektroof is a tile that produces multi-weather electricity that can be applied as a tile on the roof of a house and at the same time used as an independent power plant in each house. The power generated by the elektroof can be used to power the entire house, from fans, TVs, refrigerators to lights. So that the use of Elektroof will be able to become an alternative solution for clean, environmentally friendly energy that can reduce pollution in Indonesia and can achieve SDGs point 7, namely clean and affordable energy.

References

[1] Arinaldo D and Adiatma J C, 2019 Indonesia ' s Coal Dynamics : Toward A Just Energy Transition.
[2] Mcneil M A Karali N and Letschert V, 2019 Energy for Sustainable Development Forecasting Indonesia ' s electricity load through 2030 and peak demand reductions from appliance and lighting ef fi ciency *Energy Sustain. Dev.* **49** p. 65–77.
[3] Martins F Felgueiras C Smitkova M and Caetano N, 2019 Analysis of Fossil Fuel Energy Consumption and Environmental Impacts in European Countries † p. 1–11.
[4] Owusu P A and Asumadu-sarkodie S, 2016 sustainability issues and climate change mitigation A review of renewable energy sources , sustainability issues and climate change mitigation *Cogent Eng.* **15**, 1.
[5] Tim Sekretaris Jenderal Dewan Energi Nasional, 2019 Indonesia Energy Out Look 2019 *J. Chem. Inf. Model.* **53**, 9 p. 1689–1699.
[6] Greenpeace, Apr-2016, Kita, Batubara & Polusi Udara, April, p. 1–16.
[7] esdm, 2018, Cadangan Batubara Indonesia Sebesar 26 Miliar Ton, *www.esdm.go.id.* .
[8] Chen Z Wang Y Zhang X and Xu J, 2020 The energy-saving mechanism of coal- fi red power plant with S e CO 2 cycle compared to steam-Rankine cycle *Energy* **195** p. 116965.
[9] Wahyu Sabubu T A, 2020 Pengaturan Pembangkit Listrik Tenaga Uap Batubara Di Indonesia Prespektif Hak Atas Lingkungan Yang Baik Dan Sehat *J. Lex Renaiss.* **5**, 1 p. 72–90.
[10] Taqwa A, 2019 Higher Education Role in Supporting Indonesian Government Policy in Developing Renewable Energy Higher Education Role in Supporting Indonesian Government Policy in Developing Renewable Energy.
[11] Suharmanto P Fitria A N and Ghaliyah S, 2015 Indonesian Geothermal Energy Potential as Source of Alternative Energy Power Plant **1** p. 119–124.
[12] Martosaputro S and Murti N, 2014 Blowing the Wind Energy in Indonesia *Energy Procedia* **47** p. 273–282.
[13] Zikra M, 2017 Preliminary Assessment of Wave Energy Potential around Indonesia Sea **862** p. 55–60.

[14] Muhammad A Utama N A and Kristianto M A, 2014 A Technical and Economic Potential of Solar Energy Application with Feed-in Tariff Policy in Indonesia *Procedia Environ. Sci.* **20** p. 89–96.
[15] Eduardo C *et al.*, 2015 Performance of monocrystalline and polycrystalline solar panels in a water pumping system in Brazil *Renew. Sustain. Energy Rev.* **51** p. 1610–1616.
[16] Taşçioğlu A Taşkin O and Vardar A, 2016 A Power Case Study for Monocrystalline and Polycrystalline Solar Panels in Bursa City, Turkey *Int. J. Photoenergy* **2016**.
[17] Ueberschlag P, 2001 Features PVDF piezoelectric polymer **21**, 2 p. 118–125.
[18] DIE WERBTÄTIGEN, Polyvinylidene Fluoride (PVDF). .
[19] Wilcox G . and Gabe D ., 1993 Electrodeposited Zinc Alloy Coatings **35**, X p. 1251–1258.
[20] Hariyati R Qosim M N and Hasanah A W, 2019 Konsep Fotovoltaik Terintegrasi On Grid dengan Gedung STT-PLN Energi dan Kelistrikan : Jurnal Ilmiah **11**, 1 p. 17–26.
[21] Windarta J Sinurya E W Abidin A Z Setyawan A E and Kusuma A, 2019 Penerapan Teknologi Pembangkit Listrik Tenaga Surya (PLTS) di SMA Negeri 6 Surakarta sebagai Sekolah Hemat Energi dan Ramah Lingkungan 79 p. 215–227.
[22] Almanda D Dermawan E Ramadhan A I Diniardi E and Fajar A N, 2015 Analisis Desain Optimum Model Piezoelektrik Pvdf Untuk Sumber Pembangkit Listrik Air Hujan Berskala Mini *Semin. Nas. Sains dan Teknol. 2015* November 2015 p. 1–5.
[23] Simanjuntak Y M *et al.*, 2017 Studi Perencanaan Modul Praktikum Pembangkit Listrik Tenaga Surya (PLTS).
[24] Schmid F and Behrendt F, 2021 Optimal sizing of Solar Home Systems: Charge controller technology and its influence on system design *Sustain. Energy Technol. Assessments* **45**, October 2020 p. 101198.
[25] Voon-Kean Wong J-H H and Ai-Bao Chai, 2017 Performance of a piezoelectric energy harvester in actual rain *Energy* energy p. 8.
[26] Ketjoy N Chamsa-ard W and Mensin P, 2021 Analysis of factors affecting efficiency of inverters: Case study grid-connected PV systems in lower northern region of Thailand *Energy Reports* **7** p. 3857–3868.
[27] Zubi G Adhikari R S Sánchez N E and Acuña-Bravo W, 2020 Lithium-ion battery-packs for solar home systems: Layout, cost and implementation perspectives *J. Energy Storage* **32**, August.
[28] Malagó A *et al.*, 2021 An analytical framework to assess SDG targets within the context of WEFE nexus in the Mediterranean region *Resour. Conserv. Recycl.* **164**, April 2020.

An overview on carbon nanotubes as innovative absorbent for marine oil spill

L A S Arum[1], Y E Pawestri[1], M Zaki1, M H W Mahendratha[1], N Awaliya[1] and M M A Pratama[1]

[1]Department of Civil Engineering, Faculty of Engineering, Universitas Negeri Malang, Indonesia

mirza.abdillah.ft@um.ac.id

Abstract. Marine oil spills due to drilling activities is a serious problem worldwide because it poses a major risk to marine ecosystem. So far, researchers in the world continues to innovate materials to clean up marine oil spills. Among them, based on carbon nanotubes. Carbon nanotubes are known have perfect ability to absorb oil and neutralize substance. This study review provides the various types of carbon nanotubes that have best quality to absorb marine oil spills. Continue with various types of carbon nanotubes technology, which technology should be developed in the future by considering each advantage possessed by of carbon nanotubes in order to create product innovation that efficient and environmentally friendly because does not make some new pollution. Therefore, carbon nanotubes is a good choice material to clean up marine oil spills.

1. Introduction

The preservation of the marine environment and its biodiversity is one thing that is important for the survival of human life, either directly or indirectly. This is also supported in the Sustainable Development Goals (SDGs) program number 9 where underwater life is one of the seventeen focuses of attention. The sea has become a part of human life by distributing its existence as much as 97% on earth. On the other hand, the presence of offshore activities often contributes to marine pollution, for example oil spills. The activity of spilling a mixture of complex hydrocarbon compounds, both unintentionally and intentionally, has become one of the biggest contributors to hazardous materials for marine ecosystems [1]. In 2020 it was recorded that around 1000 tons of oil spills occurred with 3 medium scale incidents (7 – 700 tons) and zero cases with large scale (> 700 tons). It is recorded that from 1970 to 2020 more than 5 million tons of oil spills contributed to marine pollution [2]. Although there has been a decline in number of oil spill cases since the 1970s (shown in Figure 1), this does not justify that oil spill cases will not occur. This indicates that an oil spill is an unavoidable incident if offshore activities are still taking place. In the span of five decades, it was recorded that 50% of cases of major oil spills occurred in ship shipping activities on the high seas and 58% were caused by accidents and ship grounding activities. [2]. Not only these two things, but the threat of oil spills also comes from high seas oil exploration [3]. Threats to marine life will continue to occur as long as the oil spill occurs, such as the inhibition of growth and development, digestion, respiration, and reproduction in fish and marine mammals as well as possible swelling of their vital organs. [4]. Damage to coral reefs is also unavoidable if the oil spill continues to spread deeper and deeper. Oil spills have a crucial impact on the aquaculture sector, such as the narrowing of fishing grounds for fishermen, thus having a substantial impact on their economic sector. [5].

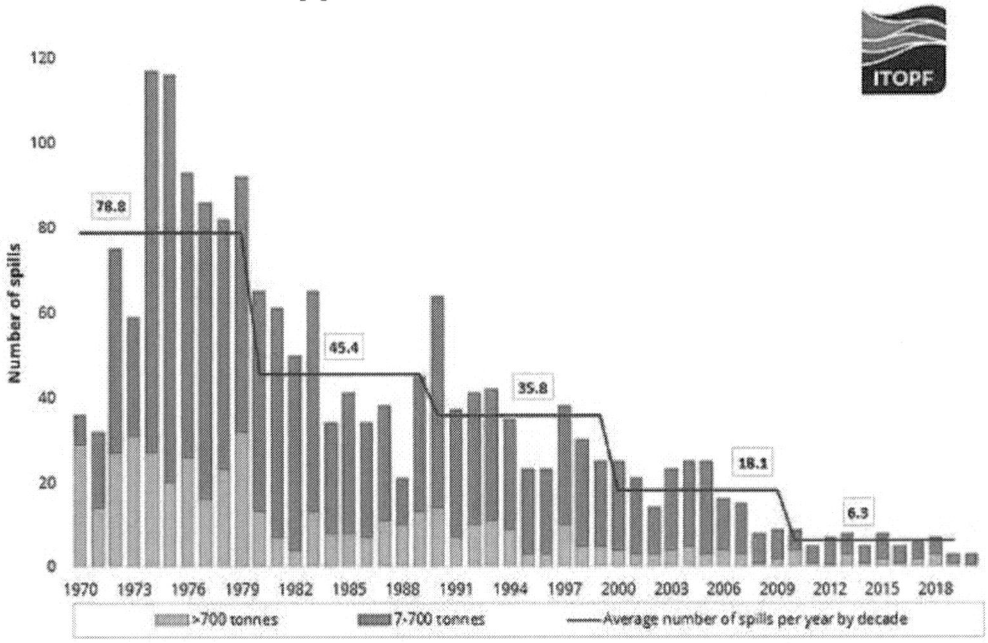

Figure 1. Number of oil spills (medium scale and large scale) from 1970 – 2020 [2]

Based on SDG point number nine, namely innovation and infrastructure, this paper will provide an explanation of product innovation to overcome oil spills. The innovation product in question is the use of nanotechnology-based products, carbon nanotubes (CNTs). CNTs are known as hydrophobic materials, but have excellent oil absorption capabilities [6]. This advanced innovation of oil absorbent has greater porosity and absorption surface so that the absorption of oil will be greater than conventional methods and is more environmentally friendly and cost effective. Reflecting on the mitigation efforts that have been implemented so far, in situ burning, for example, the activity of burning the remaining oil spills in the affected area actually endangers the ecosystem and pollutes marine life and requires large costs. [7]. With the presence of carbon nanotubes, it is expected to be able to advance industrial independence in the field of innovation and infrastructure in terms of mitigating marine pollution, in this case oil spills.

2. Method of research

The literature review method was used to gather data on carbon nanotubes and marine oil spills for this paper. The literature used comes from pre-existing journals. Research methods contain a literature review, data collecting, data analysis, idea development, and conclusion. The flowchart of the research methods used is shown in Figure 2.

As shown in Figure 2, the first stage is to review related journals that have been circulated before. To strengthen ideas, planning requires data that is by the related concepts. The data used in this paper is secondary data from valid sources. The sources are obtained from international scientific journal articles, ministerial regulation, proceedings, and SNI regulation. To make it easier to find the journal want to search, a search is performed by including keywords in the search field. Keywords used when surfing, namely carbon nanotubes, marine oil spill, offshore oil drilling, oil transporting by vessel, and SDGs. The data will be analyzed according to the ideas proposed after obtaining the necessary data. Furthermore, ideas are carried out development and innovation from the results of the analysis that has been obtained. After the development of ideas and innovations, it can be concluded that carbon nanotubes can be an alternative to overcome oil spills at sea.

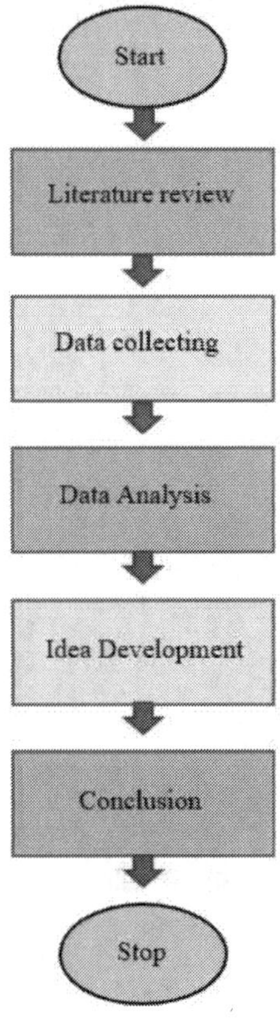

Figure 2. Flowchart of research methods

3. Oil transporting by vessel

3.1. *General information about oil transporting by vessel*

The demand for oil and other products made mainly of oil for industrial purposes is quite increased from year to year, therefore, exploration for the process of taking crude oil and its distribution becomes very important and becomes the key to the successful fulfilment of oil products. In this case, the discussion was conducted by exploring the vessel as a means of marine transportation that will be used to transport oil products. Sea transportation that is the basis of transportation mode for oil distribution has increased a lot, both in terms of quality and quantity. Sea transportation is also one of the most popular modes of transportation for intercontinental products and world trade [8]. According to the International Tanker Owners Pollution Federation (ITOPF), there has been a doubling of oil transport from 1970-1980 [2]. With the increase in sea transportation, it is certain that there will be an increase in the number of ships and the density of sea traffic lanes, which will increase the risk of accident rates at sea [9].

3.2. *Standard Operating Procedure during oil exploration transporting*

Vessels, including cargo ships used for oil unemployment, have a high fuel consumption rate of almost 78% of the total [10]. The consumption of energy used can cause damage to environmental impacts, such as damage to greenhouse gas emissions, and also impact the economy. Standard Operational Procedure (SOP) is created as a standardized work and management system for workers to reduce the risk of work accidents. In addition to worker safety, the standard operational procedure is also mentioned about an effective environmental management system to help organizations to control environmental damage caused by oil transporting work to help improve sustainable environmental performance [11].

3.3. *Fuel ship transport permit*

The permit to transport fuel into sea transportation for the distribution of oil is stipulated in the regulation of the Minister of Transportation of the Republic of Indonesia number 65 of 2019 concerning companies and ship agencies. Every ship that does work at sea in Indonesia must show a ship agency license [12].

3.4. *Causes of tanker ship accident and Oil transport ship track*

Reported from ITOPF 2020, the number of shipwrecks that caused severe oil spills (>700 tons) has decreased from 1970 by 24.5% to 1.8% in 2010. However, the oil spill will still be a serious problem for environmental and maritime industry issues [13]. It shown that there is a development in the maritime industry, which is proven by the increase in sea transportation lines with a decrease in shipwrecks that cause severe oil spills. With the rapid growth of sea transportation, it will be directly proportional to accidents between ships, both on a small and large scale, which can result in death, injury, economic problems, and environmental problems [9]. In addition to the increase in the ship population, the cause of shipwrecks can also stem from ship characteristics, type of operation, weather conditions and type of waterways [14]. The path passed by oil carriers in the world is quite wide, one of which is the world trade route where the line is the centre.

4. Offshore oil drilling

4.1. *General information of offshore oil drilling*

Offshore oil drilling is one of the energy resources utilization in filling the oil needs. Gasoline, petroleum, diesel, vehicle lubricants, liquefied petroleum gas (LPG), paraffin, and asphalt, as crude oil products, have been used in a massif for the welfare of human life. That operation utilizes a giant structure, a platform, as a place where drilling machines and oil products placed has created a lot of job fields, offered promising tax income for governments, boosted economic growth, and reduced dependence on oil import [15,16]. In addition to these benefits, some have questioned the risk of marine pollution and the impacts on marine biodiversity, in this case, the risk of oil spills resulted from offshore oil drilling [17]. Therefore, it is necessary to know how the overview of offshore oil drilling is carried out and what are the provisions need to be considered in carrying out such operations.

4.2. *How is the offshore oil drilling carried out?*

Offshore oil drilling has the same drilling principle as onshore, namely drilling through rocks which identified that the sedimentation of organic materials that have accumulated for a long time then due to high heat and pressure those organic materials are converted into oil and gas exist. [18]. The only difference is the geographical location so that additional elements are required in offshore oil drilling, there are marine riser, subsea wellhead, and subsea blow-out preventer (BOP) [19]. Offshore oil operations commonly start with drilled phases. Drilling starts when the drill bit is spudded into the rig floor carrying steel pipe behind until the drill assembly has already injected and the wellhead is right above the seafloor. The drill bit continues drilling along with the drill string (drilling system consists of drill pipe, stabilizer, drill collar, bottom hole assembly, and a bit) to the further sediments. At the same time, "mud" is pumped down into the drill pipe then through the drill bit to lift the drill cuttings (rock fragments have been drilled together) through the space between the wellbore wall and the drill string and keep the bit and well lubricated [17]. In this process, vibration usually occurs due to pressure between the drilling string and the drilling mud, the contact of the drilling string with the wellbore, and the bit with the wellbore which can damage the drilling equipment, drilling string also its stabilizer [20].

After the entire assembly has settled and the surface casing has paired through the wellbore, cementation should be done to preserve a permanently secure casing place [21]. The deeper the drilling process is carried out, the higher the hydrostatic pressure from rocks received, thus, the BOP and the riser are locked on the wellhead. As the pressure controller to prevent uncontrolled "kick" or blow-out from formation fluid flow, the BOP, a system that has a chain of valves controlling the well, has a critical role in securing all employees, equipment, and work environment [22]. Following the drilling process would be continued downward the part of another rock accompanied by those looping processes as shown in Figure 3.

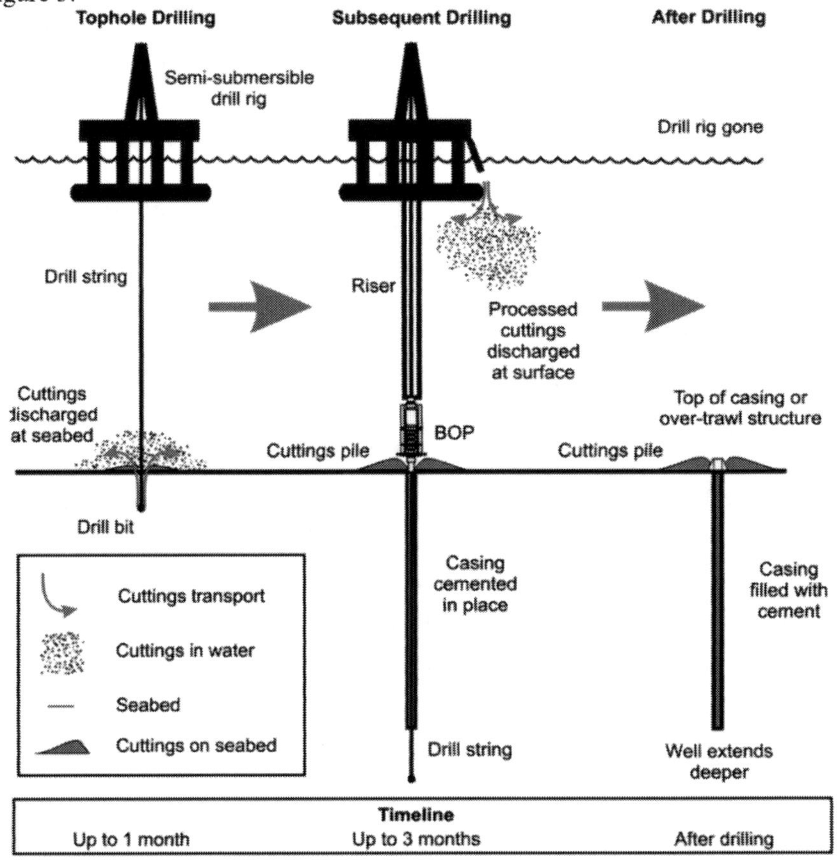

Figure 3. Sediment drilling process on the seabed [17]

4.3. *What kind of platform that should be considered?*

There are two types of units for drilling rigs, fixed units and mobile units. Fixed units (shown in Figure 5) in which the deck is supported by a giant steel structure known as *jacket* are supposed to steady upon their location and commonly used for moderate water depth with longer-term production from the reservoir. Meanwhile, mobile units are easily relocated and supposed to be used in moderate to very deep-water depth (500 ft to 10.000 ft) as the well geographical situation is considered [19]. There are several types in mobile units, namely *swamp/drilling barges, jack-up rigs, drillships (floater), and semisubmersible rigs*. Each of them has the terms of specific uses. For offshore oil drilling, *jack-up rigs* (shown in Figure 4) are suitable for the 500 ft seafloor under the surface as it have restricted leg anchored to the sea floor and can be dragged to the spot location [19]. On the other hand, *semisubmersible drilling rigs* are more flexible as they have to deal with unpredictable sea conditions at 10,000 ft above the seabed with the help of structures such as pontoons supporting the deck area, in this case, the principle of buoyancy is needed for stabilization of the platform rig legs at the unstable sea [19,23]. Similar to *semisubmersible drilling rigs*, *drillships* are designed to drill at very deep depths (2000 – 10000 ft) and face severe weather conditions. A mooring system and dynamic positioning system (DPS) are integral in drilling success. In shallower water conditions, the *drillships* are moored on the seabed with 6 to 12 anchors. After the water depth gets deeper, *drillships* rely on DPS as a positioning adjustment. This system in the form of a thruster connected to a satellite is useful for controlling the angular movement of the ship so as to prevent movement due to the agitated waves [24]. *Semisubmersible drigs* and *drillships* are shown in Figure 5

Figure 4. *Jack-up* rigs [25]

Figure 5. (a) Fixed units, (b) Semisubmersible units, (c) *Drillships* [19]

5. Marine oil spill and its mitigation

5.1. *Causes of oil marine spill*

Currently, spilled oil poses a serious problem at sea. Not a small amount of oil spilled onto the surface of the waters. Oil spills can be caused by two factors, namely natural factors and human activities. Natural factors occur due to the release of petroleum from petroleum-producing rocks at the bottom of the ocean. In addition, another thing that causes the spill of oil is the result of human activity itself. Spills can occur for various reasons during the process of exploration, extraction, and transportation, such as overpressure, mechanical failure, pipe corrosion, ship collision [26].

5.2. *Impact an oil spill on the area*

The spill of petroleum into the sea has an impact on the surrounding ecosystem, such as coral reefs, fish, organisms, and seabirds in the vicinity. The impact that occurs can be direct or indirect. In addition to having an impact on marine life, the spill of oil can have an impact on the economy. Spilled oil at sea causes enormous losses, including cleaning costs of the spilled area, equipment costs, spilled oil losses, etc. The impact can be seen in the Table 1.

Table 1. The impact of the oil spill on the surrounding ecosystem

Marine biota	Effects that occur	Reference
Fish	1. Exposure to high concentrations of polycyclic aromatic hydrocarbon (PAH) in water columns. 2. Inhibit the growth and development of fish	[27]
Coral reefs	1. Heavy oil will cover focus, and cause death or sub-lethal effects on algae 2. Damage the spawning place of the fish 3. Fish become inedible because it is dangerous for the body	[28]
Seabirds	1. Causes death 2. Damage organs when entering the body 3. Oil can damage feathers on the body of birds	[26]
Planktons	1. Oil exposure can cause immediate mortality, acute toxicity, decreased feeding, delayed egg production, slow hatching rates, reduced swimming speeds, reproductive abnormalities in planktonic copepods.	[29]

5.3. *Technology currently used to deal with oil spills and the effectiveness*

Currently, oil spills can be addressed by physical methods and chemical methods. Handling of petroleum spills in the sea, namely by using Oil booms and in situ techniques. Judging from the handling done, there are no very powerful techniques to overcome marine oil spills. The treatment is done just to hold the oil so that it does not spread more widely. The technology used today, namely:

5.3.1. Oil booms

Oil booms are technologies that are currently widely used. This technology works to hold spilled oil from spreading more widely. This technology is like a floating wall and is controlled by a tow ship. After that, a skimmer is needed to suck up the petroleum spilled in the sea. However, this technology is not too flexible to follow the pattern of the sea. This technology is less effective when the ocean waves are too high.

5.3.2. In situ Burning

In situ combustion is the burning of a portion of oil that is carried out in the reservoir to obtain heat. The heat resulted in a decrease in viscosity value, development, and distillation of oil with the effect of gas drive and solvent extraction so that the spread of oil spills did not expand. The handling that is still being used to tackle petroleum spills is still not effective. The handling of the treatment causes new pollution and more use of costs. Now, innovations are needed that are more flexible and cost less. The use of innovations from Carbon nanotubes can be the right solution in handling petroleum spills.

6. SDGs number 9: industry, innovation, and infrastructure

6.1. *General information about Sustainable Development Goals (SDGs)*

In general, Sustainable Development Goals (SDGs) are global action plans agreed upon by world leaders, including Indonesia, to end poverty, protect the environment and reduce inequality. SDGs contain 17 goals and 169 targets that are expected to be achieved by 2030 as shown in the Figure 6

Figure 6. Sustainable Development Goals, 17 Points of Sustainable Development Goals [30]

On this topic, we will focus on SDGs number 9 which to build and develop a resilient infrastructure, support inclusive and sustainable industrialization, help development, and create innovation.

6.2. *Why focus on SDGs number 9? And what relevance between SDGs number 9 with marine oil spill?*

Because it is reviewed from the objectives of SDGs number 9, the success of industry, the marine oil spill in Indonesia will hinder achievement of these goals. In addition, marine oil spill in Indonesia will cause a lot of adverse effects if left in the short or long term. The mass death of fish due to lack of oxygen and carbon dioxide poisoning, as well as damage to mangrove ecosystems and marine ecosystems are examples of the adverse effects of marine oil spill. The government and oil companies are still lacking in innovation tackling marine oil spill, so the underwater life preservation system cannot be achieve the Sustainable Developments Goals (SDGs).

6.3. *Analysis/ information the impact of using carbon nanotubes to resolve oil spills on the development of infrastructure, industry, and information.*

Marine oil spill is a problem that is considered a serious threat to the environment around the world, including Indonesia. This will hinder the development stage of industry and infrastructure, it also pose a potential hazard to marine ecosystems and aquatic life. So, we need to create an innovation that can overcome these problems. The government itself has used several ways to deal with marine oil spill, including the use of oil booms or oil pumping equipment, and in situ burning.

In the current era, there are many new innovations that are more useful for tackling marine oil spill in Indonesia. Carbon nanotubes are one of the innovative materials that are able to bind oil twice more effectively than oil absorbents and can also neutralize toxins carried by oil spill. Modified carbon nanotubes can absorb up to 180 times their weight of oil, which is certainly more effective and efficient than commercially available adsorbents for cleaning marine oil spill in Indonesia [31,32]. In addition, another advantage of carbon nanotubes is that in the future it will have a bright future because it has many benefits in terms of effective and efficient separation of oil or water, removal of contaminants and organic dyes, contaminants emerging from industrial wastewater, oil absorption, and photo catalysis to keep surface water free from pollution and contamination.

The presence of carbon nanotubes is expected to advance the independence of the industry in terms of innovation in the field of environmental pollution management, in this case marine oil spill. With the use of carbon nanotubes to overcome marine oil spill in Indonesia, it will help industry owners and also the government in tackling marine oil spill. It is also expected to produce environmentally friendly materials for industries that have a direct impact on marine ecosystems.

7. Carbon nanotubes

7.1. *General information about carbon nanotubes*

Carbon nanotubes are micrometre-scale graphene sheets that rolled into cylinders and covered with circular fullerene [33]. Carbon nanotubes have been categorized into 3 groups based on their type of the walls. As reported by Jiajia [34]. Carbon nanotubes are classified into single-walled carbon nanotubes, double-walled carbon nanotubes, and multi-walled carbon nanotubes. The specific gravity multi-walled carbon nanotubes are higher than the other nanomaterials such as fullerene. Fullerene specific gravity is 1.7 g/cm^3 and the carbon nanotubes specific gravity is 1.8 g/cm^3 [33].

Due the reason that carbon nanotubes have a high number of specific gravity with complex porous cavity structure, than carbon nanotubes are the best absorbent material that can be applied in contaminant-polluted environment. In addition, carbon nanotubes also good absorbent to surmount polycyclic aromatic carbon, dioxin, and benzene [33]. Thus, the use of carbon nanotubes to clean up marine oil spills is a correct and wise solution because it is effective and environmentally friendly. Carbon nanotubes can be modified to aerogels or sponge to clean up the oil spills [35].

7.2. *Various type of carbon nanotubes for the new technology as oil absorbent and ability of various type carbon nanotubes used for removal oil spill*

The major focus in this research is to development nano-materials that from one of the broad benefits of carbon nanotubes that can clean up marine oil spills, carbon nanotubes has various type that can used for clean up the oil spills. In the allotment of cleaning up marine oil spills, each type of carbon nanotubes has different quality and absorption capacities. Further information about types od carbon nanotubes shown in Table 2.

Table 2. Types and capabilities of each carbon nanotubes in clean up oil spills

No	Type	Mass (g)	Density (mg Cm^-1)	Sorbate	Absorption time (minutes)	Absorption capacity (g g^-1)	Desorption	Reference
1	Carbon Nanotubes microfiber sponge		45	Heptane and pump oil		43	Distillation or squeezing	[36]
2	Multi-Wall Carbon Nanotubes	0.2		Gasoline Kerosene and Diesel no 1 and no 2	120	10.5	Heating treatment	[37]
3	Carbon Fiber Fabric	0.018	7.5	Diesel and crude oil	15	92.3		[38]
4	Carbon aerogels	1	18.5	Gasoline, olive oil, and pump oil	Less than 1	26	Squeezing and distillation	[39]

Figure 7. Carbon Nanotubes Microfiber Sponge [36]

Figure 8. The SEM of Multi-Walled Carbon Nanotubes [37]

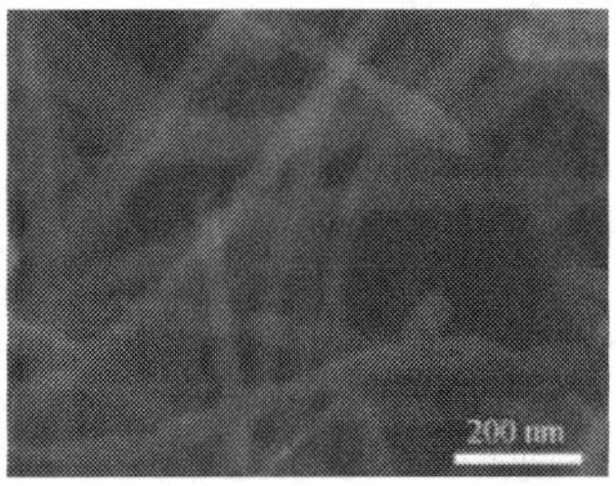

Figure 9. SEM ofCarbon Fiber Fabric[38]

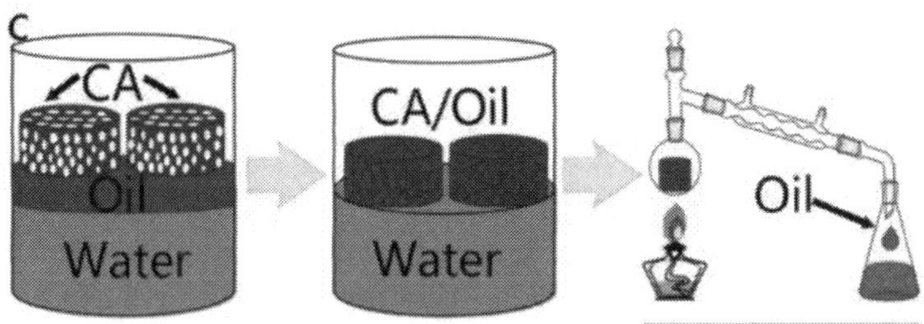

Figure 10. Process of Carbon Aerogels in absorbing and desorbing oil [39]

The physical shape of the carbon nanotubes microfiber in Figure 7 is irregular with the fibers sticking out. It has good quality in absorption capacity and can desorb by distillation and squeezing method. For multi-walled carbon nanotubes, the SEM results in Figure 8 show it has quite a lot of gap which will full when it absorb oil [37]. However, the SEM results of carbon fiber fabric in Figure 9 show that there are more gap, this indicates the absorption capacity of this material is greater than other types. It has longer absorption time than the carbon aerogel in Figure 10. Carbon aerogels are the fastest on the absorption time because of the physical shape. An illustration of the absorbing until desorbing process of the carbon aerogels can be seen in Figure 10

7.3. *Conventional technology used to deal with marine oil spill and excellence of carbon nanotubes sorption technology toward oil spills removal*

Prior to advances in the use of carbon nanotubes to clean up marine oil spills, oil spills was cleaned with nano porous sorbent, such as graphene [40], magnetic nano particles [41], bioremediation with active bacteria [42], photochemical oxidation [43], and in situ burning [44]. The problem, each technology has a limit of ability that still causes bad side effects. These limit of ability need some innovation for fast, effective, sustainable, and ecosystem friendly to avoid the danger effects by marine oil spills [45].

Among the inorganic materials such as active carbon [46] , there are several organic material such as hucks, cotton [47], and bagasse [48]. Organic materials certainly has a low impact on the marine ecosystem, but it has significant drawbacks such as low buoyancy, the oil spills not correctly absorbed, and low hydrophobic quality so it affects to the absorption capacity [49]. Over the past decade, the technology used to deal with marine oil spills has been appropriate because it focuses on limiting the spread of oil spills, the success rate of oil recovery, and how it affects the marine ecosystem. It needs the newest innovation of the material that continuous develop [50].

As we know, specific gravity of graphene is under carbon nanotubes. Even though graphene contains carbon, the absorption is not good as carbon nanotubes. As well as in situ burning, in situ burning can clean up the oil spill but this technology creates a new pollution to the environment. As shown in the table 1, carbon nanotubes-based materials such as carbon nanotubes microfiber sponge, multi-wall carbon nanotubes, carbon fibre fabric, and carbon aerogels can be used for oil absorbent materials that efficient, effective and environmentally friendly because its reusable.

8. Project implementation concept

8.1. *Main concept which equipped with illustration picture*

The concept that the author brings is to realize the use of carbon nanotube-based products as the main material used to deal with marine oil spills. For initial processing, carbon nanotube powder is processed into fibres and then the fibres will be reprocessed into carbon nanotube fabric sheets. The cloth made from carbon nanotubes is then cut to a predetermined size and dimension and then linked to a buoy, this is intended so that oil spills that occur in the area can be handled effectively and efficiently and maintain the sustainability of marine ecosystems with carbon properties. owned by the product. With this concept, the oil produced in carbon nanotube sheets can be reused by related companies so that losses due to oil spills can be minimized. Illustrations showing the detail of carbon nanotubes-based fabrics, the manufacture of carbon nanotubes-based fabrics, and the application of oil-absorbing products from carbon nanotubes can be seen in Figure 11, Figure 12, and Figure 13.

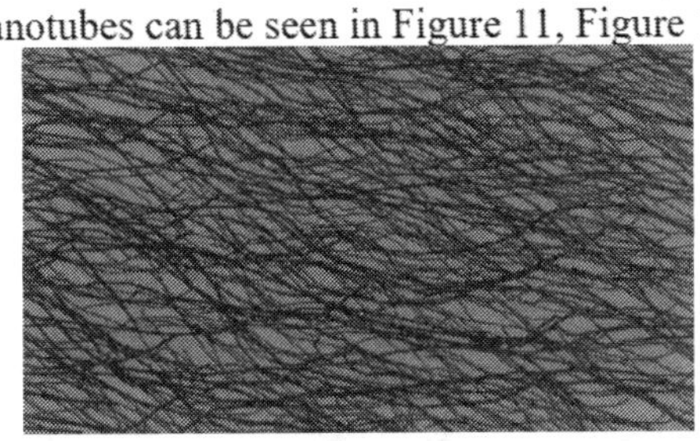

Figure 11. Carbon nanotube fabric details

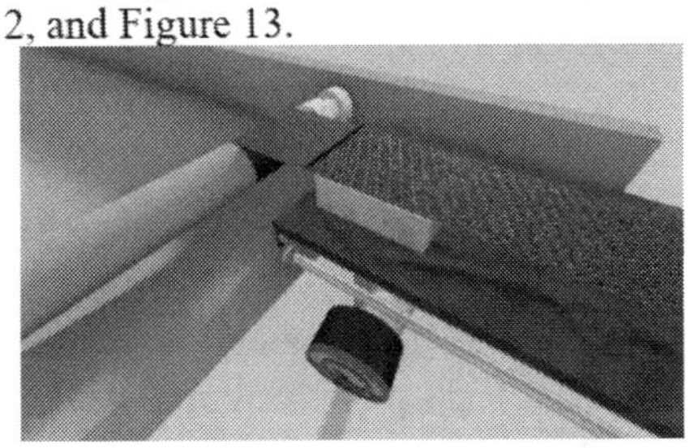

Figure 12. Manufacture of fabric sheets made from carbon nanotubes

Figure 13. Application of products based on carbon nanotubes to absorb oil

8.2. *Best impact for the environment by using carbon nanotubes as the oil absorbent*

Resolution to the problem of water contamination is considered important so as not to damage marine ecosystems. Features that make physical remediation as an act of water contamination, especially in the sea are of particular concern to environmental activists as well as the government. The cause of this contamination is due to oil exploration, sea transportation, and industrial operations in the world [51]. Tackling oil spills efficiently and being able to recover a considerable quantity of oil with minimally damaging marine ecosystems or not at all is indispensable for this case [51]. The use of carbon nanotubes-based products as the handling of oil spills at sea due to oil exploration is believed to be a new idea that is more effective and efficient in handling oil spills so as to reduce the effects of damage to marine ecosystems. With the use of carbon nanotubes-based products, it will cope with oil spills in a superior way, namely with highly effective performance in the absorption and recovery of organic oil and solvent products [35].

8.3. *Target community who get benefit by using this innovation*

8.3.1. Government.

The area of the sea based on United Nation Conversation the Law of the Sea (UNCLOS) on December 10, 1982 which reached 6,400.00 km2. If carbon nanotubes-based products are realized and applied to cope with oil spills in the sea, then the government will benefit such as a maintained marine ecosystem, so as to support the tourism sector that can contribute to the country's income and impact the economic sector.

8.3.2. Entrepreneurs involved in oil exploration.

The realization of carbon nanotubes to overcome oil spills at sea, will certainly bring benefits for entrepreneurs who move in the field of oil exploration. This is because in addition to absorbing oil spills in the sea, carbon nanotubes-based products can also recover oil spills at sea to be reprocessed by related companies. In addition, carbon nanotubes are also a recyclable material so that they can be used to the maximum.

8.3.3. Fisherman.

With a livelihood in the sea, of course marine oil spill will have a big influence on marine ecosystems, which will certainly affect the income of fishermen. If it is left unchecked, then the marine ecosystem will be damaged and the fish that will later be caught have poor quality, can even die. Therefore, with the realization of carbon nanotubes to overcome marine oil spill in the Sea of Indonesia, fishermen will benefit in the form of marine ecosystems that are maintained and avoid pollution due to marine oil spill, so that the fish produced will be better quality.

8.3.4. Parties who can help implement this oil absorbent innovation.

Related parties that will succeed the formation of oil absorbent products based on carbon nanotubes, including the government of the Republic of Indonesia, especially the Ministry of Marine Affairs and Fisheries, the Ministry of Transportation, and the Ministry of Health conduct cooperation to realize the idea. Furthermore, experts in the field of oil and gas aimed at designing plans, making products, as well as the operation of ideas, and lastly, the company involved in the oil exploration project, in this case related to the purchase and management of shares.

9. Conclusion

Oil exploration in Indonesia causes the risk of marine oil spill in the Indonesian sea to increase. In addition, offshore oil drilling also presents considerable risks to marine pollution and its impact on biodiversity. The adverse impact is the destruction of mangrove and marine ecosystems, such as fish, coral reefs, and others. The handling that is still applied by the government is considered ineffective because it causes new pollution and requires a large cost. We need a solution that can solve the problem effectively and not cause any new damage. Innovation of carbon nanotubes as a product that will be used to overcome marine oil spill in Indonesia can be the right solution.

Carbon nanotubes are excellent adsorbents when applied in contaminant-contaminated environments. Carbon nanotubes in the form of powders are processed into fibers and then into sheets. Fabric-based carbon nanotubes are then cut to such dimensions and associated with buoys. With this concept, oil produced in carbon nanotubes sheets can be reprocessed by the relevant company so as to minimize losses due to oil spills. Another advantage of carbon nanotubes is that it does not cause pollution and binds more oil so that it is more efficient than other methods.

The occurrence of marine oil spill hinders the achievement of SDGs number 9 goal which is to succeed the industry, develop infrastructure, and create innovation. The government and oil companies are still less innovating in tackling marine oil spill so the underwater life preservation system cannot be said to meet the Sustainable Development Goals (SDGs). The presence of carbon nanotubes is expected to advance the independence of the industry in terms of innovation in the field of environmental pollution management, in this case marine oil spill. With the use of carbon nanotubes to overcome marine oil spill in Indonesia, it will help industry owners and also the government in tackling marine oil spill. It is also expected to produce environmentally friendly materials for industries that have a direct impact on marine ecosystems.

Related parties that will succeed the establishment of oil absorbent products based on carbon nanotubes are the government of the Republic of Indonesia, especially the ministry of marine affairs and fisheries, the ministry of transportation, and the ministry of health to work together to realize the idea. Furthermore, oil and gas experts aimed at designing, creating products, and operating ideas, and finally, companies involved in oil exploration projects, in this case related to the purchase and management of shares. While the parties who benefited with the realization of carbon nanotubes as a solution to overcome the marine oil spill are the government, entrepreneurs involved in oil exploration, and fishermen.

Acknowledgement

The paper was supported by PKM Center Universitas Negeri Malang, Universitas Negeri Malang, and Directorate General of Higher Educational, Ministry of Education, Culture, Research, and Technology of the Republic of Indonesia, who has provided an opportunity in the Student Creativity Program (PKM) for the 2021 funding year.

References

[1] Singh H, Bhardwaj N, Arya S K and Khatri M 2020 Environmental impacts of oil spills and their remediation by magnetic nanomaterials *Environ. Nanotechnology, Monit. Manag.* **14** 100305
[2] ITOPF 2020 Statistics - ITOPF *Int. Tanker Owners Pollut. Fed.*
[3] Mrozowska A 2021 Formal Risk Assessment of the risk of major accidents affecting natural

environment and human life, occurring as a result of offshore drilling and production operations based on the provisions of Directive 2013/30/EU *Saf. Sci.* **134** 105007
[4] NOAA 2018 How does oil impact marine life? *US Dep. Commer. Natl. Ocean. Atmos. Adm.*
[5] Pascoe S and Innes J P 2018 Economic Impacts of the Development of an Offshore Oil and Gas Industry on Fishing Industries: A Review of Experiences and Assessment Methods *Rev. Fish. Sci. Aquac.* **26** 350–70
[6] Ong C C, Sundera Murthe S, Mohamed N M, Perumal V and Mohamed Saheed M S 2018 Nanoscaled Surface Modification of Poly(dimethylsiloxane) Using Carbon Nanotubes for Enhanced Oil and Organic Solvent Absorption *ACS Omega* **3** 15907–15
[7] Liang G, Li Y, Zhang Y, Zi C, Zhao W and Zhang D 2019 Progress on Magnetically Responsive Oil Absorption Materials *Cailiao Daobao/Materials Reports* **33**
[8] Celik M and Topcu Y I 2009 Use of an ANP to prioritize managerial responsibilities of maritime stakeholders in environmental incidents: An oil spill case *Transp. Res. Part D Transp. Environ.* **14** 502–6
[9] Ye X, Chen B, Li P, Jing L and Zeng G 2019 A simulation-based multi-agent particle swarm optimization approach for supporting dynamic decision making in marine oil spill responses *Ocean Coast. Manag.* **172** 128–36
[10] Çetin O and Ziya Sogut M 2021 A new strategic approach of energy management onboard ships supported by exergy and economic criteria: A case study of a cargo ship *Ocean Eng.* **219** 108137
[11] Indonesia S N and Nasional B S 2005 Sistem manajemen lingkungan – Panduan umum tentang prinsip, sistem dan teknik pendukung
[12] Menteri PUPR 2020 Peraturan Menteri Perhubungan Republik Indonesia *Mentri Perhub. Republik Indones.* **2011** 13
[13] Cakir E, Sevgili C and Fiskin R 2021 An analysis of severity of oil spill caused by vessel accidents *Transp. Res. Part D Transp. Environ.* **90** 102662
[14] Talley W K, Jin D and Kite-Powell H 2005 Post OPA-90 vessel oil transfer spill prevention: The effectiveness of coast guard enforcement *Environ. Resour. Econ.* **30** 93–114
[15] Mukherjee D and Rahman M A 2016 To drill or not to drill? An econometric analysis of US public opinion *Energy Policy* **91** 341–51
[16] Chen M and Martens P 2021 Coastal residents' attitudes toward offshore oil and gas drilling in China *Extr. Ind. Soc.* 100942
[17] Cordes E E, Jones D O B, Schlacher T A, Amon D J, Bernardino A F, Brooke S, Carney R, DeLeo D M, Dunlop K M, Escobar-Briones E G, Gates A R, Génio L, Gobin J, Henry L A, Herrera S, Hoyt S, Joye M, Kark S, Mestre N C, Metaxas A, Pfeifer S, Sink K, Sweetman A K and Witte U 2016 Environmental impacts of the deep-water oil and gas industry: A review to guide management strategies *Front. Environ. Sci.* **4**
[18] Alisson E and Mandler B 2018 What Determines the Location of a Well ? 2016–9
[19] Zulqarnain M and Fike R 2017 Overview of Offshore Drilling Technologies *Encycl. Marit. Offshore Eng.* 1–12
[20] Afriansyah A and Basri H 2019 a Review Dynamic Vibration on Drill String in Drilling Operation *Sebatik* **23** 1–8
[21] Fang H and Duan M 2014 Offshore Oil and Gas Drilling Engineering and Equipment
[22] Alexandri A, Nurbayanah S and Palilu J M 2015 Blow Out Preventer Test Sebagai Bagian Dari Pemeriksaan Rutin **05**
[23] Simpson D A 2017 Well-Bore Construction (Drilling and Completions)
[24] Ma K T, Luo Y, Kwan T and Wu Y 2019 Mooring system engineering for offshore structures *Mooring Syst. Eng. Offshore Struct.* 1–350
[25] Ma K Y, Kim J H, Park J S, Lee J M and Seo J K 2020 A study on collision strength assessment of a jack-up rig with attendant vessel *Int. J. Nav. Archit. Ocean Eng.* **12** 241–57
[26] Zhang B, Matchinski E J, Chen B, Ye X, Jing L and Lee K 2018 *Marine oil spills-oil pollution,*

sources and effects (Elsevier Ltd.)
[27] Langangen, Olsen E, Stige L C, Ohlberger J, Yaragina N A, Vikebø F B, Bogstad B, Stenseth N C and Hjermann D 2017 The effects of oil spills on marine fish: Implications of spatial variation in natural mortality *Mar. Pollut. Bull.* **119** 102–9
[28] Michel J 2015 Chapter 7 Oil Spills : Causes , Consequences , Prevention , and Countermeasures
[29] Cohen J H, McCormick L R and Burkhardt S M 2014 Effects of dispersant and oil on survival and swimming activity in a marine copepod *Bull. Environ. Contam. Toxicol.* **92** 381–7
[30] United Nations 2018 THE 17 GOALS | Sustainable Development *Dep. Econ. Soc. Aff.*
[31] Gui X, Cao A, Wei J, Li H, Jia Y, Li Z, Fan L, Wang K, Zhu H and Wu D 2010 Soft, highly conductive nanotube sponges and composites with controlled compressibility *ACS Nano* **4** 2320–6
[32] Gui X, Wei J, Wang K, Cao A, Zhu H, Jia Y, Shu Q and Wu D 2010 Carbon nanotube sponges *Adv. Mater.* **22** 617–21
[33] Lee J 2018 Carbon Nanotube-Based Membranes for Water Purification (Elsevier Inc.)
[34] Sun J, Li H, Feng L, Jia Y, Song Q and Li K 2017 A novel treatment of carbon fibers with improving tensile strength to synthesize evenly distributed carbon nanotubes on their surface *Appl. Surf. Sci.* **403** 95–102
[35] Kukkar D, Rani A, Kumar V, Younis S A, Zhang M, Lee S S, Tsang D C W and Kim K H 2020 Recent advances in carbon nanotube sponge–based sorption technologies for mitigation of marine oil spills *J. Colloid Interface Sci.* **570** 411–22
[36] Song J, Wang H, Li Z, Long Y, Liu Z, Wang H, Li X, Fang M, Li B and Wu H 2018 Large-scale blow spinning of carbon microfiber sponge as efficient and recyclable oil sorbent *Chem. Eng. J.* **343** 638–44
[37] Liu T, Chen S and Liu H 2015 Oil adsorption and reuse performance of multi-walled carbon nanotubes *Procedia Eng.* **102** 1896–902
[38] Zhu K, Shang Y Y, Sun P Z, Li Z, Li X M, Wei J Q, Wang K L, Wu D H, Cao A Y and Zhu H W 2013 Oil spill cleanup from sea water by carbon nanotube sponges *Front. Mater. Sci.* **7** 170–6
[39] Han S, Sun Q, Zheng H, Li J and Jin C 2016 Green and facile fabrication of carbon aerogels from cellulose-based waste newspaper for solving organic pollution *Carbohydr. Polym.* **136** 95–100
[40] Diraki A, Mackey H, McKay G and Abdala A A 2018 Removal of oil from oil–water emulsions using thermally reduced graphene and graphene nanoplatelets *Chem. Eng. Res. Des.* **137** 47–59
[41] Cardona D S, Debs K B, Lemos S G, Vitale G, Nassar N N, Carrilho E N V M, Semensatto D and Labuto G 2019 A comparison study of cleanup techniques for oil spill treatment using magnetic nanomaterials *J. Environ. Manage.* **242** 362–71
[42] Atlas R M and Hazen T C 2011 Oil biodegradation and bioremediation: A tale of the two worst spills in U.S. history *Environ. Sci. Technol.* **45** 6709–15
[43] Shivaraju H P, Muzakkira N and Shahmoradi B 2016 Photocatalytic treatment of oil and grease spills in wastewater using coated N-doped TiO2 polyscales under sunlight as an alternative driving energy *Int. J. Environ. Sci. Technol.* **13** 2293–302
[44] Sulistyono 2013 Dampak Tumpahan Minyak (Oil Spill) di Perairan Laut Pada Kegiatan Industri Migas dan Metode Penanggulangannya *Forum Teknol.* **3** 49–57
[45] Alaa El-Din G, Amer A A, Malsh G and Hussein M 2018 Study on the use of banana peels for oil spill removal *Alexandria Eng. J.* **57** 2061–8
[46] Parthasarathy P and Narayanan S K 2014 Effect of Hydrothermal Carbonization Reaction Parameters on *Environ. Prog. Sustain. Energy* **33** 676–80
[47] Doshi B, Sillanpää M and Kalliola S 2018 A review of bio-based materials for oil spill treatment *Water Res.* **135** 262–77
[48] Youssefian L, Vahidnezhad H, Raoofian R, Greene W and Heidari M 2012 Potential functions of

the human homeobox TGIFLX/Y genes in normal and abnormal development *Eur. J. Clin. Med. Oncol.* **4** 23–6
[49] Liu H, Geng B, Chen Y and Wang H 2017 Review on the Aerogel-Type Oil Sorbents Derived from Nanocellulose *ACS Sustain. Chem. Eng.* **5** 49–66
[50] Dave D and Ghaly A E 2011 Remediation technologies for marine oil spills: A critical review and comparative analysis *Am. J. Environ. Sci.* **7** 424–40
[51] Dhaka A and Chattopadhyay P 2021 A review on physical remediation techniques for treatment of marine oil spills *J. Environ. Manage.* **288** 112428

IOP Conf. Series: Earth and Environmental Science **847** (2021) 012035 doi:10.1088/1755-1315/847/1/012035

An overview of the permeable reactive barrier as part of water remediation system in tropical countries

I N Ibrahim[1], M A A Q Rahman[1], M A Hannandya[1], M H Avicenna[1], M D Bathista[1] and M M A Pratama[1]

[1]Department of Civil Engineering, Faculty of Engineering, Universitas Negeri Malang, Indonesia

mirza.abdillah.ft@um.ac.id

Abstract. Natural resources are widely used for daily life, but due to poor utilization it can cause natural disasters. Disasters can occur due to several circumstances, such as flooding, which is caused by a malfunction of the drainage system. Floods may contain some bad compounds; therefore, proper solutions are needed to prevent soil pollution. Permeable reactive barrier or PRB can be a solution to this problem. There are various ingredients and benefits of Permeable Reactive Barriers, such as the addition of carbon particles, which will filter water containing petroleum when water passes through PRB. With the help of porous concrete, the water will flow into the permeable reactive barrier and then flow into the bio pores, so that it can make the surrounding soil more fertile. This paper review aims to discuss the use of permeable reactive barriers for soil remediation in tropical countries.

1. Introduction

The environment provides natural resources for us to use. However, sometimes humans use it excessively so that it often causes environmental damage. Environmental conditions change every year either quickly or slowly, this can happen due to many factors and can cause many problems. One of the problems caused by unwise environmental management is flooding. Flood is a natural phenomenon in the form of abnormal overflow of water that can drown a large area so that it can have an impact on the daily life conditions of the people and the environment affected by it [1–4]. Floods every year hit Indonesia, especially metropolitan areas such as Jabodetabek. The rapid rate of urbanization and the ineffectiveness of urban spatial control have resulted in environmental damage. In addition, the rapid population growth that is in line with the increasing need for vacant land encourages the convention of land which was originally used as a water catchment area to become watertight land, this is the main cause of flooding in urban areas [5,6].

The tropical climate of Indonesia makes Indonesia has high rainfall. The tropical climate of Indonesia makes Indonesia has high rainfall. For example, the results of the analysis of rainfall in February in the Jabodetabek area are more than 300 mm per month where the frequency is included in the category of high monthly rainfall or in the category of above normal rainfall (AN) with a comparison of the characteristic value of rain around 116-150% [7]. The high rainfall is caused by La Nina or a decrease in sea surface temperature which can increase the potential for flooding. BMKG data in February 2021 states that the potential for flooding in the Jabodetabek area has a medium level, which means that it can potentially lead to flooding [8].

High rainfall can lead to reduced availability of clean water. This can happen because the incoming flood can be polluted, so it can cause pollution and water resistance. With the various problems above, efficient solutions are needed to be able to achieve the Sustainable Development Goals (SDGs) targets, namely industry goals, innovation and infrastructure, and life on land goals. The use of a permeable reactive barrier combined with bio pore holes and porous concrete as a rainwater remediation solution in residential areas with tropical climates can be a solution in accelerating the infiltration process to reduce the volume of puddles that can potentially become flooded. Utilization of bio pore holes can help in fertilizing plants around roads so that they can help in creating green open spaces [9]. On the other hand, the use of porous concrete can be a solution in reducing the number of accidents, especially when it rains because the rough texture of the concrete makes the roads not slippery.

This concept is to apply porous concrete to the road in order to increase the soil infiltration process, then the water will be channelled into a permeable reactive barrier (PRB) which functions to remediate rainwater after it is channelled into bio pore holes as a place for water infiltration and to fertilize plants and increase groundwater volume.

2. Research method

This paper using the literature review method to collect the data. The data in this paper also cited from trusted sources. The sources of the data were taken from scientific journal articles and trusted website articles. The flowchart of research methods can be seen in Figure 1.

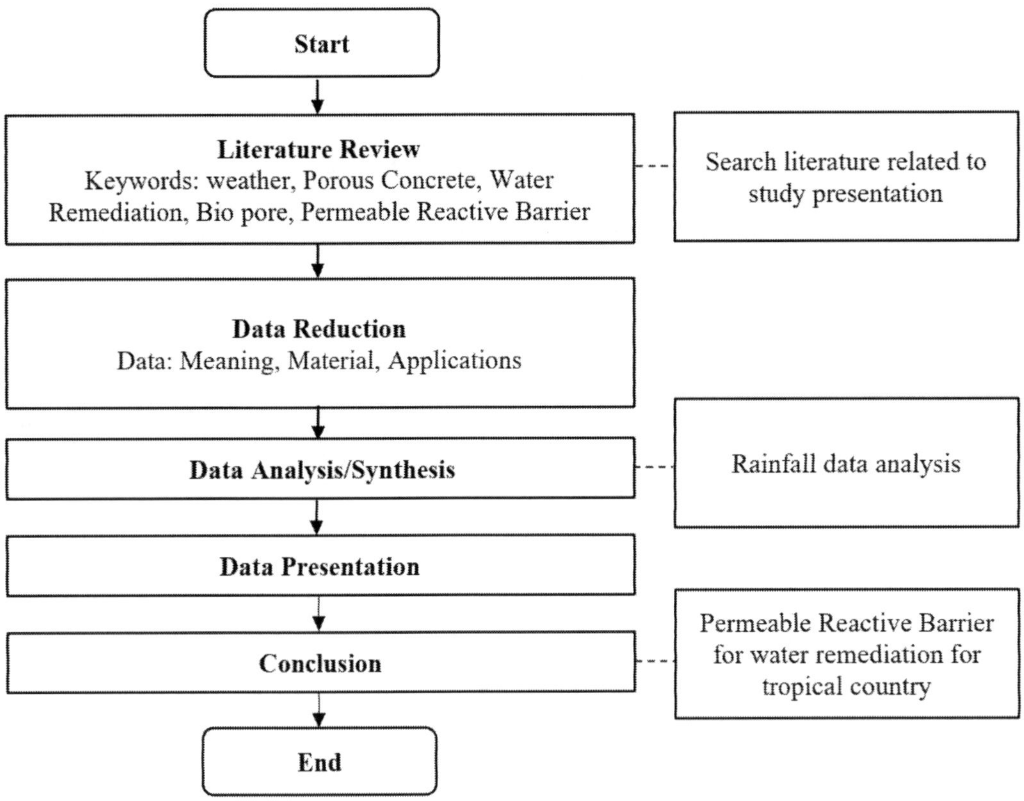

Figure 1. Flowchart of research method

Related to Figure 1, in compiling this paper, it begins by searching for literature related to papers such as weather, porous concrete, water remediation, bio pores, and permeable reactive barriers and continues by reducing the data obtained previously, which focuses on meaning, materials and applications. After the data has been reduced, proceed to the stage of analysing the data and proceed to its presentation in this article. The last method is to conclude the data obtained and conclude permeable reactive barriers for water remediation for tropical countries.

3. Porous concrete and its application

Porous concrete or pervious concrete is a special type of cementitious material composed of gap graded aggregates, combine with a thin layer of cement paste. Porous concrete is made by continuous voids that will make it into concrete. The gaps that exist in the concrete occur due to the size of the voids [10].

The material that made porous concrete are coarse aggregates, Ordinary Portland cement (OPC), admixtures, and water its similar to the common concrete, but porous concrete just uses coarse aggregates from 9.5 mm to 19 mm various of the size [11]. The strength of the porous concrete is affected by the gap of the voids, aggregate gradation, and the paste aggregate ratio is smaller [12–14]. The aggregate gradation also affected the permeability [14]. Figure 2 show a schematic diagram of porous concrete. The figure shows the gap that occur because the size of the voids that make the concrete [10]. The schematic diagram of porous concrete method can be seen in Figure 2.

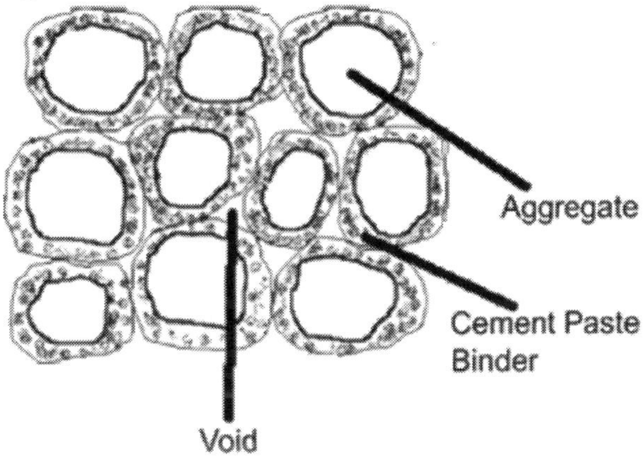

Figure 2. Schematic diagram of porous concrete method [15]

The porous concrete is normally used without any corrosion reinforcement due to the gap in its structure. Porous concrete is usually considered an friendly environmentally pavement [16]. The applications of porous concrete are rigid drainage layers under exterior areas, greenhouse floors, structural walls for better thermal insulation characteristics of lightweight, or both [17]. The water permittivity test can be seen in Figure 3. Figure 3 show the application of porous concrete as interlocking pavement. The interlocking pavement with porous concrete is shows no occurring puddles on the surface. The water is directly absorbed to the blocks whereas in contrast to conventional concrete water will puddles on the surface as seen on Figure 4 [18,19].

Figure 3. Water permittivity test [18]

Figure 4. Porous concrete sample in the passage of sample [19]

4. Current technology on water remediation

One of the biggest challenges of the twenty-first century is to provide safe, clean, and low-cost water. The fast growth of industrialization, increasing population, and changing people's lifestyles have led to increased production. This change leads to improper disposal of hazardous industrial waste in water systems, such as dyes, heavy metals, pesticides, pharmaceuticals, and residues [20]. The demand for clean water continues to increase along with the growth of the human population, but there is water quality deterioration [21]. These contaminants are micro and nano plastic in size [22]. Several attempts have been made to remediate contaminated water by local and international organizations [23]. One of the efforts made for water remediation is water remediation using biosorbents obtained from agriculture and fruits waste [24]. Agricultural waste materials used include bagasse powder (SBP), rice husk ash (RHA), powdered fruit peel waste (WFPP), and powder mixture (PB) were used as fluoride absorbent materials. Bagasse and rice husk ash were obtained from local farmers and rice mills respectively. Skin fruits are waste from the local juice shop. The steps of water remediation using biosorbents obtained from agriculture and fruits waste can be seen in Figure 5.

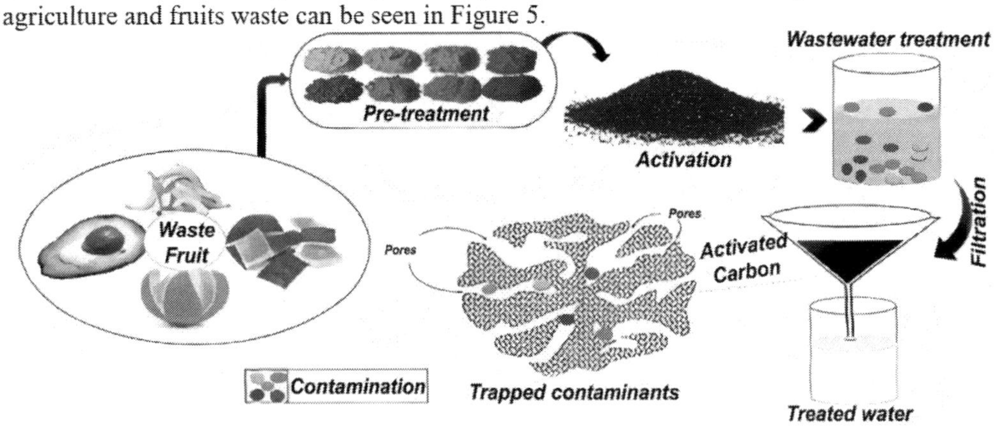

Figure 5. Water remediation using biosorbents obtained from agriculture and fruits waste [25]

Figure 5 show the steps water remediation using biosorbents obtained from agriculture and fruits waste. All ingredients except rice husk ash cut into small pieces, dried in the sun for 72 hours to remove the moisture content, and then crushed in the grinder separately to become a fine powder. Then the prepared bio-adsorbent powder is stored in a tight container to protect it from moisture. To take notes the pH in the solution that has been treated with bio adsorbent is used as a Microprocessor-Based (Labtronics) pH meter. The experiment was carried out by shaking the fluoride solution with a known fluoride concentration bio adsorbent in a shaker. Then the result is % fluoride removal. After analysis, it can be concluded that the use of agricultural waste to remove fluoride ions is effective for use in pH range of 4-6 with maximum dose PB to remove 85% of fluoride ions i.e. 1.2 g per 100 mL solution. However, remediation wastewater containing various pollutants will require more research. The suitability of other biomaterials as bio adsorbents should be explored to overcome water pollution [24].

5. Bio pore infiltration process

Bio pores are a method to overcome the problem of decreased water absorption. The bio pore form resembles a synaptic burrow located in the ground and branched and effective for channelling water and air into the soil. In order for bio pore holes to function optimally, the organic matter must be routinely added so that in the hole the process of aerobic composting will still take place by soil microorganisms [26].

Ingredients contained in bio pores are organic waste such as vegetable waste, animal food waste, paper, twig pieces, and grass. The utilization of these microbes can accelerate the process of waste degradation by adding a combined inoculum of four microorganisms such as Saccharomyces, Lactobacillus, Bacillus, and Acetobacter. The combined inoculum has effectiveness in degrading organic waste into compost [26]. There are also several types of bio pores, including root bio pores and earthworm bio pores. Most subsoil carbon (C) turnover occurs in bio pore hotspots such as root channels and earthworm burrows. Bio pores allocate large C amounts into the subsoil, where a vast capacity for long-term C sequestration is predicted. Sufficient content of C inputs creates several years lasting hotspots for preferential root growth and nutrient mobilisation in the subsoil. Concluded that root- and earthworm-derived bio pores are vertical pathways for plant C from the soil surface into the subsoil and for intensive processing of litter C and sequestration of microbial necromass [27].

The use of holes in porous pipes has the aim to improve the infiltration process and help organic waste to have direct contact with air so that air can go in and out along the pipe as well as assist the decomposition process by microorganisms [28]. The benefits produced by the use of bio pores, namely increasing the absorbency of water, converting organic waste into compost, utilizing the role of soil fauna and plant roots, and as a carbon sink to help prevent global warming [26]. The bio pore system methods can be seen in Figure 6.

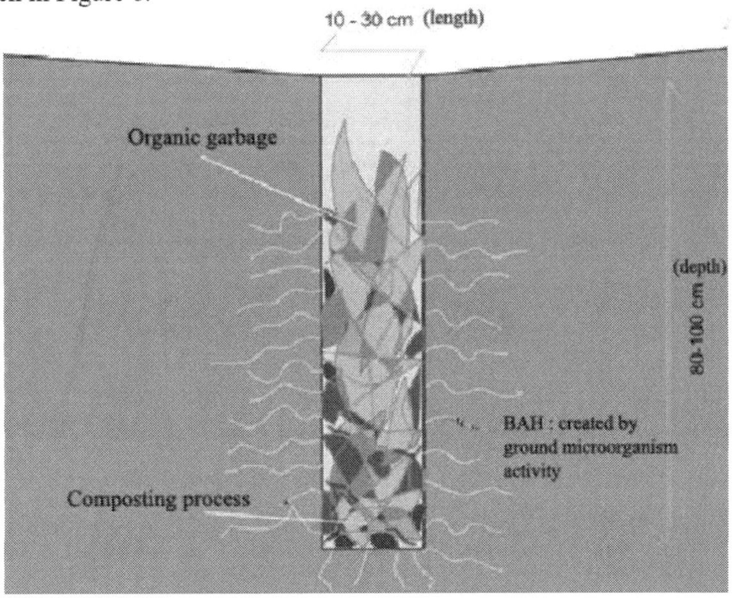

Figure 6. Bio pore system [29]

Bio pore made by PVC pipes that have been given hole on each side. Holes in bio pore pipes have a function to increase the rate of infiltration in the soil, it is influenced by aeration factors determined by porosity and water content of materials. Holes in the pipe can help organic waste to contact directly with the air so that air can go in and out along the bio pore pipe and help the decomposition process by microorganisms [28]. Organic waste applied to bio pores has a function to make the soil more fertile because the waste will undergo a decomposition process by microorganisms and can turn into compost that serves to fertilize the soil [26].

6. Permeable reactive barrier

Permeable Reactive Barrier or what we usually call PRB is a water remediation technology in the form of a vertical wall as a barrier layer that is used to minimize water contamination before being channelled to groundwater [30]. The water will be filtered through the pores on the wall before being channelled to the bio pore so that the water used to fertilize the soil is more hygienic and will then be forwarded to the soil [31].

In the manufacture of Permeable reactive barrier (PRB) there are several materials that can be used which have their respective functions and roles. The materials used include iron, limestone, soil, clay, carbon, and mulch. The material used has a role as a media for filtration, one example is limestone which plays an active role in depositing lead and copper solutions in PRB.

The first step, we will dig a trench for the placement of PRB. Usually, PRBs are placed no more than 50 feet below the ground. Then the trench will be filled with reactive materials such as iron, carbon, mulch, and limestone. These materials can be mixed with sand to make the walls more permeable. The side walls are made of clay to make it watertight so that it is easier to channel water to PRB [31].

When it rains, the water will flow through the ditch to the PRB. Contaminated rainwater will be absorbed (stick) to the surface of the reactive material. For example, carbon particles will filter water containing petroleum when water passes through PRB. Limestone will precipitate metals dissolved in water and cause dissolved lead and copper to settle in PRB [31].

PRB has previously been applied to clean groundwater at a former semiconductor manufacturing site in Sunnyvale, California 1995 [31]. The funnel and gate configuration of PRB can be seen in Figure 7.

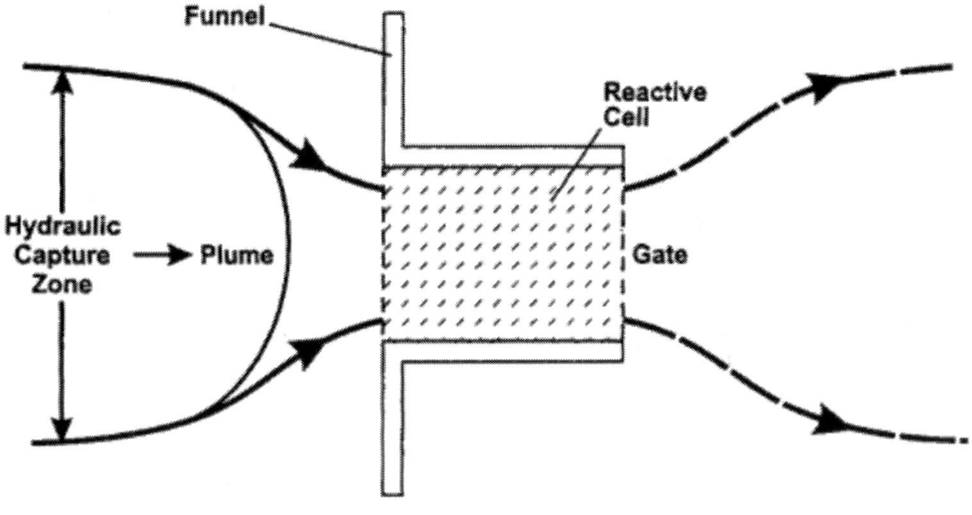

Figure 7. Funnel and gate configuration of PRB [32]

When the water will be in the underground channel, before being directed to the bio pore, the water will first pass through the PRB in an arrangement like the picture above where the funnel is made of clay or water-resistant material so that water does not penetrate through the funnel wall and can be channelled into the PRB to filter contaminated water.

7. Project implementation concept

7.1. Layout plan
The layout of the Permeable Reactive Barriers is a planning strategy for design element arrangement relate to the implementation of Permeable Reactive Barriers concept in the road in tropical countries. It is supposed to form an appropriate arrangement and can maximize the work function of the plan elements. The design of permeable reactive barriers and bio pores system can be seen in Figure 8.

Figure 8. Layout plan

Figure 8 shows the layout plan of the installation of PRB and the bio pore system. Porous concrete is installed under the asphalt layer so that rainwater can be absorbed and flowed into the pipe to PRB. After going through PRB, rainwater will go through the pipe again and enter the bio pore system before being distributed to the surrounding soil. Bio pores are made using PVC pipes and covers that have been given holes on each side.

7.2. Detail of permeable reactive barriers and bio pores system
The water will be filtered through the pores on the wall before being channeled to the bio pore so that the water used to fertilize the soil is more hygienic and will then be forwarded to the soil. After that, the water will enter the bio pore system. The use of holes in porous pipes has the aim to improve the infiltration process and help organic waste to have direct contact with air so that air can go in and out along the pipe as well as assist the decomposition process by microorganisms. The detail of permeable reactive barriers and bio pores system can be seen in Figure 9.

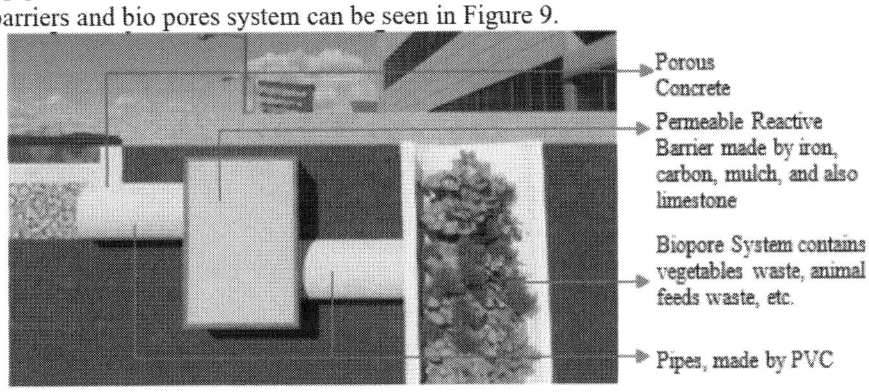

Figure 9. Detail of PRB and bio pores system

Figure 9 shows the detail of porous concrete, PRB, and bio pore system. The material that made porous concrete are coarse aggregates, ordinary portland cement (OPC), admixtures, and water its similar to the common concrete, but porous concrete just uses coarse aggregates from 9.5 mm to 19 mm various of the size. PRB are placed under the pavement filled with reactive materials such as iron, carbon, mulch, and limestone. Ingredients contained in bio pores are organic waste such as vegetable waste, animal food waste, paper, twig pieces, and grass. The utilization of these microbes can accelerate the process of waste degradation by adding a combined inoculum of four microorganisms such as Saccharomyces, Lactobacillus, Bacillus, and Acetobacter.

7.3. The stages of "Permeable Reactive Barriers" implementation

The stage of "Permeable Reactive Barriers" implementation can be seen on Figure 10 and explained as bellows.

- Starting with planning for the implementation of permeable reactive barriers and bio pore systems.
- Water pollution socialization. The first stage is giving socialization for local communities "Permeable Reactive Barriers" discussing the water pollution and how to solve it.
- "Permeable Reactive Barriers" socialization, socialization of permeable reactive barriers, providing an understanding to local communities about permeable reactive barriers and their benefits. Persuade them to help take care after being built.
- License, done by asking a license to government, so that "Permeable Reactive Barriers" can be implemented in the streets of tropical countries.
- Agreement, done by making agreement and signing Memorandum of Understanding (MoU) with government and local communities.
- DED (Detailed Engineering Design) manufacture by collaborating with related departments.
- Study of land for procurement and construction of "Permeable Reactive Barriers".
- Inauguration by the government and local communities.

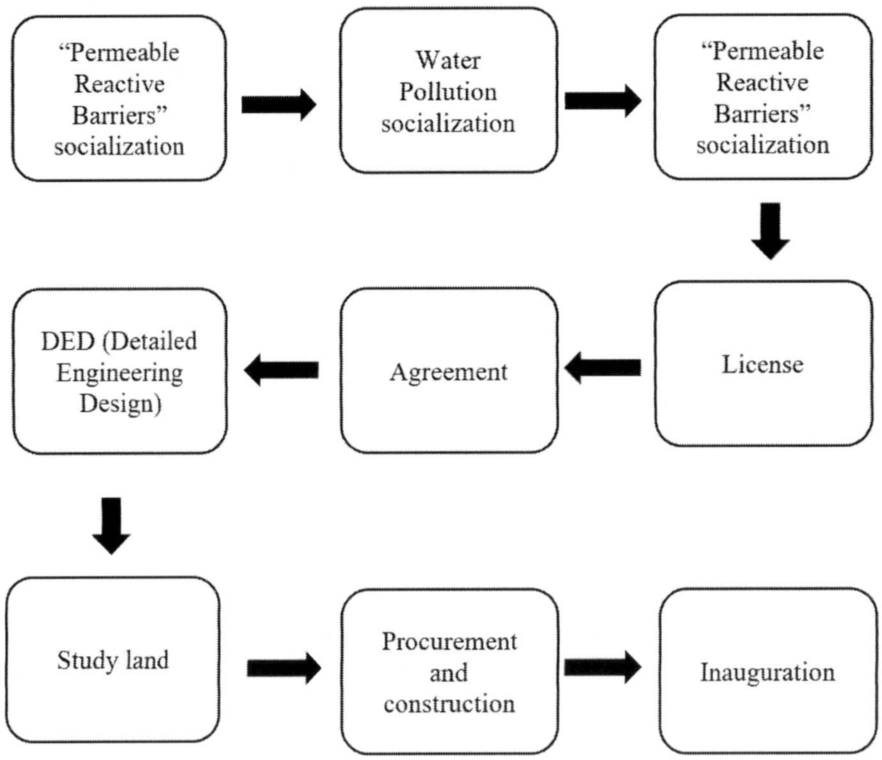

Figure 10. The permeable reactive barriers implementation stage

7.4. The Analysis of Effect from Implementing "Permeable Reactive Barriers" for Tropical Countries

The development of public transportation facilities must have an effect. The analysis of the effects of the construction of Permeable Reactive Barriers as Part of Water Remediation System in Tropical Countries can be seen in Table 1.

Table 1. Effect from implementing "Permeable Reactive Barriers" for tropical countries.

No	Impacted Areas	Explanation
1	Environment	The roads are protected from flooding due to puddles of rain. In addition, water enters the canal to PRB and is channelled back to the ground making the soil healthier because of reduced water pollution. Plants also become healthier due to improved water quality.
2	Economics	The community's economy increases because the flood barriers on the highway are reduced. After all, rainwater is absorbed by the pore concrete.

8. Conclusion

By stating this idea dealing with problems caused by the lack of water infiltration and the lack of application of water remediation in industrial areas with high rainfall is the use of Permeable Reactive Barriers as a solution for rainwater remediation on the surface of industrial tropical climates. This program has several parts for its application so that it can be carried out optimally. This section consists of, 1) Use of Pervious Concrete for water absorption on roads, 2) Permeable Reactive Barrier as a filter for rainwater contaminated with pathogenic microorganisms, decaying organic waste, toxic chemicals, and hazardous substances from the atmosphere carried by rain, 3) Use of Bio pore as the final reservoir of water which aims to make the soil around the road more fertile, because in the bio pore process the pollutants in the water are reduced so that it can reduce soil pollution.

Acknowledgment

This paper was supported by PKM Center Universitas Negeri Malang, Universitas Negeri Malang, and Directorate General of Higher Education, Ministry of Education, Culture, Research, and Technology of the Republic of Indonesia, who has provided an opportunity in the Student Creativity Program (PKM) for the 2021 funding year.

References

[1] El-Zein A, Ahmed T and Tonmoy F 2021 Geophysical and social vulnerability to floods at municipal scale under climate change: The case of an inner-city suburb of Sydney *Ecol. Indic.* **121** 106988

[2] Huang H, Chen Y, Clinton N, Wang J, Wang X, Liu C, Gong P, Yang J, Bai Y, Zheng Y and Zhu Z 2017 Mapping major land cover dynamics in Beijing using all Landsat images in Google Earth Engine *Remote Sens. Environ.* **202** 166–76

[3] Mehebub S, Raihan A, Nuhul H and Haroon S 2015 Assessing flood inundation extent and landscape vulnerability to flood using geospatial technology: A study of Malda district of West Bengal, India *Forum Geogr.* **XIV** 156–63

[4] Tarasova L, Merz R, Kiss A, Basso S, Blöschl G, Merz B, Viglione A, Plötner S, Guse B, Schumann A, Fischer S, Ahrens B, Anwar F, Bárdossy A, Bühler P, Haberlandt U, Kreibich H, Krug A, Lun D, Müller-Thomy H, Pidoto R, Primo C, Seidel J, Vorogushyn S and Wietzke L 2019 Causative classification of river flood events *WIREs Water* **6** 1–23

[5] Abolade O, Muili A B and Ikotun S A 2013 Impacts of flood disaster in Agege local government area Lagos , Nigeria *Int. J. Dev. Sustain.* **2** 2354–67

[6] Tingsanchali T 2012 Urban flood disaster management *Procedia Eng.* **32** 25–37

[7] Anon Buletin Hujan Bulanan di Indonesia - Updated Februari 2021 | BMKG

[8] BMKG 2021 Prakiraan Daerah Potensi Banjir Update 10 April 2020 1–61

[9] Yohana C, Griandini D and Muzambeq S 2017 Penerapan Pembuatan Teknik Lubang Biopori Resapan Sebagai Upaya Pengendalian Banjir *J. Pemberdaya. Masy. Madani* **1** 296–308

[10] Ramadhansyah P J, Mohd Ibrahim M Y, Mohd Rosli H and Mohd Haziman W I 2014 A review of porous concrete pavement: Applications and engineering properties *Appl. Mech. Mater.* **554** 37–41

[11] Guntakal S N and Selvan S 2017 Application of pervious concrete for pavements: A review *Rasayan J. Chem.* **10** 32–6

[12] Lian C, Zhuge Y and Beecham S 2011 The relationship between porosity and strength for porous concrete *Constr. Build. Mater.* **25** 4294–8

[13] Park S B and Tia M 2004 An experimental study on the water-purification properties of porous concrete *Cem. Concr. Res.*

[14] Xu G, Shen W, Huo X, Yang Z, Wang J, Zhang W and Ji X 2018 Investigation on the properties of porous concrete as road base material *Constr. Build. Mater.* **158** 141–8

[15] Yang J and Jiang G 2003 Experimental study on properties of pervious concrete pavement materials *Cem. Concr. Res.*

[16] Zhang Y, Li H, Abdelhady A, Yang J and Wang H 2021 Effects of specimen shape and size on the permeability and mechanical properties of porous concrete *Constr. Build. Mater.* **266** 121074

[17] Sonebi M, Bassuoni M and Yahia A 2016 Pervious Concrete: Mix Design, Properties and Applications *RILEM Tech. Lett.* **1** 109

[18] Utomo A B, Hidayatiningrum L F and Dhanardono B 2021 Application of porous concrete to resolveflood on the roads *IOP Conf. Ser. Earth Environ. Sci.* **708**

[19] Pishdadakhgari M 2018 Pervious Concrete : Potential Advantages in Regions with High Precipitation Pervious Concrete : Potential Advantages in Regions with High Precipitation 1–5

[20] Leonel A G, Mansur A A P and Mansur H S 2021 Advanced Functional Nanostructures based on Magnetic Iron Oxide Nanomaterials for Water Remediation: A Review *Water Res.* **190** 116693

[21] Lu F and Astruc D 2020 Nanocatalysts and other nanomaterials for water remediation from organic pollutants *Coord. Chem. Rev.* **408** 213180

[22] Bolade O P, Adeniyi K O, Williams A B and Benson N U 2021 Remediation and optimization of petroleum hydrocarbons degradation in contaminated water using alkaline activated persulphate *J. Environ. Chem. Eng.* **9** 105801

[23] Sarcletti M, Park H, Wirth J, Englisch S, Eigen A, Drobek D, Vivod D, Friedrich B, Tietze R, Alexiou C, Zahn D, Apeleo Zubiri B, Spiecker E and Halik M 2021 The remediation of nano-/microplastics from water *Mater. Today* **xxx** 1–9

[24] Garg R, Garg R, Thakur A and Arif S M 2021 Water remediation using biosorbent obtained from agricultural and fruit waste *Mater. Today Proc.*

[25] Solangi N H, Kumar J, Mazari S A, Ahmed S, Fatima N and Mubarak N M 2021 Development of fruit waste derived bio-adsorbents for wastewater treatment: A review *J. Hazard. Mater.*

[26] Lubis S W M A Y and Komala dan O 2015 Aktivitas Degradasi Sampah Organik Dalam Biopori *J. Ekol.* **15** 1–5

[27] Banfield C C, Pausch J, Kuzyakov Y and Dippold M A 2018 Microbial processing of plant residues in the subsoil – The role of bio pores *Soil Biol. Biochem.*

[28] Juliandari M 2013 Efektivitas Lubang Resapan Biopori Terhadap Laju Resapan (Infiltrasi) *J. Teknol. Lingkung. Lahan Basah* **1** 1–10

[29] Yulianur A, Fauzi A and Humaira Z 2018 The difference of drainage channel dimensions at Kopelma Darussalam on the land with and without the use of Bio pores *IOP Conf. Ser. Mater. Sci. Eng.* **352**

[30] Powell R M, Puls R W, Blowes D W, Vogan J L, Gillham R W, Powell P D, Schultz D, Sivavee T and Landis R 1998 *Permeable Reactive Barrier Technologies for Contaminant Remediation*

IOP Conf. Series: Earth and Environmental Science **847** (2021) 012035 doi:10.1088/1755-1315/847/1/012035

[31] Painter B D M 2004 Reactive barriers: Hydraulic performance and design enhancements *Ground Water* **42** 609–17
[32] He Q, Si S, Yang J and Tu X 2019 Application of permeable reactive barrier in groundwater remediation *E3S Web Conf.* **136** 1–5

Life cycle analysis of Guanxi freeway pavement maintenance interval

P A Paramitha[1], Shih-Huang Chen[1] and C P Dewi[2]

[1]National Central University, No. 300 Zhongda Rd., Zhongli, Taoyuan 32001, Taiwan
[2]Universitas Negeri Malang, Jl. Semarang no.5, Malang, Indonesia

cynthia.dewi.ft@um.ac.id

Abstract. Life Cycle Analysis (LCA) is widely used as an environmental impact over the full life cycle of a system. This paper incorporates LCA including impacts that occur throughout the construction, operation, and, maintenance of pavement. The historical freeway database was collected to analyze the pavement life cycle responsible for substantial energy and resource consumption. The analysis was done for Guanxi section as part of Taiwan National Freeway no.3 to compare a long-cycled and short-cycle maintenances. Calibrated HDM-4 and GaBi used to simulate total life cycle conditions. Road network, climate, vehicle fleet, and traffic data were collected and analyzed. Two aspects including Pavement Condition Index (PCI) and International Roughness Index (IRI) were considered for the simulations. The result shows that short-cycle is more acceptable for maintenance in Guanxi section. Environmental analysis by Global Warming Potential (GWP) shows that short-cycle maintenance contributed the smallest GWP with 0.02% up to 0.67% lower than another maintenance type. However, Taiwan focus on increase the level of service of the freeway by setting the minimum IRI value 1.7 m/km. Likewise, this study validates that recent maintenance use in Guanxi section is better than the previous long-cycle, while reduce the maintenance frequency to 1.5 years' might could increase the effectivity of short-cycle maintenance.

1. Introduction

Taiwan freeway was built in 1971 and completed in 1974, which was adopted the Interstate Highway System of the United States design with the length over 1054 km [1]. 95% of the roads are paved with asphalt [1]. The growing number of road in Taiwan was followed by some maintenance issues like low time interval of maintenance, lack of large-scale maintenance, and budget limitation. Those problems resulted in low pavement condition on the freeway and increase the environmental effect. High traffic volume and low pavement condition found increase the oil consumption which affect the environmental [2]. Traffic consumes more energy than road construction over the same period, depending on whether the traffic [3]. However, since 2013 Taiwan no longer focused on freeway construction and intended to focus on increase the pavement condition standard. It intended to concentrate on short-cycle maintenance instead of long-cycle maintenance. Those two maintenance types have different basic concept on maintenance timing and treatment. Short-cycle maintenance is a principal treatments used to extend pavement life, slow, or reverse pavement deterioration with small amount of pavement material and close maintenance time interval (every 0.5-2 years). It only can be applied for the pavement with minimal deterioration. While long-cycle maintenance extended the pavement life with longer maintenance time interval and greater amount of pavement material. Applied short-cycle maintenance indicates better road performance. However, it might increase the environmental effect due to the construction during maintenance. It becomes essential to embracing sustainability as an effective tool to assist in the selection of maintenance treatments for the roads. As an important feature of sustainability, life cycle assessment (LCA) emphasizing environmental factors in the decision-making with more systematic and organized [4]

The Ecological Energy saving Waste reduction Health (EEWH), Taiwan Green Building Rating System serves as a scorecard for the design, construction, and operation of green buildings [5]. Pavement management on its own can contribute directly to some EEWH credits. Increase road density in Taiwan should require a better maintenance and improvement management that considered on environmental effect to minimize total life cycle energy use and vehicle emissions. LCA can be used to identify and quantify the energy consumption and emissions associated with the pavement life cycle [6]. The life cycle assessment starts with a definition of the aim and scope of the study. Life Cycle Inventory (LCI) covers all of the significant environmental burdens from the pavement lifetime and will be quantified and compiled [7]. It followed by Life Cycle Impact Assessment (LCIA) which calculating and presenting the result that supports comparison or further analysis. The concept and working phases of LCA are described in the ISO14040 [7].

This study focuses on Guanxi section (Sta. 42+00 km - Sta. 110+00 km) as part of Taiwan freeway no. 3 and located in the north part of Taiwan. Special case of network level on Sta. 83+00 km - Sta. 89+00 km lane 3 was taken for worst case example in this section. This analysis will include lanes with flexible pavements and not includes some area with rigid pavement. A historical database will be collected and be used for long-cycle and short-cycle maintenance analysis which applied in Taiwan before and after 2013. It has some difficulties in the data collecting and analyzing process. Lack of data available for entire year or at least 4 years, resulted in poor calibration and analysis. Some of the data are available difficult to analyze due to the random pattern. Therefore, forecasting model will be used to fill the empty data and predict data in the future.

HDM-4 used in this simulation analysis to predicts the life cycle pavement conditions. It provides simulation feature for environmental analysis and emissions resulted from the traffic and maintenance process, although rule out the possibility to use another life cycle program like GaBi for more detail analysis.

2. Methods

2.1. Life cycle inventory database

Freeway database includes road network database, traffic database, vehicle fleet database, road work database, and climate database. The data were collected from the beginning of Guanxi section open to traffic up to 2016. It was taken from various sources, including long-term pavement performance database from Taiwan National Freeway Bureau, Electronic Toll Collection (ETC) database, and Central Weather Bureau of Taiwan. Pavement condition and pavement roughness was obtained by conducting road survey by automatic monitoring vehicle. Pavements are responding to the influences of environmental, 2016 climate data from Taiwan Central Weather Bureau use in this analysis.

The geometric design of roads is one of engineering concerned with the positioning of the physical elements such as slope, profile, cross section, and altitude according to standards and constraints. The pavement cross section illustrated in Figure. 1. Five types of vehicles were use in this analysis, includes passenger car, SUV, light truck, heavy truck, and trailer. The traffic data were collected for both northbound and southbound base on hourly, daily, weekly, monthly pattern, also holiday patterns. Traffic composition, volume and growth rates specified as required by the purpose of the study being performed. Annual average daily traffic (AADT) was then count to predict pavement condition progression, fuel consumption and environmental impact on HDM-4 program.

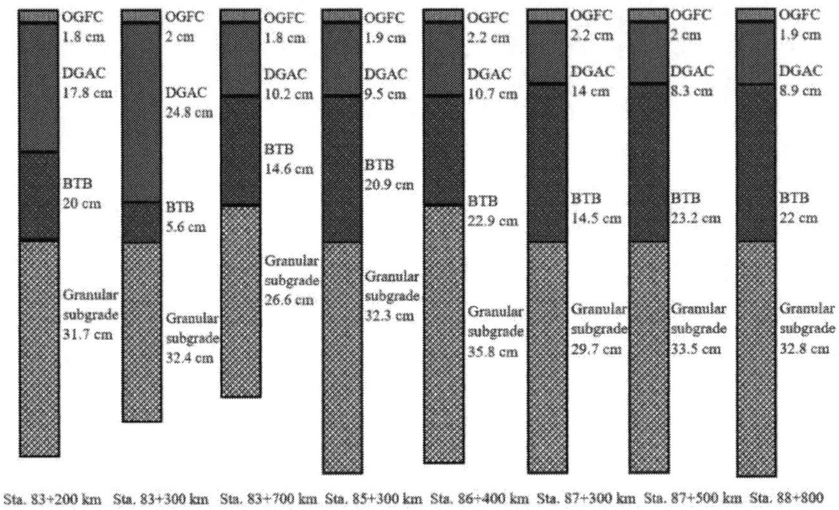

Figure 1. Pavement structure in Guanxi section.

Note: Open Graded Friction Course (OGFC); Dense Grade Asphalt Concrete (DGAC), Bitumen Treated Based (BTB).

2.2. Data manipulation

Statistical analyses were conducted in this study, including clustering, correlation, and regression analysis. Clustering was used to separate the traffic peak hour and traffic volume based on the time. Correlation and regression analysis were used to find the variables affecting average daily traffic and checked the relationship between two variables, included prediction result and observed data.

2.3. Pavement Condition Prediction Model

Pavement condition affect by geometric, traffic, pavement strength, historical maintenance and environmental condition. Pavement historic data, pavement deterioration, Pavement condition Index (PCI) and International Roughness Index (IRI) use as a parameter to decide the pavement maintenance interval. The simulation analysis uses 4 types of maintenance for 6 km road section (mill 5 cm, repave DGAC 5 cm, PAC 2 cm; mill 11 cm, repave DGAC 10 cm, PAC 2 cm; mill 5 cm, repave DGAC 3.5 cm, PAC 2 cm; repave PAC 2 cm) HDM-4 predicted the life cycle pavement conditions user-specified scenario of circumstances. It will be automatically analyzed the equivalent of two or more project alternatives for each of the road sections. A set of default data will be provided with the system, but users will be able to modify it to reflect Taiwan condition. This study collects 23 years' data from the beginning of open traffic until 2016. Trial and error used to calibrate the model, due to many variables and few available data. As mentioned by Rohde, prediction can be done by compare the result with actual observed performance and HDM-4 calibration coefficient can be adjust manually to minimize different between predicted and measured distress. Important current network condition that important as monitoring indicators are follows [8]:

- Network length, network utilization distribution and surface type.
- Average network roughness divided by km and vehicle-km
- Network distribution charts
- Road network monitoring indicators.

2.4. Environmental life cycle

Life cycle in pavement included many components from material phase, construction phase, use phase, maintenance phase. It should be analyzed by collect the inventory data that compose by many data includes quantifying energy and raw material requirements, atmospheric emissions, waterborne emissions, solid wastes and other releases for the entire pavement life cycle [9]. Life Cycle Inventory (LCI) database formed into a flowchart model of processes shows inflows and outflows of the item. The second step is calculating the global warming potential (GWP) to understand their environmental significances. These processes divided into some parameters includes resources, deposited goods, emissions to air, emission to fresh water, emission to sea water, emission to agricultural soil, and emissions to industrial, which describe their impact to the atmosphere. The third steps measure how much Greenhouse gasses (GHG) trapped in the atmosphere. The analysis cycle decides to be 40 years. GaBi and HDM-4 software were used to simulate LCA. The LCA model in GaBi consisted of data from production and construction phase for all of pavement maintenance types, while LCA model in operating phase was done in HDM-4. The environmental analysis framework illustrated in Figure. 2.

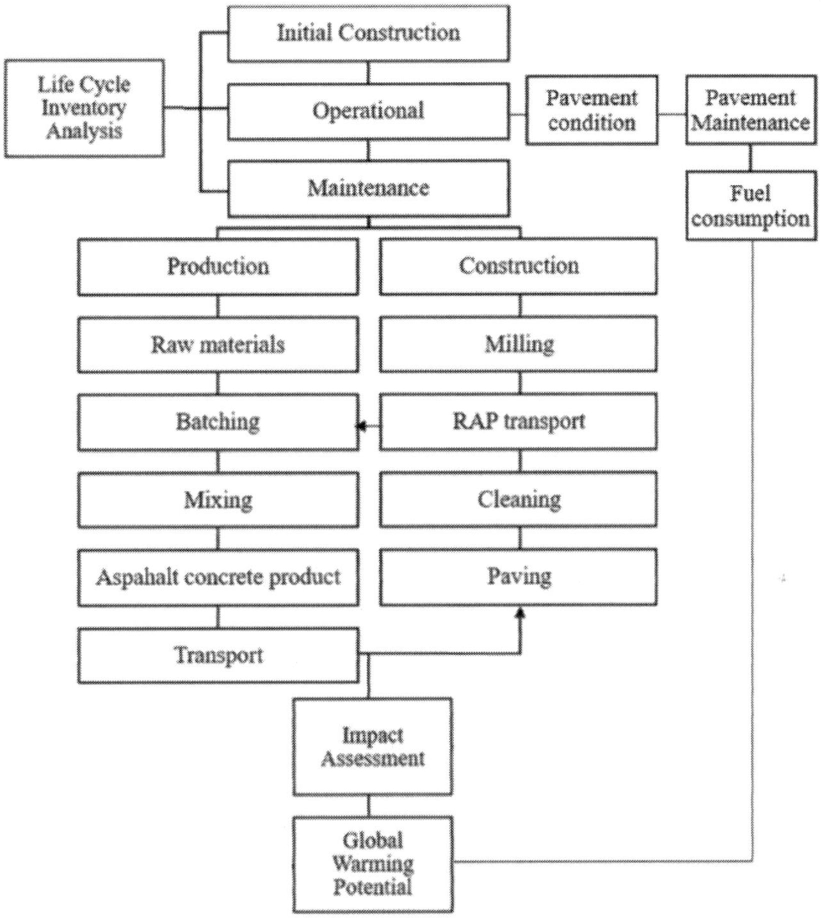

Figure 2. Environmental analysis framework.

3. Results and discussion

3.1. The prediction results of pavement condition index

Pavement Condition Index (PCI) is one of an important pavement condition parameter in Taiwan PMS. The pavement distress data collected two times per year or more with manual and Charge-coupled Device (CCD) automatic vehicle. Distress type and the severity are recorded, was then calculate the PCI. Standard PCI 85-100 used to control the pavement condition to stay in excellent condition. The PCI data in this study use two years' data in a row from May 2015 to May 2017. Figure. 3 shows the PCI progression based on five types of distress progression (cracking, pothole, rutting, raveling, edge break). All of distresses count as light distress, since Taiwan freeway always stay in excellent condition. However, this result has different progression with the survey result with R-squared 0.3 thus maybe cause by the distress type use for the analysis. The real shows 10 kind of distresses exist in Guanxi section with patching dominated, while HDM-4 only analyzed 5 types of distresses without patching. This factor finds as the HDM-4 software limitation.

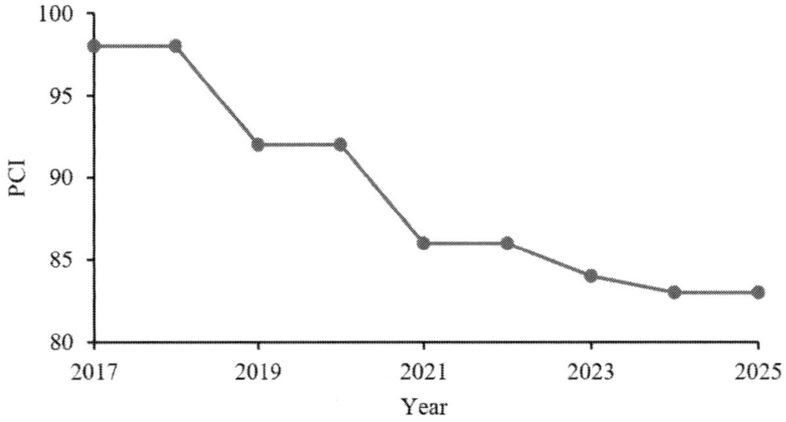

Figure 3. PCI progression based on distress progression

IRI is the most important pavement condition indicator to decide maintenance both in Taiwan and HDM-4. Taiwan freeway increase the IRI threshold requirement year by year, and recently applied 1.75 m/km IRI as a standard minimum the pavement to stay in excellent condition. The Figure. 4 bellow shows the long-cycle and short-cycle of maintenance before and after 2013. Long-cycle shows done every few years with IRI threshold 3.5 m/km, while short-cycle has been done every year or half year to make the pavement condition stay around 1.75 m/km.

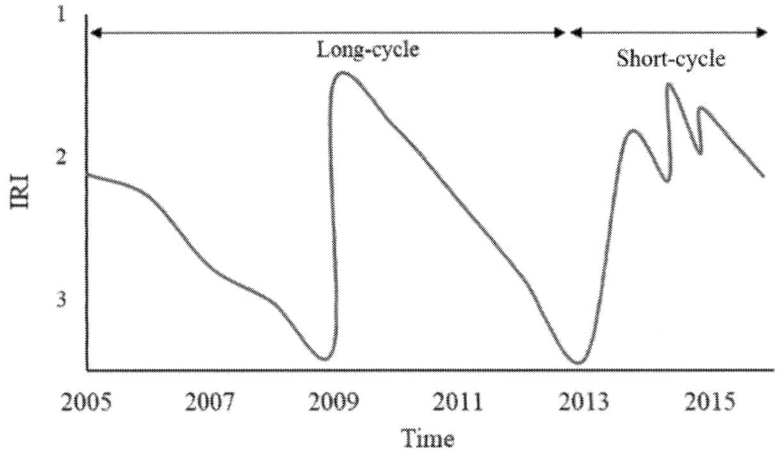

Figure 4. Average IRI.

Calibration was done by compare the prediction of HDM-4 and actual observed performance. Calibration factor adjusted to minimize the difference between the predicted and observed distress and IRI. Trial and error for calibration factor was use to adjust the simulation model. Figure. 5 shows the IRI progression for 4 types of maintenance applied in Guanxi section, includes long-cycle and short-cycle (repave PAC 2 cm). The linear regression between observed IRI and predicted IRI from HDM-4 after calibration shows R^2 0.88 which means the model fits the data. It convincing that calibration need to do to make the results become closer with the real condition. The simulation result shows after 20 years use short-cycle maintenance; the pavement condition become not stable due to aging of the under layer. The pavement become worst faster than original condition. This condition similar with maintenance with DGAC 5 and DGAC 3.5. To control this condition, major rehabilitation need to improve the road condition after 20 years. As simple alternative, mill and repave with 10cm of DGAC and 2cm PAC (or another pavement types) also could be apply in this section. The conditions with rehabilitation after 20 years' cycle can be simulate like in Figure. 5.

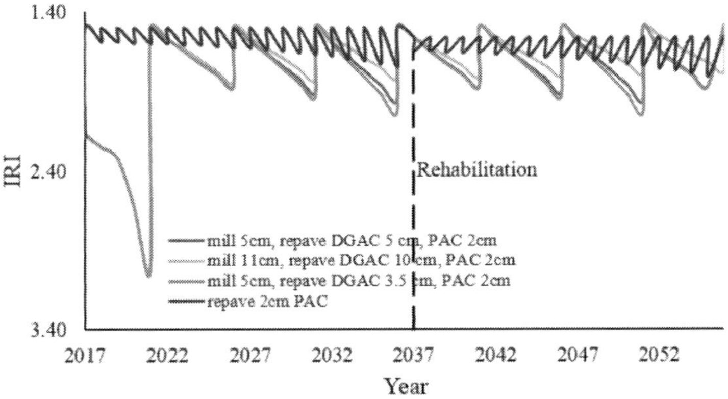

Figure 5. IRI progression with rehabilitation alternative.

3.2. Impact analysis of Guanxi section

3.2.1. Material.

This analysis uses four types of surface materials for 6 km road section, with the typical road section in Guanxi section. The analysis conduct per 1 km of road length for one lane. Flexible pavement consists of fine aggregate, coarse aggregate, and bitumen. Road construction involves a huge amount of energy and non-renewable materials. To construct and maintain flexible roads, a considerable amount material that include bitumen, aggregates and additives are needed which consumes a significant amount of energy for the material production, transportation and paving of roads. Emissions from these activities impact air, water and soil pollution, loss of biodiversity and threat on human life. Since Taiwan data only available for 1 ton of aggregate (4.72 kg CO2-e) [10], another coefficient uses the default coefficient value from GaBi software.

3.2.2. Transportation.

HDM-4 determines effects of emissions and noise, predicts numbers of road accidents and amounts of energy consumption with features to calculate emissions by the road work. Traffic information used to calculate energy used by vehicles. The energy use related to vehicle size, weight, design, age, road characteristics and condition, traffic characteristic, fuel consumption, lubricating oil, tire wear, and vehicle part consumption. Vehicle emissions output includes hydrocarbon, carbon monoxide, nitrous oxide, Sulphur oxide, carbon dioxide, and some particulates. Those pollutants gave effect to the air quality, health, damage the environmental and result on global warming [11].

Project alternative in this research or Guanxi section use the real maintenance type and compare long-term and short-cycle. The real data in this section available in more detail.

3.2.3. Asphalt Production (Maintenance).

LCI study focused on asphalt pavements was initiated in 1998 by European Asphalt Pavement Association (EAPA) and Eurobitume [12]. Meanwhile, Eurobitume carried out a partial LCI study on bitumen covering the life period from crude oil extraction to refinery deposit [13]. During the production of asphalt mixes, there are several factors that may affect energy consumption such as moisture content of aggregate, ambient temperature, as well as binder type. Moisture Content and Gradation Increased aggregate moisture content increases the amount of energy used to dry the aggregate in the asphalt plant especially for finer aggregates [14]. Thus, temperature and aggregate size influenced the amount of fuel needed to dry/heat.

3.2.4. Construction.
The construction phase includes mixture paving and compacting. In this case equipment data was taken from many companies that related to the mixture paving and compacting. Engine power and fuel consumption were obtained. In the life cycle inventory calculations of asphalt pavers only the direct fuel consumption and the production of the corresponding amount of fuel have been taken into account based on material weight and total layer (5 cm per layer). Compaction process apply based on density and requirement void by performing rolling patterns on a test strip. There is no standard for determining the number of passes and coverage location by the compactor instead use experience. However, the default compaction setting from GaBi used in this analysis with consider the layer thickens.

Impact analysis includes total flows and emission for maintenance and operation phase. It includes milling, production, laying, compaction, and transportation. Asphalt milling flow result show that road section which applied 11 cm milling contribute more emission flow due to double milling process and more transport flow to discard the disposal material. Flow result for asphalt laying and compaction shows much shorter than another maintenance activity, since it is only use some vehicles on the location.

Asphalt production phase simulation result present that asphalt production requires a lot of natural resources and stockpile goods. It is required average 50.8 % of the total flow. Despite of material needed, this process shows contribute much effect to the air, fresh water and sea water. Emission to air dominated by CO_2 that contribute 50.5% from the emission to air followed by NO, CO, NO_2, SO_2, CH_4 and another particle. This process also affects the water by the needed of water on production process and for cleaning the equipment. Moreover, pavement also found contaminate the storm water and groundwater [15]. This study found that pavement production contributes in emission to water around 47% and 0.15% to sea water.

Every kilometer of road contribute different emission based on material use and maintenance type use on the road. Sta. 83+00 shows contribute bigger emission compare with another mileage caused it to repave the road more with 10 cm DGAC, while another road dominate with repave 5 cm or 3.5 cm DGAC. Total flow for short-cycle analysis for asphalt production much lower than total flow on long-cycle maintenance. It worth to do to reduce emission per year analysis. However, Figure. 6 shows different result for the flows of long-cycle and short-cycle in 40-years analysis. Short-cycle maintenance that conduct every year will contribute more emission compare with another types of maintenance with longer cycle. It shows asphalt production flow of short-cycle maintenance is two to three times more than asphalt production flow of long-cycle.

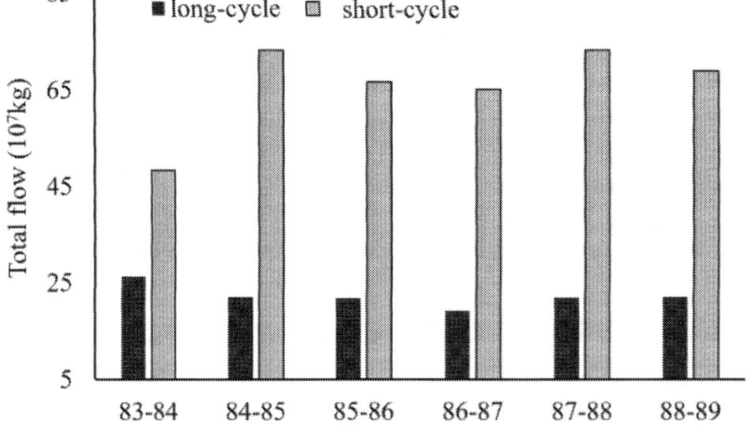

Figure 6. Total flows for asphalt production activities in 40 years.

3.2.5. Operation Phase.

This analysis considers the transportation during operation phase of Guanxi section. Analysis conducted by HDM-4 which considered AADT, equivalent single axle load (ESAL), road geometric and pavement conditions. The fuel consumption carbon footprint value in Taiwan for gasoline and diesel marked to be 0.671 and 0.752 respectively [10]. Operation phase emitted more GHG compared to another phases due to high volume of traffic. Pavement roughness are identified give influence to the vehicles' oil consumption. This analysis includes HC, CO, NO, SO_2, CO_2, Pb, and particle. Figure. 7 presented the GHG emission in Guanxi section. The GHG quantities shows different per kilometer road affected by road length, AADT, ESAL, road geometric, rise and fall. Every lane has different GHG emission rate based on the vehicle proportion. However, the result per lane shows almost similar proportion. Only lane 1 and 2 have more Pb that emitted from small vehicle that use gasoline.

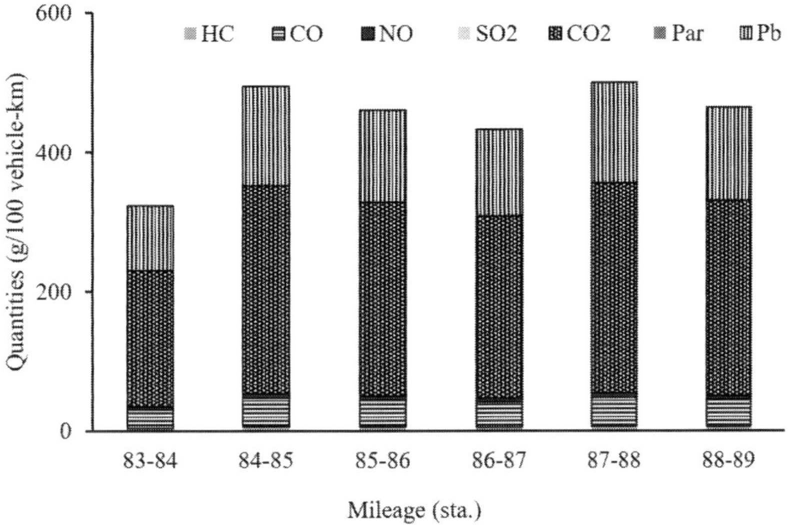

Figure 7. GHG emission on Guanxi section

3.3. GWP of Guanxi section

The GWP for milling process in six km road of Guanxi section depend on the section length. While transportation GWP depend on the layer thickness. Figure. 8 illustrated the total GWP for production phase in six kilometers of Guanxi section. It shows that 80% GWP for production comes from transportation of truck when take the materials from asphalt plant followed by fine aggregate production, mixing process, bitumen refinery, and coarse aggregate production. Transportation is a sector where measures need to be taken in order to reduce the energy demand and environmental impact.

GWP for construction phase consist of GWP for pavement laying and pavement compaction presents in Figure 8. In this result can be seen that the GWP is mostly affected by laying process. GWP for laying affected by total materials use for the construction, while compaction affected by the thickness of construction.

Laying and compaction process was design for every 5-cm layer. Repave for 5 cm and 3.5 cm shared same energy, however repave with 10 cm layer need two times energy which reaches 125.00 kg CO_2-e. This analysis shows that different types of maintenance proportion and road length resulted in different GWP.

This session focus on environmental impact of different maintenance alternatives on Guanxi section as part of Taiwan National Freeway No. 3. This simulation was conducted on GaBi and HDM-4 software for life cycle phases, especially production and construction (maintenance) phase and use phase. It conducted LCI, LCIA and calculate GWP to evaluate the different maintenance alternatives and maintenance cycle. Comparison analysis between long-cycle maintenance and short-cycle maintenance in 40 years shows that average short-cycle give slightly higher environmental impact compare with long-cycle as shown in Figure. 10, since it is conduct in every year. However, when it combined with the vehicle operation, short-cycle maintenance contributed the smaller GWP with 60.8% lower than another maintenance type. Maintenance interval, material consumption and pavement roughness are identified give influence to the environmental. Moreover, further analysis need to make sure the short-cycle worthy to conduct in Taiwan freeway to replace long-cycle maintenance.

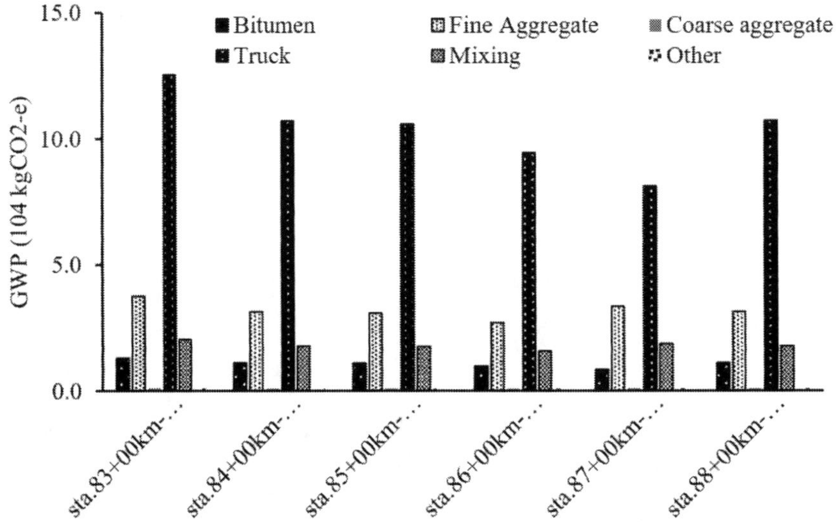

Figure 8. Each km GWP for production phase in Guanxi section.

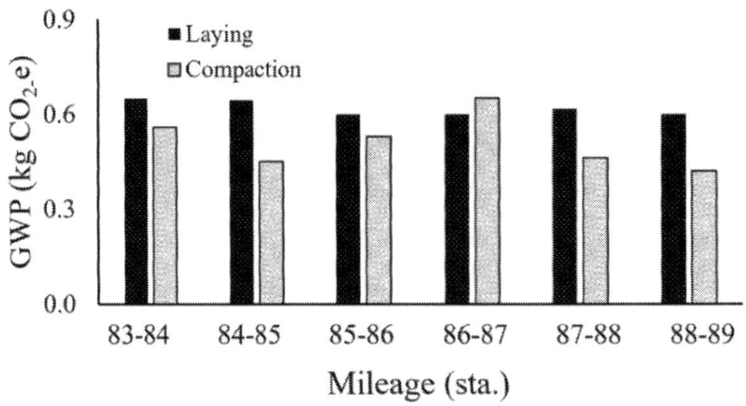

Figure 9. Each km GWP for construction in Guanxi section.

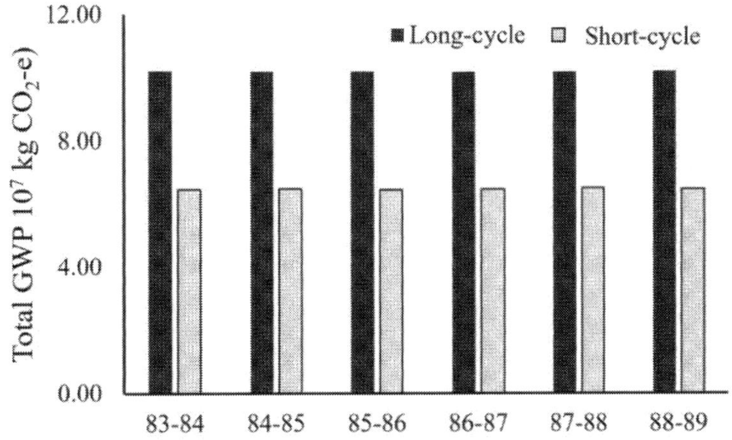

Figure 10. Total GWP for long-cycle and short-cycle.

4. Conclusion

This research aims to collect historical pavement life cycle database and predict pavement life cycle of Guanxi section. Two different maintenance interval (short-cycle and long-cycle) used to predict road network performance in different maintenance standards. Environmental effect due to pavement maintenance process and increase in vehicle number was then considered and analyzed by LCA. In accordance with policy objectives within environmental constraints. This research concludes:

- Transportation or operation phase emitted more GHG than the asphalt production or maintenance. Pavement roughness are identified give influence to the vehicles' oil consumption.
- Short-cycle is more acceptable for maintenance in Guanxi section, which resulted in better IRI value and pavement condition.
- Environmental analysis by GWP shows that short-cycle maintenance contributed the smaller GWP with 60.8% lower than another maintenance type. Maintenance interval, material consumption and pavement roughness are identified to influence to the environmental.
- Reduce the maintenance interval to 1.5 years might decrease the cost of short-cycle maintenance with similar pavement condition. This maintenance method could be used as future maintenance strategy in another freeway sections in Taiwan.
- When inquiring about life cycle analysis, more precisely define the big continuous local database in continuous time is required to maintain the better calibration and simulations.

Acknowledgments

We gratefully acknowledge the Prof. Lin, Jyh-Dong's Lab in National Central University and Taiwan Freeway Bureau, MOTC for providing the data and Taiwan Ministry of Science and Technology, MOST for providing the financial support.

References

[1] MOTC 2016 Statistical Yearbook of Republic of China Statistical Department
[2] Christopher R.B., and Greenwood I.D. 2001 Modelling road user and environmental effects in HDM-4. *Highway Development and Management* **7** 401.
[3] Michel C. and Julian B. 2003 Energy consumption and greenhouse gas emissions. *Sustainable Developmentethe Environmental Road of the Future* 47.
[4] U.S Environmental Protection Agency 2000 Hot Mix Asphalt Plants Emission Assessment Report 46.
[5] EEWH Taiwan 2017 Green building.

[6] Baumann H. and Tillman A-M. 2004 The hitch hiker's guide to LCA (Life Cycle Assessment): an orientation in Life Cycle Assessment methodology and application.
[7] ISO 14040 1997 Environmental Management – Life Cycle Assessment – Principles and Framework American National Standards.
[8] Henry G.R.K. and Odoki J.B. 2000 Analytical Framework & Model Description Part E *HDM-4 Highway Development and Management*.
[9] N. a. E. Santero, "Life-cycle assessment of pavements. Part I: Critical Review," ScienceDirect, vol. 55, no. Resources, Conservation and Recycling, pp. 801-809, 2011.
[10] T. EPA Taiwan 2010 Product Carbon Footprint.
[11] Henry G. R. K., Odoki J.B. and Stannard E.E. 2000 Overview of HDM-4. The highway development and management series 4 part F
[12] Johannes S. 2005 Climate Change After the International: Rethinking Security.
[13] Timo B., et al. 1999 Partial life cycle inventory or "eco-profile" for paving grade bitumen. 99.007.
[14] David S. 2010 Moisture in Asphalt Production-The importance of accurate moisture measurement and control. *Quarry Management* 37 1 19.
[15] Nemeth A. F., Ward D.A. and Woodington W.G. 2010 The effect of asphalt pavement on stormwater contamination. 79

AUTHOR INDEX